Technology and Change

Technology and Change

A Courses by Newspaper Reader

Edited by

John G. Burke

and

Marshall C. Eakin

Courses by Newspaper is a project of
University Extension, University of California, San Diego
Funded by the National Endowment for the Humanities

Boyd & Fraser Publishing Company
San Francisco

John G. Burke and Marshall C. Eakin, editors

TECHNOLOGY AND CHANGE
A Courses by Newspaper Reader

Published by Boyd & Fraser Publishing Company
3627 Sacramento Street, San Francisco, CA 94118

Library of Congress Cataloging in Publication Data:

Burke, John G
    Technology and change.

    "A Courses by Newspaper reader."
    1. Technology—Social aspects.   2. Technological innovations.
I. Eakin, Marshall, joint author.   II. Courses by Newspaper (Project)
III. Title.
T14.5.B87   301.24'3   79-14346
ISBN 0-87835-082-9
ISBN 0-87835-083-7 pbk.

1 2 3 4 5 · 2 1 0 9

Courses by Newspaper
Connections: Technology and Change

Contents

Part Three: Conditions of Technological Development

Part Four: Sources of Technological Change

Part Five: Retrospect and Prospect

Preface

This is the eleventh in a series of books developed for Courses by Newspaper (CbN). A national program, originated and administered by University Extension, University of California, San Diego, and funded by the National Endowment for the Humanities, Courses by Newspaper develops materials for college-level courses that are presented to the general public through the nationwide cooperation of newspapers and participating colleges and universities.

The program offers three levels of participation: readers interested in self-learning can follow a series of weekly newspaper articles; they can pursue the subjects further in the supplementary anthology (or Reader) and study guide; and they can enroll for credit at one of the 300 participating colleges or universities. In addition, many community organizations offer local forums and discussion groups based on the Courses by Newspaper series.

This volume supplements the fifteen newspaper articles written by noted experts from around the country especially for the eleventh Course by Newspaper, "Connections: Technology and Change." The weekly series was produced for newspapers throughout the nation, with a starting date of September, 1979.

We would like to acknowledge some of the many people whose efforts have made Courses by Newspaper possible. Hundreds of newspaper editors and publishers have contributed valuable space to bring the newspaper series to their readers; and the faculties and administrations of the many colleges and universities participating in the program have cooperated to make credit available on a nationwide basis.

Deserving special mention at the University of California, San Diego, are Paul D. Saltman, vice chancellor for academic affairs and professor of biology, who has chaired the faculty committee and guided the project since its inception, in addition to serving as the first academic coordinator in 1973; Caleb A. Lewis of University Extension, who originated the idea of Courses by Newspaper; and the members of the Courses by Newspaper staff—Project Director George Colburn, and Yvonne Hancher, Stephanie Giel, Elliot Wager,

Bethany Gardella, Beverly Barry, and Sharon Porter, who have been critical to the success of this year's program.

Many thanks are due to Jane Scheiber for her invaluable editorial advice and for expediting the completion of the project. The CbN faculty committee and National Board made important contributions to the conception of this course.

We also with to thank the authors of the newspaper articles—Kingsley Davis, Peter F. Drucker, A. Hunter Dupree, Eugene S. Ferguson, Joseph Gies, Clarence Glacken, G. Allen Greb, Melvin Kranzberg, Edwin T. Layton, Jr., Bertram Morris, Robert P. Multhauf, Derek de Solla Price, Nathan Rosenberg, Lynn White, Jr., and Herbert F. York—for their bibliographic suggestions for this volume.

Finally, we wish to express our gratitude to our funding agency, the National Endowment for the Humanities. The Endowment, a federal agency created in 1965 to support education, research, and public activities in the humanities, has generously supported this nationwide program from its beginning. We wish particularly to acknowledge the support and advice of James Kraft, Director of the Endowment's Office of Special Projects.

Although Courses by Newspaper is a project of the University of California, San Diego, and is supported by the National Endowment for the Humanities, the views expressed in course materials are those of the authors only and do not necessarily reflect those of the funding agency or of the University of California.

Introduction

World War II was the first conflict in which scientists were effectively mobilized and scientific knowledge was systematically applied to aid the war effort. With the deployment of radar by the British, an effective system which gave timely warning of the approach of enemy aircraft came into existence. During the course of the war, the Germans perfected jet aircraft engines and long-range rockets. Blood plasma, penicillin, and new pesticides saved the lives of countless soldiers and civilians. And the employment of the atomic bomb by the United States marked the end of the bloodiest and most tragic confrontation among the world's major nations.

Although it is a terrifyingly destructive weapon, the significance of the atomic bomb lies in other areas. It became, first, a symbol of the power and ability of science and technology to shape the course of events in the modern world. Further, the bomb was a signal that humans now possessed the awesome capability to destroy civilization on this planet and to poison the environment for centuries, possibly leading to the extermination of the human species.

These consequences stimulated two kinds of reaction. On the one hand, the major nations became actively engaged in promoting scientific research and technological development in both the military and civilian sectors by massive expenditures, which dwarfed the monies allocated prior to World War II. Examples of the results of these efforts are the intercontinental ballistic missiles, the nuclear submarines, the manned space flights to the moon, and the support of medical research on cancer, heart disease, and stroke.

On the other hand, many concerned people began to have second thoughts and to raise important questions about science and technology. Where are science and technology leading us? Do we have the wisdom to control the awesome powers that have emerged from modern science and technology? How can we ensure that science and technology are used for humane purposes? There appeared, in other words, critics who voiced serious misgivings about the course and conduct of modern science and technology, and whose arguments were much more cogent than any advanced in the past. Their number increased substantially during the sixties when the environmental and consumer movements gained recognition and took on momentum.

We have, then, at the present time two major camps, although differences are apparent among the adherents of each. One group seeks actively to encourage technological change, the other to stem its advance. It is this broad question of technology and change that will be addressed in this book. In order to gain a perspective on our present situation, however, issues and problems will be presented in a historical context. This approach will not only help to explain how the current state of affairs came about, it will also give some knowledge of past relationships between technology and society.

Most readers are aware of the immediate problems stemming from technological advance; for example, air and water pollution or the question of the safety of nuclear power plants. While analyses and discussions of such immediate problems are very important, Part One will concentrate on more difficult and more significant issues: the sources of, or the responsibility for, the deficiencies of our technological society, and the question as to whether technology is now completely out of human control.

Part Two will focus on the long-range effects of technological advance. From historical accounts we know that certain technological innovations have spawned social crises and changes that led in some societies to the establishment of novel institutions and that altered customs, habits, and thought. Further, advancing technology over time appears to have created more complexity and to have resulted in the appearance of classes of experts, conditions which some modern critics of technology deplore. Also, technological change has always changed the ways in which goods are produced and work is performed. Considering these effects will shed some light on the questions as to whether the changes we are currently experiencing represent a sharp and radical break with the past and whether technological change has been mainly beneficial or harmful.

Some societies, we know, have not experienced the rapid technological changes characteristic of the West, which leads us to a consideration of the conditions of technological development. The most important of these are the natural environment—climate, geography, and resources; the social environment or the culture of a society; and the size and distribution of the population. Technological advance rests on these three conditions, which are themselves intimately related. In turn, the environment, the culture, and the size and distribution of the population may be affected by technological change. Attention will be given to these three conditions in Part Three, introducing the reader to the complex problems involved in global technological development.

Given the appropriate conditions, however, there still must be a stimulus or an agency that produces technological innovation and change. Part Four considers five of the most important sources of technological change. They are: economic, scientific, engineering, military, and government. With the exception of science, these sources have been chiefly responsible for technological innovation since the beginnings of civilization about seven millennia ago. The application of scientific knowledge to create new technologies, on the other hand, has been a factor for only about the past two centuries. Critics of technology have singled out one or another of these sources as being primarily responsible for the unwanted effects of technological innovations; for example, the emphasis on economic growth or the machinations of the military-industrial complex. The reading selections will explore some of these concerns.

Part Five confronts three important problem areas: the nature of technological innovation, the major ethical issues involved in technological change, and current efforts to monitor innovations. All of these areas are receiving close attention at the present time, and the well-being of future generations will surely be affected by decisions that are now being made relative to these problems.

Readers of this book will not find in it any pat solutions to the problems created by advancing technology. The arguments of the critics of technology that are presented herein are just as articulate and persuasive as those of technology's advocates. What should be realized is the fact that the problem of technological change is extraordinarily complex, and that any proposed simple solution may actually imperil the future. Further, it should be recognized that each and every one of us is involved in the process of technological change because of the choices we make—where we live, what we eat and drink, how we dress, how we get to work, how we spend our spare time. Complexity does not preclude study, analysis, and understanding. It is this type of activity with respect to the problem of technology and change that this book seeks to encourage.

John G. Burke

PART ONE

Technology On Trial

Introduction

Critics and proponents of technology make a bewildering variety of complaints, assertions, rebuttals, and predictions. It is therefore difficult for any thoughtful person to decide whether the advance of science and technology is beneficial or harmful, whether the belief in progress through technology is realistic or misguided, and whether we should view the future with hope or with despair. A close look at statements about technology reveals three levels of discussion or areas of concern. Thinking about technology in these ways permits a better assessment of our situation.

The first level addresses immediate or urgent problems, which are attributed to unchecked technological advance. One major problem area in this cluster is our physical environment. There is concern about the quality of air and water; about the endangerment or extinction of species of fish, animals, or birds; and about possible future harmful climatic changes and destruction of the ozone layer, which protects us from harmful ultraviolet radiation. There are numerous other immediate or urgent problem areas—the exploitation of the consumer, hazardous working conditions, the use and misuse of computers, nuclear reactors and radioactive wastes, to name a few.

Legislative or regulatory action by governments on the local, state, or federal level has already solved or alleviated some of these immediate problems. Bitter controversy still exists about others. Current federal clean air standards, for example, have eliminated about 80 percent of the noxious automobile exhaust products. Auto manufacturers claim that the stricter standards to be enforced in a few years, which would reduce pollutants by another 10 percent, are unnecessary. Present standards, they claim, are sufficient; they have increased the cost of an average car by about $400, and the stricter standards would add at least another $350.

Controversy in these immediate problem areas, then, often boils down to whether the incremental further benefit to be gained in quality or safety justifies the necessary additional costs. It becomes a question of judgment. How safe is safe? Should we insist on absolute safety or permit probabilities to enter the picture and reduce costs?

The second level of concern and discussion involves attempts to track down the *sources* of immediate and urgent problems, which are spawned by technological advance. In some instances, critics attempt to place blame on a conspiracy. More often, there is an honest effort to determine responsibility or to pinpoint deficiencies in the structure of our economic, political, legal, or societal institutions or customs, which permitted the problems to arise in the first place.

Some view the capitalist system as chiefly responsible. Inasmuch as the central goal of major corporations is profit, their managers, it is said, have little if any concern about any undesirable environmental or social effects of the new technologies they introduce. The fact that pollution is occurring in socialist or communist nations, however, undermines this argument. On this level, the "military-industrial complex" and the "scientific-technological elite," both of which President Eisenhower warned against, are subject to searching analysis.

Another possible source of our problems, frequently cited, is the growth ethic, which pervades our society, and which many economists view as necessary for affluence.

Institutions and organizations have made some attempts to respond to concerns voiced on this level. The federal Environmental Protection Agency requires detailed studies on the environmental impact of projects proposed either by private industries or by agencies of the federal government. Congress has created an Office of Technology Assessment, whose present mandate is to forewarn government leaders of possible harmful side effects of newly conceived technologies in the civilian sector of the federal government. Also, some major corporations have made their own assessment studies, and others stress energy and resource conservation in their advertisements. Scientific and engineering societies have established new codes of ethics for their members. Despite this activity, debates on this level, concerned with the institutional or societal sources of our problems, still continue.

The third level of concern is both philosophical and ethical, having to do with the very nature of technology and what effects its development has on human beings. Proponents of technology claim that although technology is a necessary human activity, it is only one among many in which people engage. Others repudiate this description of technology. Technology, they say, is now autonomous—completely out of human control. It gives us abundance but destroys our freedom, because it shapes and directs all aspects of human life.

Reference will be made throughout this book to such first-level problems as environmental pollution. The selections that follow illustrate discussions occurring on the *second* and *third* levels. Barry Commoner views our problems as emanating directly from scientifically based technology. He voices the opinion that complexity has reached such a point that even the experts may no longer be in control.

Lewis Mumford believes that what he terms the "megatechnic system" or the "Power Complex" is the source of our troubles. Rampant technology, he writes, results from the decisions of anonymous technocrats—scientists, engineers, corporation and advertising executives. They compose the "system," which seeks to gain complete power and to extend its authority into all areas of human life.

Jacques Ellul writes almost exclusively on the third level. In fact, he thinks discussion of, or concern about, such first-level difficulties as environmental pollution or urban congestion is absurd, since to him they are "fake" problems, which Technique will soon overcome. In his view, a technological society, which is incompatible with true civilization, already dominates humankind. Although Ellul does not explicitly lay the blame for our plight on scientists and engineers, he calls Technique "an ensemble of rational and efficient practices," which of course describes the chief goals of science and engineering.

Samuel Florman, a proponent of human progress through science and engineering, attempts to refute all of the arguments advanced by Mumford, Ellul, and others writing in the same vein. There are defects, he admits, but these are due primarily to man's imperfection, "the irrepressible human will."

These selections, then, mirror the widespread diversity of prevalent opinions today concerned with technology and cultural change. Some of these will be voiced again in later selections, and new views on these three levels of discussion will be expressed. The reader should attempt to follow the logic of these various arguments, and to come to conclusions that can be supported.

BARRY COMMONER

Are We Really In Control?

*This article by biologist Barry Commoner is representative of a growing senti-
ment among scientists and others that we should no longer have implicit faith
that scientific discoveries and technological progress will bring us only benefits.
Commoner believes that the 1965 New York City blackout—an event that was
repeated in New York in 1977 and throughout France in 1978—demonstrated
that engineers are now creating systems that they do not fully understand. Similarly, nuclear tests in Nevada in the 1950s had consequences that scientists were
unable to foresee. Commoner does not reject science and technology, but he is
deeply concerned about their increasingly unpredictable effects. He wonders
whether science and technology have gotten out of hand.*

The age of innocent faith in science and technology may be over. We were
given a spectacular signal of this change on a night in November 1965.
On that night all electric power in an 80,000-square-mile area of the north-
eastern United States and Canada failed. The breakdown was a total surprise.
For hours engineers and power officials were unable to turn the lights on
again; for days no one could explain why they went out; even now no one can
promise that it won't happen again.

The failure knocked out a huge network which was supposed to shift electric power from areas with excess generating capacity to those facing a heavy
drain. But on that night the power grid worked against its intended purpose.

Instead of counteracting a local power failure, it spread the trouble out of control until the whole system was engulfed and dozens of cities were dark.

The trouble began with the failure of a relay which controlled the flow of electricity from the Sir Adam Beck No. 2 power plant in Queenston, Ontario, into one of its feeder lines. The remaining lines, unable to carry the extra load, shut down their own safety switches. With these normal exits blocked the plant's full power flowed back along the lines that tied the Queenston generators into the U.S.–Canadian grid. This sudden surge of power, traveling across New England, quickly tripped safety switches in a series of local power plants, shutting them down. As a result the New England region, which until then had been feeding excess electricity into the Consolidated Edison system in New York, drained power away from that city; under this strain the New York generators were quickly overloaded and their safety switches shut off. The blackout was then complete. The system had been betrayed by the very links that were intended to save local power plants from failure.

In one of the magazine reports of the great blackout, there is a photograph that tells the story with beautiful simplicity. It shows a scene in Consolidated Edison's Energy Control Center. Stretched purposefully across the photograph is an operational diagram of the New York power system; an intricate but neat network of connections, meters, and indicators symbolizing the calculated competence of this powerful machine. In the foreground, dwarfed by

Consolidated Edison's Energy Control Center

the diagrammatic system and in curious contrast to its firm and positive lines, is a group of very puzzled engineers. . . .

One man, however, if he had lived to see it, would not have been surprised by the great blackout—Norbert Wiener, the mathematician who did so much to develop cybernetics, the science which guides the design of complex electrical grids and their computerized controls. Cybernetics has produced electronic brains and all the other marvelous machines that now operate everything from election reports to steel plants; that have made the robot no longer a cartoon but a reality; that made the U.S.–Canadian power grid feasible.

Just six years before the blackout Dr. Wiener reviewed a decade of remarkable progress in the science which he helped to create. He reported at that time on the development of a new kind of automatic machine, a computer that had been programmed to play checkers. Engineers built into the electronic circuits a correct understanding of the rules of checkers and also a way of judging what moves were most likely to beat the computer's opponents. The computer made a record of its opponent's moves in the current and previous games. Then, at great speed, it calculated its opponent's most likely moves in any given situation and, having figured those out, adjusted its own game, move by move, to give itself the best chance of winning. The engineers designed a machine that not only knew how to play checkers but could learn from experience and actually improve its own game.

Dr. Wiener described the first results of the checkers tournaments between the computer and its programmers. The machine started out playing an accurate but uninspired game which was easy to beat. But after about ten or twenty hours of practice the machine got the hang of it, and from then on the human player usually lost and the machine won.

Dr. Wiener emphasized this point: Here was a machine designed by a man who built into it everything that it could do. Yet, because it could calculate complicated probabilities faster than the man could, the machine learned to play checkers against the man better than he could against the machine. Dr. Wiener concluded that it had become technically possible to build automatic machines that "most definitely escape from the complete effective control of the man who has made them."

The U.S.–Canadian power grid is just such a machine. By following the rules built into its design, the machine acted—before the engineers had time to understand and countermand it—in a way that went against their real wishes.

One month after the great blackout, there occurred in Salt Lake City, Utah, a little-noticed event that can take its place beside the power failure as a monument to the blunders which have begun to mar the accomplishments of modern science and technology. There, nine children from Washington

County, Utah, entered a hospital for tests to determine whether abnormal nodules in their thyroid glands were an indication of possible thyroid disease: nontoxic goiter, inflammation, benign or malignant tumors. Fifteen years earlier these children had been exposed to radioactive iodine produced by fallout from the nearby Nevada atomic test site.

It will be some time before anyone can tell whether the incidence of thyroid nodules in this group of children is statistically significant, and if so, whether the nodules are really due to fallout. But regardless of the outcome, the mere fact that health authorities felt compelled to look for an effect of fallout on the health of these children is itself a surprise.

The chain of events which brought the children into the hospital began in the 1950s when the AEC started a long series of nuclear explosions at its Nevada test site in the conviction that "... these explosives created no immediate or long-range hazard to human health outside the proving ground." But among the radioactive particles of the fallout clouds that occasionally escaped into the surrounding territory was the isotope iodine-131. As these clouds passed over the Utah pastures, iodine-131 was deposited on the grass; being widely spread, it caused no alarming readings on outdoor radiation meters. But dairy cows grazed these fields. As a result, iodine-131, generated in the mushroom cloud, drifted to Utah farms, was foraged by cows, passed to children in milk, and was gathered in high concentration in the children's thyroid glands. Here in a period of a few weeks the iodine-131 released its radiation. If sufficiently intense, such radiation passing through the thyroid cells may set off subtle changes which, though quiescent and hidden for years, eventually give rise to disease.

Like the Northeast blackout, this too is a chain reaction. Where the blackout reaction chain took minutes, the iodine-131 chain took days and in a sense years. But in both cases the process was over and the damage done before we understood what had happened.

Modern science, and the huge technological enterprises which it produces, represent the full flowering of man's understanding of nature. Scientific knowledge is our best guide to controlling natural forces. In this it has been magnificently successful; it is this success which has given us the marvels of modern electricity, and the tremendous power of nuclear bombs.

The power blackout and the Utah thyroid problem have cast a shadow—small, but deeply troubling—over the brilliance of these scientific successes. Is it possible that we do not know the full consequences of the new power grids and the new bombs? Are we really in control of the vast new powers that science has given us, or is there a danger that science is getting out of hand?

LEWIS MUMFORD

The Technique of Total Control

Social critic Lewis Mumford agrees with Commoner that current science and technology present real dangers, but he views the problem quite differently. For Mumford, technology is increasingly depriving man of freedom and of the possibility of self-realization. Science and technology, he warns, have created a "megatechnic system"—a massive technological complex in which control not only over nature but ultimately over man himself are the chief goals. There is no such thing as technological inevitability, Mumford concludes, but we must resort to cultural inventions to rid ourselves of the system.

Though we are too close to it to make a completely objective judgement, it has become obvious that our own culture has fallen into a dangerously unbalanced state, and is now producing warped and unbalanced minds. One part of our civilization—that dedicated to technology—has usurped authority over all the other components, geographical, biological, anthropological: indeed, the most frenetic advocates of this process are now proclaiming that the whole biological world is now being supplanted by technology, and that man will either become a willing creature of this technology or cease to exist.

Not merely does technology claim priority in human affairs: it places the demand for constant technological change above any considerations of its own efficiency, its own continuity, or even, ironically enough, its own capacity to survive. To maintain such a system, whose postulates contradict those that underlie all living organisms, it requires for self-protection absolute conformity by the human community; and to achieve that conformity it proposes to institute a system of total control, starting with the human organism itself, even before conception has taken place. The means for establishing this control is the ultimate gift of the megamachine; and without submergence in the subjective "myth of the machine," as omnipotent, omniscient, and omnicompetent, it would not already have advanced to the point it has now reached. . . .

The business of creating a limited, docile, scientifically conditioned human animal, completely adjusted to a purely technological environment, has kept pace with the rapid transformation of that environment itself: partly this has been effected . . . by re-enforcing conformity with tangible rewards, partly by denying any real opportunities for choices outside the range of the megatechnic system. American children, who, on statistical evidence, spend from three to six hours a day absorbing the contents of television, whose nursery songs are advertisements, and whose sense of reality is blunted by a world dominated by daily intercourse with Superman, Batman, and their monstrous relatives, will be able only by heroic effort to disengage themselves from this system sufficiently to recover some measure of autonomy. The megamachine has them under its remote control, conditioned to its stereotypes, far more effectively than the most authoritative parent. No wonder the first generation brought up under this tutelage faces an "identity crisis.". . .

In the final stage of technical development, as various science-fiction writers have been quick to perceive, the organized sciences will attempt to do directly, mainly by physical and chemical devices, what other human institutions—religion, morals, law—sought to do more indirectly, with only partial success, by exhortation, persuasion, or warning threat: namely, to transform the nature of man. Science confidently proposes to alter his potentialities at the source through genetic intervention and through further programming his existence so as to permit no unforeseen departures or rebellions. Radical alterations that kings and priests never succeeded in performing except by evisceration scientists now confidently propose to do on the living corpse by surgical alteration, chemotherapy, and electronic control. . . .

The decisive factor of safety in human development lies in the fact that man's many specific experimental errors and subjective aberrations have *not* been deliberately fixed in the genes. To an extent that no other species enjoys, each fresh generation shakes the genetic dice and rolls out fresh combinations, leaving it open to new human factors to repair past errors and embark on fresh experiments. Many mistakes have been made in the development of every known culture, and some of them, like war, slavery, and class exploitation, have seriously crippled human development. Yet none of these aberrations is so deeply embedded in the flesh that it is unalterable or immortal. If in future fresh human possibilities should be closed off, it would be because the dominant power system had deliberately closed them, in the very fashion that technocratic spokesmen advocate.

In so far as the illusion of technological inevitability is taken for an inescapable reality—*e.g.*, *"genetic control is bound to occur"*—this attitude only adds an inner compulsiveness to the many external compulsions imposed by the Power Complex. Such beliefs often prove self-fulfilling prophecies, and they make more probable the riveting together of a planetary megamachine. This superimposed power system, with its insistent compulsiveness and automatism, may prove in the end the gravest menace to man's own development. While the cultural inheritance is partly re-programmed from generation to generation, from culture to culture, and is modified even from hour to hour by the plans and acts of individual minds, genetic control might program man out of existence, and create a substitute homunculus: the fixed component of a humanly vacuous automatic system. By its cultural inventions the human species has up till now avoided such a fatal arrest.

JACQUES ELLUL

The Technological Order

Jacques Ellul, a French sociologist, takes the logic of the anti-technology stance to its ultimate conclusion. Ellul considers technology as now completely autonomous and self-determining, entirely out of human control. Technology, he writes, destroys human freedom and makes the search for ethical and spiritual values meaningless. The problem now, Ellul thinks, is not how to control technology, since that is impossible, but how best to live with it.

Technique has become the new and specific *milieu* in which man is required to exist, one which has supplanted the old *milieu,* viz., that of nature. This new technical *milieu* has the following characteristics:

1. It is artificial;
2. It is autonomous with respect to values, ideas, and the state;
3. It is self-determining in a closed circle. Like nature, it is a closed organization which permits it to be self-determinative independently of all human intervention;
4. It grows according to a process which is causal but not directed to ends;
5. It is formed by an accumulation of means which have established primacy over ends;
6. All its parts are mutually implicated to such a degree that it is impossible to separate them or to settle any technical problem in isolation....

Jacques Ellul, "The Technological Order," from *Technology and Culture* 3 (1962). Published by the University of Chicago Press, Chicago 60637. © 1962 by the Society for the History of Technology. All rights reserved. Reprinted by permission.

13

Since Technique has become the new *milieu,* all social phenomena are situated in it. It is incorrect to say that economics, politics, and the sphere of the cultural are influenced or modified *by* Technique; they are rather situated *in* it, a novel situation modifying all traditional social concepts. Politics, for example, is not modified by Technique as one factor among others which operate upon it; the political world is today *defined* through its relation to the technological society. Traditionally, politics formed a part of a larger social whole; at the present the converse is the case.

Technique comprises organizational and psycho-sociological techniques. It is useless to hope that the use of techniques of organization will succeed in compensating for the effects of techniques in general; or that the use of psycho-sociological techniques will assure mankind ascendancy over the technical phenomenon. In the former case, we will doubtless succeed in averting certain technically induced crises, disorders, and serious social disequilibrations; but this will but confirm the fact that Technique constitutes a closed circle. In the latter case, we will secure human psychic equilibrium in the technological *milieu* by avoiding the psycho-biologic pathology resulting from the individual techniques taken singly and thereby attain a certain happiness. But these results will come about through the *adaptation of human beings to the technical milieu.* Psycho-social techniques result in the *modification* of men in order to render them happily subordinate to their new environment, and by no means imply any kind of human domination over Technique.

The ideas, judgments, beliefs, and myths of the man of today have already been essentially modified by his technical *milieu.* It is no longer possible to reflect that on the one hand, there are techniques which may or may not have an effect on the human being; and, on the other, there is the human being himself who is to attempt to invent means to master his techniques and subordinate them to his own ends by *making a choice* among them. Choices and ends are both based on beliefs, sociological presuppositions, and myths which are a function of the technological society. Modern man's state of mind is completely dominated by technical values, and his goals are represented only by such progress and happiness as is to be achieved through techniques. Modern man in choosing is already incorporated within the technical process and modified in his nature by it. He is no longer in his traditional state of freedom with respect to judgment and choice.

To understand the problem posed to us, it is first of all requisite to disembarrass ourselves of certain fake problems.

We make too much of the disagreeable features of technical development, for example, urban over-crowding, nervous tension, air pollution, and so forth. I am convinced that all such inconveniences will be done away with by the ongoing evolution of Technique itself, and indeed, that it is only by means of such evolution that this can happen. The inconveniences we emphasize are always dependent on technical solutions, and it is only by means of techniques that they can be solved. This fact leads to the following two considerations:

1. Every solution to some technical inconvenience is able only to reinforce the system of techniques *in their ensemble;*

2. Enmeshed in a process of technical development like our own, the possibilities of human survival are better served by more technique than less, a fact which contributes nothing, however, to the resolution of the basic problem. . . .

What, then, is the real problem posed to men by the development of the technological society? It comprises two parts: 1. Is man able to remain master in a world of means? 2. Can a new civilization appear inclusive of Technique?

The answer to the first question, and the one most often encountered, seems obvious: Man, who exploits the ensemble of means, *is* the master of them. Unfortunately, this manner of viewing matters is purely theoretical and superficial. We must remember the autonomous character of Technique. We must likewise not lose sight of the fact that the human individual himself is to an ever greater degree the *object* of certain techniques and their procedures. He is the object of pedagogical techniques, psychotechniques, vocational guidance testing, personality and intelligence testing, industrial and group aptitude testing, and so on. In these cases (and in countless others) most men are treated as a collection of objects. But, it might be objected, these techniques are exploited by other men, and the exploiters at least remain masters. In a certain sense this is true; the exploiters *are* masters of the particular techniques they exploit. But, they, too, are subjected to the action of yet other techniques, as, for example, propaganda. Above all, they are spiritually taken over by the technological society; they believe in what they do; they are the most fervent adepts of that society. They themselves have been profoundly technicized. They never in any way affect to despise Technique, which to them is a thing good in itself. They never pretend to assign values to Technique, which to them is in itself an entity working out its own ends. They never claim to subordinate it to any value because for them Technique *is* value.

It may be objected that these individual techniques have as their end the best adaptation of the individual, the best utilization of his abilities, and, in the long run, his happiness. This, in effect, is the objective and the justification of all techniques. (One ought not, of course, to confound man's "happiness" with capacity for mastery with, say, freedom.) If the first of all values is happiness, it is likely that man, thanks to his techniques, will be in a position to attain to a certain state of this good. But happiness does not contain everything it is thought to contain, and *the absolute disparity between happiness and freedom* remains an ever real theme for our reflections. To say that man should remain *subject* rather than *object* in the technological society means two things, viz., that he be capable of giving direction and orientation to Technique, and that, to this end, he be able to master it. . . .

We must ask ourselves realistically and concretely just who is in a position to choose the values which give Technique its justification and to exert mastery over it. . . .

Can the *technician* himself assume mastery over Technique? The trouble here is that the technician is *always* a specialist and cannot make the slightest claim to have mastered any technique but his own. Those for whom Technique bears its meaning in itself will scarcely discover the values which lend meaning to what they are doing. They will not even look for them. The only thing they can do is to apply their technical specialty and assist in its refinement. They cannot *in principle* dominate the totality of the technical problem or envisage it in its global dimensions. *Ergo,* they are completely incapable of mastering it.

Can the *scientist* do it? There, if anywhere, is the greatest hope. Does not the scientist dominate our techniques? Is he not an intellectual inclined and fit to put basic questions? Unfortunately, we are obliged to re-examine our hopes here when we look at things as they are. We see quickly enough that the scientist is as specialized as the technician, as incapable of general ideas, and as much out of commission as the philosopher. Think of the scientists who, on one tack or another, have addressed themselves to the technical phenomenon: Einstein, Oppenheimer, Carrel.* It is only too clear that the ideas these gentlemen have advanced in the sphere of the philosophic or the spiritual are vague, superficial, and contradictory *in extremis.* They really ought to stick to warnings and proclamations, for as soon as they assay anything else, the other scientists and the technicians rightly refuse to take them seriously, and they even run the risk of losing their reputations as scientists.

Can the *politician* bring it off? In the democracies the politicians are subject to the wishes of their constituents who are primarily concerned with the happiness and well-being which they think Technique assures them. Moreover, the further we get on, the more a conflict shapes up between the politicians and the technicians. We cannot here go into the matter which is just beginning to be the object of serious study. But it would appear that the power of the politician is being (and will continue to be) outclassed by the power of the technician in modern states. Only dictatorships can impose their will on technical evolution. But, on the one hand, human freedom would gain nothing thereby, and, on the other, a dictatorship thirsty for power has no

---

*Editors' note: Alexis Carrel (1873–1944), a French biologist and surgeon, won the Nobel prize in medicine for his work in organ and tissue transplants and in blood-vessel surgery.

recourse at all but to push toward an excessive development of various techniques at its disposal.

Any of us? An individual can doubtless seek the soundest attitude to dominate the techniques at his disposal. He can inquire after the values to impose on techniques in his use of them, and search out the way to follow in order to remain a man in the fullest sense of the word within a technological society. All this is extremely difficult, but it is far from being useless, since it is apparently the only solution presently possible. But the individual's efforts are powerless to resolve in any way the technical problem in its universality; to accomplish this would mean that *all* men adopt the same values and the same behavior.

The second real problem posed by the technological society is whether or not a new civilization can appear which is inclusive of Technique. The elements of this question are as difficult as those of the first. It would obviously be vain to deny all the things that can contribute something useful to a new civilization: security, ease of living, social solidarity, shortening of the work week, social security, and so forth. But a civilization in the strictest sense of the term is not brought into being by all these things.

A threefold contradiction resides between civilization and Technique of which we must be aware if we are to approach the problem correctly:

1. The technical world is the world of material things; it is put together out of material things and with respect to them. When Technique displays any interest in man, it does so by converting him into a material object. The supreme and final authority in the technological society is fact, at once ground and evidence. And when we think on man as he exists in this society it can only be as a being immersed in a universe of objects, machines, and innumerable material things. Technique indeed guarantees him such material happiness as material objects can. But, the technical society is not, and cannot be, a genuinely humanist society since it puts in first place not man but material things. It can only act on man by lessening him and putting him in the way of the quantitative. The radical contradiction referred to exists between technical perfection and human development because such perfection is only to be achieved through quantitative development and necessarily aims exclusively at what is measurable. Human excellence, on the contrary, is of the domain of the qualitative and aims at what is not measurable. Space is lacking here to argue the point that spiritual values cannot evolve as a function of material improvement. The transition from the technically quantitative to the humanly qualitative is an impossible one. In our times, technical growth monopolizes all human forces, passions, intelligences, and virtues in such a way that it is in practice nigh impossible to seek and find anywhere any distinctively human

excellence. And if this search is impossible, there cannot be any civilization in the proper sense of the term.

2. Technical growth leads to a growth of power in the sense of technical means incomparably more effective than anything ever before invented, power which has as its object only power, in the widest sense of the word. The possibility of action becomes limitless and absolute. For example, we are confronted for the first time with the possibility of the annihilation of all life on earth, since we have the means to accomplish it. In *every* sphere of action we are faced with just such absolute possibilities. Again, by way of example, governmental techniques, which amalgamate organizational, psychological, and police techniques, tend to lend to government absolute powers. And here I must emphasize a great law which I believe to be essential to the comprehension of the world in which we live, viz., that when power becomes absolute, values disappear. When man is able to accomplish anything at all, there is no value which can be proposed to him; when the means of action are absolute, no goal of action is imaginable. Power eliminates, in proportion to its growth, the boundary between good and evil, between the just and the unjust. We are familiar enough with this phenomenon in totalitarian societies. The distinction between good and evil disappears beginning with the moment that the ground of action (for example the *raison d'état,* or the instinct of the proletariat) claims to have absolute power and thus to incorporate *ipso facto* all value. Thus it is that the growth of technical means tending to absolutism forbids the appearance of values, and condemns to sterility our search for the ethical and the spiritual. Again, where Technique has place, there is the implication of the impossibility of the evolution of civilization.

3. The third and final contradiction is that Technique can never engender freedom. Of course, Technique frees mankind from a whole collection of ancient constraints. It is evident, for example, that it liberates him from the limits imposed on him by time and space; that man, through its agency, is free (or at least tending to become free) from famine, excessive heat and cold, the rhythms of the seasons, and from the gloom of night; that the race is freed from certain social constraints through its commerce with the universe, and from its intellectual limitations through its accumulation of information. But is this what it means really to be free? Other constraints as oppressive and rigorous as the traditional ones are imposed on the human being in today's technological society through the agency of Technique. New limits and technical oppressions have taken the place of the older, natural constraints, and we certainly cannot aver that much has been gained. The problem is deeper— the operation of Technique is the contrary of freedom, an operation of determinism and necessity. Technique is an ensemble of rational and efficient

practices; a collection of orders, schemas, and mechanisms. All of this expresses very well a necessary order and a determinate process, but one into which freedom, unorthodoxy, and the sphere of the gratuitous and spontaneous cannot penetrate. All that these last could possibly introduce is discord and disorder. The more technical actions increase in society, the more human autonomy and initiative diminish. The more the human being comes to exist in a world of ever increasing demands (fortified with technical apparatus possessing its own laws to meet these demands), the more he loses any possibility of free choice and individuality in action. This loss is greatly magnified by Technique's character of self-determination, which makes its appearance among us as a kind of fatality and as a species of perpetually exaggerated necessity. But where freedom is excluded in this way, an authentic civilization has little chance. Confronted in this way by the problem, it is clear to us that no solution can exist, in spite of the writings of all the authors who have concerned themselves with it. They all make an unacceptable premise, viz., rejection of Technique and return to a pre-technical society. One may well regret that some value or other of the past, some social or moral form, has disappeared; but, when one attacks the problem of the technical society, one can scarcely make the serious claim to be able to revive the past, a procedure which, in any case, scarcely seems to have been, globally speaking, much of an improvement over the human situation of today. All we know with certainty is that it was different, that the human being confronted other dangers, errors, difficulties, and temptations. Our duty is to occupy ourselves with the dangers, errors, difficulties, and temptations of modern man in the modern world. All regret for the past is vain; every desire to revert to a former social stage is unreal. There is no possibility of turning back, of annulling, or even of arresting technical progress. What is done is done. It is our duty to find our place in our present situation and in no other. Nostalgia has no survival value in the modern world and can only be considered a flight into dreamland.

Unbounded faith in American science and engineering, 1952. This cartoon from the
Chicago Tribune *reflected the optimism that marked the centennial of the first national*
engineering society in the United States, founded in 1852.

SAMUEL C. FLORMAN

In Praise of Technology

Samuel C. Florman is a vigorous but not uncritical champion of technology. He takes up the charges of the anti-technologists and contests them point by point. Technology, he argues, is not a "thing" controlling man. Nor is there an elite technocratic group taking over society. Man is not being cut off from nature, nor is he being alienated by increasing industrialization and automation. For Florman, the central current problem is not the growth of technology but the accelerating demand for the benefits of technology and our inability to satisfy that demand. The solution of the problem is not to restrict technology but to make its benefits more widely available.

A generation ago most people believed, without doubt or qualification, in the beneficial effects of technological progress. Books were written hailing the coming of an age in which machines would do all the onerous work, and life would become increasingly utopian.

Today there is a growing belief that technology has escaped from human control and is making our lives intolerable. Thus do we dart from one false myth to another, ever impressed by glib and simple-minded prophets.

Hostility to technology has become such a familiar staple of our reading fare that rarely do we stop to consider how this new doctrine has so quickly and firmly gained its hold upon us. I believe that critical scrutiny of this strange and dangerous phenomenon is very much overdue.

The founding father of the contemporary antitechnological movement is Jacques Ellul, whose book, *The Technological Society,* was published in France in 1954, and in the United States ten years later. When it appeared here, Thomas Merton, writing in *Commonweal,* called it "one of the most important books of this mid-century." In *Book Week* it was labeled "an essay that will likely rank among the most important, as well as tragic, of our time."

Ellul's thesis is that "technique" has become a Frankenstein monster that cannot be controlled. By technique he means not just the use of machines, but all deliberate and rational behavior, all efficiency and organization. Man created technique in prehistoric times out of sheer necessity, but then the *bourgeoisie* developed it in order to make money, and the masses were converted because of their interest in comfort. The search for efficiency has become an end in itself, dominating man and destroying the quality of his life.

The second prominent figure to unfurl the banner of antitechnology was Lewis Mumford. His conversion was particularly significant since for many years he had been known and respected as the leading historian of technology. His massive *Myth of the Machine* appeared in 1967 (Part I: *Technics and Human Development*) and in 1970 (Part II: *The Pentagon of Power*). Each volume in turn was given front-page coverage in *The New York Times Sunday Book Review.* On the first page of *Book World* a reviewer wrote, "Hereafter it will be difficult indeed to take seriously any discussion of our industrial ills which does not draw heavily upon this wise and mighty work." The reviewer was Theodore Roszak, who, as we shall see, was soon to take his place in the movement.

The next important convert was René Dubos, a respected research biologist and author. In *So Human an Animal,* published in 1968, Dubos started with the biologist's view that man is an animal whose basic nature was formed during the course of his evolution, both physical and social. This basic nature, molded in forests and fields, is not suited to life in a technological world. Man's ability to adapt to almost any environment has been his downfall, and little by little he has accommodated himself to the physical and psychic horrors of modern life. Man must choose a different path, said Dubos, or he is doomed. This concern for the individual, living human being was just what was needed to flesh out the abstract theories of Ellul and the historical analyses of Mumford. *So Human an Animal* was awarded the Pulitzer Prize, and quickly became an important article of faith in the antitechnology crusade.

In 1970 everybody was talking about Charles A. Reich's *Greening of America.* In paperback it sold more than a million copies within a year. Reich, a law professor at Yale, spoke out on behalf of the youthful counterculture and its dedication to a liberating consciousness raising. Theodore Roszak's *Where the Wasteland Ends* appeared in 1972 and carried Reich's theme just a little further, into the realm of primitive spiritualism. Roszak, like Reich, is a college professor. Unlike *The Greening of America,* his work did not capture a mass audience. But it seemed to bring to a logical climax the antitechnological movement started by Ellul. As the reviewer in *Time* magazine said, "he has brilliantly summed up once and for all the New Arcadian criticism of what he calls 'postindustrial society.' "

There have been many other contributors to the antitechnological movement, but I think that these five—Ellul, Mumford, Dubos, Reich, and Roszak—have been pivotal. . . .

They are united in their hatred and fear of technology, and surprisingly unanimous in their treatment of several key themes. . . .

1. Technology is a "thing" or a force that has escaped from human control and is spoiling our lives.

2. Technology forces man to do work that is tedious and degrading.

3. Technology forces man to consume things that he does not really desire.

4. Technology creates an elite class of technocrats, and so disenfranchises the masses.

5. Technology cripples man by cutting him off from the natural world in which he evolved.

6. Technology provides man with technical diversions which destroy his existential sense of his own being. . . .

The antitechnologists repeatedly contrast our abysmal technocracy with three cultures that they consider preferable: the primitive tribe, the peasant community, and medieval society. . . .

Recognizing that we cannot return to earlier times, the antitechnologists nevertheless would have us attempt to recapture the satisfactions of these vanished cultures. In order to do this, what is required is nothing less than *a change in the nature of man.* The antitechnologists would probably argue that the change they seek is really a return to man's *true* nature. But a change from man's present nature is clearly their fondest hope. . . .

In the often-repeated story, Samuel Johnson and James Boswell stood talking about Berkeley's theory of the nonexistence of matter. Boswell observed that although he was satisfied that the theory was false, it was impossible to refute it. "I never shall forget," Boswell tells us, "the alacrity with which Johnson answered, striking his foot with mighty force against a large stone, till he rebounded from it—'I refute it *thus.*' "

The ideas of the antitechnologists arouse in me a mood of exasperation similar to Dr. Johnson's. Their ideas are so obviously false, and yet so persuasive and widely accepted, that I fear for the common sense of us all.

The impulse to refute this doctrine with a Johnsonian kick is diminished by the fear of appearing simplistic. So much has been written about technology by so many profound thinkers that the nonprofessional cannot help but be intimidated. Unfortunately for those who would dispute them, the antitechnologists are masters of prose and intellectual finesse. To make things worse, they display an aesthetic and moral concern that makes the defender of technology appear like something of a philistine. To make things worse yet, many defenders of technology are indeed philistines of the first order.

Yet the effort must be made. If the antitechnological argument is allowed to stand, the engineer is hard pressed to justify his existence. More important, the implications for society, should antitechnology prevail, are most disquieting. For, at the very core of antitechnology, hidden under a veneer of aesthetic sensibility and ethical concern, lies a yearning for a totalitarian society.

The first antitechnological dogma to be confronted is the treatment of technology as something that has escaped from human control. It is understandable that sometimes anxiety and frustration can make us feel this way. But sober thought reveals that technology is not an independent force, much less a thing, but merely one of the types of activities in which people engage. Furthermore, it is an activity in which people engage because they choose to do so. The choice may sometimes be foolish or unconsidered. The choice may be forced upon some members of society by others. But this is very different from the concept of technology *itself* misleading or enslaving the populace.

Philosopher Daniel Callahan has stated the case with calm clarity:

> At the very outset we have to do away with a false and misleading dualism, one which abstracts man on the one hand and technology on the other, as if the two were quite separate kinds of realities. I believe that there is no dualism inherent here. Man is by nature a technological animal; to be human is to be technological. If I am correct in that judgment, then there is no room for a dualism at all. Instead, we should recognize that when we speak of technology, this is another way of speaking about man himself in one of his manifestations.

Although to me Callahan's statement makes irrefutable good sense, and Ellul's concept of technology as being a thing-in-itself makes absolutely no sense, I recognize that this does not put an end to the matter, any more than Samuel Johnson settled the question of the nature of reality by kicking a stone.

It cannot be denied that, in the face of the excruciatingly complex problems with which we live, it seems ingenuous to say that men invent and manufacture things because they want to, or because others want them to and reward them accordingly. When men have engaged in technological activities, these activities appear to have had *consequences,* not only physical but also intellectual, psychological, and cultural. Thus, it can be argued, technology is *deterministic.* It causes other things to happen. Someone invents the automobile, for example, and it changes the way people think as well as the way they act. It changes their living patterns, their values, and their expectations in ways that were not anticipated when the automobile was first introduced. Some of the changes appear to be not only unanticipated but undesired. Nobody wanted traffic jams, accidents, and pollution. Therefore, technological advance seems to be independent of human direction. Observers of the social scene become so chagrined and frustrated by this turn of events—and its thousand equivalents—that they turn away from the old common-sense explanations, and become entranced by the demonology of the antitechnologists. . . .

In addition to confounding rational discourse, the demonology outlook of the antitechnologists discounts completely the integrity and intelligence of the ordinary person. Indeed, pity and disdain for the individual citizen is an essential aspect of antitechnology. It is central to the next two dogmas, which hold that technology forces man to do tedious and degrading work, and then forces him to consume things that he does not really desire.

Is it ingenuous, again, to say that people work, not to feed some monstrous technological machine, but, as since time immemorial, to feed themselves? We all have ambivalent feelings toward work, engineers as well as antitechnologists. We try to avoid it, and yet we seem to require it for our emotional well-being. This dichotomy is as old as civilization. A few wealthy people are bored because they are not required to work, and a lot of ordinary people grumble because they have to work hard.

The antitechnologists romanticize the work of earlier times in an attempt to make it seem more appealing than work in a technological age. But their idyllic descriptions of peasant life do not ring true. Agricultural work, for all its appeal to the intellectual in his armchair, is brutalizing in its demands. Factory and office work is not a bed of roses either. But given their choice, most people seem to prefer to escape from the drudgery of the farm. This fact fails to impress the antitechnologists, who prefer their sensibilities to the choices of real people.

As for the technological society forcing people to consume things that they do not want, how can we respond to this canard? Like the boy who said, "Look, the emperor has no clothes," one might observe that the consumers who buy cars and electric can openers could, if they chose, buy oboes and oil paints, sailboats and hiking boots, chess sets and Mozart records. Or, if they have no personal "increasing wants," in Mumford's phrase, could they not help purchase a kidney machine which would save their neighbor's life? If people are vulgar, foolish, and selfish in their choice of purchases, is it not the worst sort of cop-out to blame this on "the economy," "society," or "the suave technocracy"? Indeed, would not a man prefer being called vulgar to being told he has no will with which to make choices of his own?

Which brings us to the next tenet of antitechnology, the belief that a technocratic elite is taking over control of society. Such a view at least avoids the logical absurdity of a demon technology compelling people to act against their own interests. It does not violate our common sense to be told that certain people are taking advantage of other people. But is it logical to claim that exploitation increases as a result of the growth of technology?

Upon reflection, this claim appears to be absolutely without foundation. When camel caravans traveled across the deserts, there were a few merchant entrepreneurs and many disenfranchised camel drivers. From earliest historical times, peasants have been abused and exploited by the nobility. Bankers, merchants, landowners, kings, and assorted plunderers have had it good at the expense of the masses in practically every large social group that has ever been (not just in certain groups like pyramid-building Egypt, as Mumford contends). Perhaps in small tribes there was less exploitation than that which developed in large and complex cultures, and surely technology played a role in that transition. But since the dim, distant time of that initial transition, it simply is not true that advances in technology have been helpful to the Establishment in increasing its power over the masses.

In fact, the evidence is all the other way. In technologically advanced societies, there is more freedom for the average citizen than there was in earlier ages. There has been continuing apprehension that new technological achievements *might* make it possible for governments to tyrannize the citizenry with Big Brother techniques. But, in spite of all the newest electronic gadgetry, governments are scarcely able to prevent the antisocial actions of criminals, much less control every act of every citizen. Hijacking, technically ingenious robberies, computer-aided embezzlements, and the like, are evidence that the outlaw is able to turn technology to his own advantage, often more adroitly than the government. The FBI has admitted that young revolutionaries are almost impossible to find once they go "underground." The rebellious individual is more than holding his own.

Exploitation continues to exist. That is a fact of life. But the antitechnologists are in error when they say that it has increased in extent or intensity because of technology. In spite of their extravagant statements, they cannot help but recognize that they are mistaken, statistically, at least. Reich is wrong when he says that "decisions are made by experts, specialists, and professionals safely insulated from the feelings of the people." (Witness changes in opinion, and then in legislation, concerning abortion, divorce, and pornography.) Those who were slaves are now free. Those who were disenfranchised can now vote. Rigid class structures are giving way to frenetic mobility. The barons and abbots and merchant princes who treated their fellow humans like animals, and convinced them that they would get their reward in heaven, would be incredulous to hear the antitechnologists theorize about how technology has brought about an increase in exploitation. We need only look at the underdeveloped nations of our present era to see that exploitation is not proportionate to technological advance. If anything, the proportion is inverse. . . .

Next we must confront the charge that technology is cutting man off from his natural habitat, with catastrophic consequences. It is important to point out that if we are less in touch with nature than we were—and this can hardly be disputed—then the reason does not lie exclusively with technology. Technology could be used to put people in very close touch with nature, if that is what they want. Wealthy people could have comfortable abodes in the wilderness, could live among birds in the highest jungle treetops, or even commune with fish in the ocean depths. But they seem to prefer penthouse apartments in New York and villas on the crowded hills above Cannes. Poorer people could stay on their farms on the plains of Iowa, or in their small towns in the hills of New Hampshire, if they were willing to live the spare and simple life. But many of them seem to tire of the loneliness and the hard physical labor that goes with rusticity, and succumb to the allure of the cities.

It is Roszak's lament that "the malaise of a Chekhov play" has settled upon daily life. He ignores the fact that the famous Chekhov malaise stems in no small measure from living in the country. "Yes, old man," shouts Dr. Astrov at Uncle Vanya, "in the whole district there were only two decent, well-educated men: you and I. And in some ten years the common round of the trivial life here has swamped us, and has poisoned our life with its putrid vapours, and made us just as despicable as all the rest." There is tedium in the countryside, and sometimes squalor.

Nevertheless, I personally enjoy being in the countryside or in the woods, and so feel a certain sympathy for the antitechnologists' views on this subject. But I can see no evidence that frequent contact with nature is *essential* to human well-being, as the antitechnologists assert. Even if the human species owes much of its complexity to the diversity of the natural environment, why must man continue to commune with the landscapes in which he evolved? Millions of people, in ages past as well as present, have lived out their lives in city environs, with very little if any contact with "nature." Have they lived lives inherently inferior because of this? Who would be presumptuous enough to make such a statement? . . .

The next target of the antitechnologists is Everyman at play. It is particularly important to antitechnology that popular hobbies and pastimes be discredited, for leisure is one of the benefits generally assumed to follow in the wake of technological advances. The theme of modern man at leisure spurs the antitechnologists to derision. . . .

In their consideration of recreation activities, the antitechnologists disdain to take into account anything that an actual participant might feel. For even when the ordinary man considers himself happy—at a ball game or a vacation camp, watching television or listening to a jukebox, playing with a pinball machine or eating hot dogs—we are told that he is only being fooled into *thinking* that he is happy.

It is strategically convenient for the antitechnologists to discount the expressed feelings of the average citizen. It then follows that (1) those satisfactions which are attributed to technology are illusory, and (2) those dissatisfactions which are the fault of the individual can be blamed on technology, since the individual's choices are made under some form of hypnosis. It is a can't-lose proposition.

Under these ground rules, how can we argue the question of what constitutes the good life? . . . The antitechnologists have every right to be gloomy, and have a bounden duty to express their doubts about the direction our lives are taking. But their persistent disregard of the average person's sentiments is a crucial weakness in their argument—particularly when they ask us to consider the "real" satisfactions that they claim ordinary people experienced in other cultures of other times.

It is difficult not to be seduced by the antitechnologists' idyllic elegies for past cultures. We all are moved to reverie by talk of an arcadian golden age. But when we awaken from this reverie, we realize that the antitechnologists have diverted us with half-truths and distortions. The harmony which the antitechnologists see in primitive life, anthropologists find in only certain tribes. Others display the very anxiety and hostility that antitechnologists blame on technology—as why should they not, being almost totally vulnerable to every passing hazard of nature, beast, disease, and human enemy? As for the peasant, was he "foot-free," "sustained by physical work," with a capacity for a "nonmaterial existence"? Did he crack jokes with every passerby? Or was he brutal and brutalized, materialistic and suspicious, stoning errant women and hiding gold in his mattress? And the Middle Ages, that dimly remembered time of "moral judgment," "equilibrium," and "common aspirations." Was it not also a time of pestilence, brigandage, and public tortures? "The chroniclers themselves," admits a noted admirer of the period (J. Huizinga), tell us "of covetousness, of cruelty, of cool calculation, of well-understood self-interest. . . ." The callous brutality, the unrelievable pain, the ever-present threat of untimely death for oneself (and worse, for one's children) are the realities with which our ancestors lived and of which the antitechnologists seem totally oblivious.

It is not my intention to assert that, because we live longer and in greater physical comfort than our forebears, life today is better than it ever was. It is this sort of chamber of commerce banality that has driven so many intellectuals into the arms of the antitechnological movement. Nobody is satisfied that we are living in the best of all possible worlds.

Part of the problem is the same as it has always been. Men are imperfect, and nature is often unkind, so that unhappiness, uncertainty, and pain are perpetually present. From the beginning of recorded time we find evidence of despair, melancholy, and ennui. We find also an abundance of greed, treachery, vulgarity, and stupidity. Absorbed as we are in our own problems, we tend to forget how replete history is with wars, feuds, plagues, fires, massacres, tortures, slavery, the wasting of cities, and the destruction of libraries. As for ecology, over huge portions of the earth men have made pastures out of forests, and then deserts out of pastures. In every generation prophets, poets, and politicians have considered their contemporary situation uniquely distressing, and have looked about for something—or someone—to blame. The antitechnologists follow in this tradition, and, in the light of history, their condemnation of technology can be seen to be just about as valid as the Counter-Reformation's condemnation of witchcraft.

But it will not do to say *plus ça change plus c'est la même chose,** and let it go at that. We do have some problems that are unique in degree if not in kind, and in our society a vague, generalized discontent appears to be more widespread than it was just a generation ago. *Something* is wrong, but what? . . .

Our contemporary problem is distressingly obvious. We have too many people wanting too many things. This is not caused by technology; it is a consequence of the type of creature that man is. There are a few people holding back, like those who are willing to do without disposable bottles, a few people turning back, like the young men and women moving to the counterculture communes, and many people who have not gotten started because of crushing poverty and ignorance. But the vast majority of people in the world want to move forward, whatever the consequences. Not that they are lemmings. They are wary of revolution and anarchy. They are increasingly disturbed by crowding and pollution. Many of them recognize that "progress" is not necessarily taking them from worse to better. But whatever

---

*Editors' note: The more things change, the more they are the same.

their caution and misgivings, they are pressing on with a determination that is awesome to behold. . . .

Our blundering, pragmatic democracy may be doomed to fail. The increasing demands of the masses may overwhelm us, despite all our resilience and ingenuity. In such an event we will have no choice but to change. The Chinese have shown us that a different way of life is possible. However, we must not deceive ourselves into thinking that we can undergo such a change, or maintain such a society, without the most bloody upheavals and repressions.

We are all frightened and unsure of ourselves, in need of good counsel. But where we require clear thinking and courage, the antitechnologists offer us fantasies and despair. Where we need an increase in mutual respect, they exhibit hatred for the powerful and contempt for the weak. The times demand more citizen activism, but they tend to recommend an aloof disengagement. We surely could use a sense of humor, but they are in the grip of an unrelenting dolefulness. Nevertheless, the antitechnologists have managed to gain a reputation for kindly wisdom.

This reputation is not entirely undeserved, since they do have many inspiring and interesting things to say. Their sentiments about nature, work, art, spirituality, and many of the good things in life, are generally splendid and difficult to quarrel with. Their ecological concerns are praiseworthy, and their cries of alarm have served some useful purpose. In sum, the antitechnologists are good men, and they mean well.

But, frightened and dismayed by the unfolding of the human drama in our time, yearning for simple solutions where there can be none, and refusing to acknowledge that the true source of our problems is nothing other than the irrepressible human will, they have deluded themselves with the doctrine of antitechnology. It is a hollow doctrine, the increasing popularity of which adds the dangers inherent in self-deception to all of the other dangers we already face.

PART TWO

Technology's Effects

Introduction

Many authors who write about the immediate and urgent problems of twentieth-century civilization make certain assumptions regarding technology that are plausible but difficult to prove. One is that advancing technology is creating unprecedented cultural change. Another is that modern technology is so incredibly complex that it baffles even experts. A third is that machines are making humans obsolete.

In this section, we shall explore the bases of these assumptions by considering three major effects of technology: cultural change, the development of complexity, and the transformation of work processes.

It is possible that the influence of advancing technology on culture was viewed with mixed emotions of enthusiasm and apprehension in all ages. Drucker, Ashton, and Bronowski take up this question of the effects of innovations on culture. Drucker writes about the agricultural revolution 7,000 years ago that marked the emergence of what we now think of as civilization. At that time, advancing technology created the need for new social and political institutions, and various cultures met the challenges in different ways. Current innovations, he declares, also call for institutional and societal changes; our task is to be certain that the new institutions will be appropriate for our needs.

Moving forward in time, Ashton describes the effects of another great upheaval involving technology—the Industrial Revolution. A number of factors, he shows, combined to effect this massive change in English society, which occurred during the reign of George III (1760–1820). Customs, privileges, and traditions were swept away, and new cultural patterns came into existence. People then were aware of the changes that were occurring, as we are today, but they did not recognize their magnitude. Perhaps we are now more alert to the fact of change than Englishmen were in that period.

Bronowski sees cultural changes effected by technology not only as natural but as beneficial. A stabilized environment, he thinks, would spell intellectual stagnation. Cultural changes, he believes, cause the evolution of man, not in

35

the sense of genetic rearrangement, but instead in the increase of skills and in the achievement of more intelligent behavior.

We may also wonder whether people in every era see the appearance of novel technical devices as evidence of growing complexity, and look upon them with mixed feelings of surprise, wonder, suspicion, and fear. In the early nineteenth century, for example, organized bands of handicraft workers called Luddites pillaged factories in England and destroyed the new textile machinery that was throwing them out of work. In contrast, at the time the telephone, the electric light, and the radio were invented, people considered them as marvels. In the 1920s, General Electric's industrial research laboratory was called the "House of Magic" because of the new inventions made there.

The selection by Carlo Cipolla gives us a glimmer of understanding of the effect on people, both noble and common, of the introduction in the Middle Ages of a complex device, the weight-driven clock. We can compare and contrast the effects of this invention with one of the major innovations of our time, the computer, which Lewis Mumford and Herbert Simon examine in the following selections.

Few people, Cipolla writes, could understand the elaborate movements of such great early clocks as that erected in the Strasbourg cathedral. Gradually, further innovations made the mechanisms of clocks and watches even more intricate and improved their accuracy. People became accustomed to their complex mechanisms, which only specialists could make or repair. But the most important effects were cultural and occurred over a considerable period of time. Clocks and watches altered people's behavior by gradually changing habits and customs. Regular hours of work were established, and the clock, as a symbol of regularity, even influenced philosophical theories about the universe.

Mumford views the computer as a menace, which will enormously aid the technocratic establishment, or the "Megamachine," to invade our privacy and control our lives. Simon disputes this charge, stating that with proper precautions the computer will be no more of a threat to privacy than previously existing technology. From Simon's account, we can already see some of the immediate effects of computers in changing business practices and industrial work processes. The computer, he says, is so complex that many think about it as a kind of artificial intelligence. But should the complexity concern us as much as the cultural changes that it is certain to bring about?

Complexity necessitates specialization, which gives rise to classes of experts. Cipolla mentions the talented Bolognese technicians, who designed the complicated clockwork mechanisms of the fifteenth century. The selections by Simon Ramo, Theodore Roszak, and Victor Ferkiss treat the experts of our own times. The employment of groups of experts in a "systems approach," Ramo writes, is the only rational way, indeed the only way, to solve urgent social problems. Ramo paints a rosy picture, which Roszak repudiates entirely. The experts, Roszak asserts, are the ones creating our problems, and it would be folly to place our future in the hands of these bureaucrats, whose only goal is to consolidate their power. Ferkiss thinks that the computer presents an opportunity to deliver us from this impasse. The bureaucratic power of the experts, he writes, came into existence to protect individuals from arbitrary rule. The capabilities of the computer, properly applied, can now rid us of bureaucratic domination by permitting local autonomy and citizen participation in decision-making. Thus, Ferkiss and Mumford have entirely different views on how the computer will influence cultural change.

The final selections by Lynn White, Robert Guest, Gary Brynner and Dan Clark, and Marshall McLuhan treat the effects of technological innovation on work. White describes the new attitudes toward work that emerged from the monasteries in the Middle Ages. The new ethic incorporated the belief that manual labor was not degrading, but also that it should not be monotonous drudgery. It was in keeping with the new ethic that machinery, powered by water wheels, was introduced in the twelfth and thirteenth centuries. By reducing arduous labor, this machinery improved the worker's lot.

With the beginnings of mass production, engineers sought to increase work efficiency using time and motion studies. Guest describes the advent of scientific management and tells how its adaptation to the assembly line resulted in repetitive monotony, dehumanizing the worker. Further analyses showed the defects of scientific management and caused modification of its worst features. Guest believes, however, that the emphasis on efficiency made the regimentation of work inevitable. The selection by Brynner and Clark, two auto workers, accentuates Guest's description of the effects of the assembly line and further stresses that many individual workers now find their jobs meaningless.

Technological innovation next led to the automation of certain types of work, in which the jobs of some workers were eliminated entirely. McLuhan predicts that automation is the wave of the future, and will eventually free people from having to do any manual labor whatsoever. Automation, he thinks, will be beneficial in the long run. It will create the necessity for more advanced training and higher education, and it will permit people to enjoy more rewarding lives.

Taken together, these selections should enable the reader to begin to frame certain conclusions. Attempts should be made to answer such questions as the following: Have the innovations of our era had greater impact on our lives than those of the past did on the lives of our ancestors? Are the cultural changes that we presently perceive merely a continuation of the process of change that has occurred through history, or do they mark a radical and revolutionary break? Is Bronowski's conclusion that cultural evolution resulting from innovation is beneficial to human life sound or misguided?

PETER F. DRUCKER

The First Technological Revolution and Its Lessons

While adherents of the anti-technology movement see only tragedy and injustice stemming from technological development, others view it as a source of human and social progress. Peter Drucker, a social scientist, examines what he terms the "first technological revolution," and finds that many of our contemporary social and political institutions originated during that period. While praising the achievements brought about by technological innovation, Drucker avoids the ingenuous belief that technology brings only benefits. We must be certain, he concludes, that the social innovations that new technology brings will serve human ends.

A ware that we are living in the midst of a technological revolution, we are becoming increasingly concerned with its meaning for the individual and its impact on freedom, on society, and on our political institutions. Side by side with messianic promises of utopia to be ushered in by technology, there are the most dire warnings of man's enslavement by technology, his

By permission from Peter F. Drucker, "The First Technological Revolution and Its Lessons." *Technology and Culture*, 1966.

alienation from himself and from society, and the destruction of all human and political values.

Tremendous though today's technological explosion is, it is hardly greater than the first great revolution technology wrought in human life seven thousand years ago when the first great civilization of man, the irrigation civilization, established itelf. First in Mesopotamia, and then in Egypt and in the Indus Valley, and finally in China, there appeared a new society and a new polity: the irrigation city, which then rapidly became the irrigation empire. No other change in man's way of life and in his making a living, not even the changes under way today, so competely revolutionized human society and community. In fact, the irrigation civilizations were the beginning of history, if only because they brought writing.

The age of the irrigation civilization was pre-eminently an age of technological innovation. Not until a historical yesterday, the eighteenth century, did technological innovations emerge which were comparable in their scope and impact to those early changes in technology, tools, and processes. Indeed, the technology of man remained essentially unchanged until the eighteenth century insofar as its impact on human life and human society is concerned.

But the irrigation civilizations were not only one of the great ages of technology. They represent also mankind's greatest and most productive age of social and political innovation. The historian of ideas is prone to go back to ancient Greece, to the Old Testament prophets, or to the China of the early dynasties for the sources of the beliefs that still move men to action. But our fundamental social and political institutions antedate political philosophy by several thousand years. They all were conceived and established in the early dawn of the irrigation civilizations. Anyone interested in social and governmental institutions and in social and political processes will increasingly have to go back to those early irrigation cities. And, thanks to the work of archeologists and linguists during the last fifty years, we increasingly have the information, we increasingly know what the irrigation civilizations looked like, we increasingly can go back to them for our understanding both of antiquity and of modern society. For essentially our present-day social and political institutions, practically without exception, were then created and established. Here are a few examples.

1. The irrigation city first established government as a distinct and permanent institution. It established an impersonal government with a clear hierarchical structure in which very soon there arose a genuine bureaucracy— which is, of course, what enabled the irrigation cities to become irrigation empires.

Even more basic: the irrigation city first conceived of man as a citizen. It had to go beyond the narrow bounds of tribe and clan and had to weld people of very different origins and blood into one community. This required the first supra-tribal deity, the god of the city. It also required the first clear distinction between custom and law and the development of an impersonal, abstract, codified legal system. Indeed, practically all legal concepts, whether of criminal or of civil law, go back to the irrigation city. The first great code of law, that of Hammurabi, almost four thousand years ago, would still be applicable to a good deal of legal business in today's highly developed, industrial society.

The irrigation city also first developed a standing army—it had to. For the farmer was defenseless and vulnerable and, above all, immobile. The irrigation city which, thanks to its technology, produced a surplus, for the first time in human affairs, was a most attractive target for the barbarian outside the gates, the tribal nomads of steppe and desert. And with the army came specific fighting technology and fighting equipment: the war horse and the chariot, the lance and the shield, armor and the catapult.

2. It was in the irrigation city that social classes first developed. It needed people permanently engaged in producing the farm products on which all the city lived; it needed farmers. It needed soldiers to defend them. And it needed a governing class with knowledge, that is, originally a priestly class. Down to the end of the nineteenth century these three "estates" were still considered basic in society.

But at the same time the irrigation city went in for specialization of labor resulting in the emergence of artisans and craftsmen: potters, weavers, metal-workers, and so on; and of professional people: scribes, lawyers, judges, physicians.

And because it produced a surplus, it first engaged in organized trade which brought with it not only the merchant but money, credit, and a law that extended beyond the city to give protection, predictability, and justice to the stranger, the trader from far away. This, by the way, also made necessary international relations and international law. In fact, there is not very much difference between a nineteenth-century trade treaty and the trade treaties of the irrigation empires of antiquity.

3. The irrigation city first had knowledge, organized it, and institutionalized it. Both because it required considerable knowledge to construct and maintain the complex engineering works that regulated the vital water supply and because it had to manage complex economic transactions stretching over many years and over hundreds of miles, the irrigation city needed records, and this, of course, meant writing. It needed astronomical data, as it depended on a calendar. It needed means of navigating across sea or desert. It, therefore, had to organize both the supply of the needed information and its processing into learnable and teachable knowledge. As a result, the irrigation city developed the first schools and the first teachers. It developed the first

systematic observation of natural phenomena, indeed, the first approach to nature as something outside of and different from man and governed by its own rational and independent laws.

4. Finally, the irrigation city created the individual. Outside the city, as we can still see from those tribal communities that have survived to our days, only the tribe had existence. The individual as such was neither seen nor paid attention to. In the irrigation city of antiquity, however, the individual became, of necessity, the focal point. And with this emerged not only compassion and the concept of justice; with it emerged the arts as we know them, the poets, and eventually the world religions and the philosophers.

This is, of course, not even the barest sketch. All I wanted to suggest is the scope and magnitude of social and political innovation that underlay the rise of the irrigation civilizations. All I wanted to stress is that the irrigation city was essentially "modern," as we have understood the term, and that, until today, history largely consisted in building on the foundations laid five thousand or more years ago. In fact, one can argue that human history, in the last five thousand years, has largely been an extension of the social and political institutions of the irrigation city to larger and larger areas, that is, to all areas on the globe where water supply is adequate for the systematic tilling of the soil. In its beginnings, the irrigation city was the oasis in a tribal, nomadic world. By 1900 it was the tribal, nomadic world that had become the exception.

The irrigation civilization was based squarely upon a technological revolution. It can with justice be called a "technological polity." All its institutions were responses to opportunities and challenges that new technology offered. All its institutions were essentially aimed at making the new technology most productive. . . . The question I posed at the beginning [was] what we can learn from the first technological revolution regarding the impacts likely to result on man, his society, and his government from the new industrial revolution, the one we are living in. Does the story of the irrigation civilization show man to be determined by his technical achievements, in thrall to them, coerced by them? Or does it show him capable of using technology to his own, to human ends, and of being the master of the tools of his own devising?

The answer which the irrigation civilizations give us to this question is threefold.

1. Without a shadow of doubt, major technological change creates the need for social and political innovation. It does make obsolete existing institutional arrangements. It does require new and very different institutions of community, society, and government. To this extent there can be no doubt: technological change of a revolutionary character coerces; it *demands innovation*.

2. The second answer also implies a strong necessity. There is little doubt, one would conclude from looking at the irrigation civilizations, that specific technological changes demand equally specific social and political innovations. That the basic institutions of the irrigation cities of the Old World, despite great cultural difference, all exhibited striking similarity may not prove much. After all, there probably was a great deal of cultural diffusion (though I refuse to get into the quicksand of debating whether Mesopotamia or China was the original innovator). But the fact that the irrigation civilizations of the New World around the Lake of Mexico and in Maya Yucatan, though culturally completely independent, millennia later evolved institutions which, in fundamentals, closely resemble those of the Old World (e.g., an organized government with social classes and a permanent military, and writing) would argue strongly that the solutions to specific conditions created by new technology have to be specific and are, therefore, limited in number and scope.

In other words, one lesson to be learned from the first technological revolution is that new technology creates what a philosopher of history might call "objective reality." And objective reality has to be dealt with on *its* terms. Such a reality would, for instance, be the conversion, in the course of the first technological revolution, of human space from "habitat" into "settlement," that is, into a permanent territorial unit always to be found in the same place—unlike the migrating herds of pastoral people or the hunting grounds of primitive tribes. This alone made obsolete the tribe and demanded a permanent, impersonal, and rather powerful government.

3. But the irrigation civilizations can teach us also that the new objective reality determines only the gross parameters of the solutions. It determines where, and in respect to what, new institutions are needed. It does not make anything "inevitable." It leaves wide open *how* the new problems are to be tackled, what the purposes and values of the new institutions are to be.

In the irrigation civilizations of the New World the individual, for instance, failed to make his appearance. Never, as far as we know, did these civilizations get around to separating law from custom nor, despite a highly developed trade, did they invent money.

Even within the Old World, where one irrigation civilization could learn from the others, there were very great differences. They were far from homogeneous even though all had similar tasks to accomplish and developed similar institutions for these tasks. The different specific answers expressed above all different views regarding man, his position in the universe, and his society—different purposes and greatly differing values.

Impersonal bureaucratic government had to arise in all these civilizations; without it they could not have functioned. But in the Near East it was seen at a very early stage that such a government could serve equally to exploit and hold down the common man and to establish justice for all and protection for the weak. From the beginning the Near East saw an ethical decision as crucial to government. In Egypt, however, this decision was never seen. The question of the purpose of government was never asked. And the central quest of government in China was not justice but harmony.

It was in Egypt that the individual first emerged, as witness the many statues, portraits, and writings of professional men, such as scribes and administrators, that have come down to us—most of them superbly aware of the uniqueness of the individual and clearly asserting his primacy. It is early Egypt, for instance, which records the names of architects who built the great pyramids. We have no names for the equally great architects of the castles and palaces of Assur or Babylon, let alone for the early architects of China. But Egypt suppressed the individual after a fairly short period during which he flowered. . . . There is no individual left in the records of the Middle and New Kingdoms, which perhaps explains their relative sterility.

In the other areas two entirely different basic approaches emerged. One, that of Mesopotamia and of the Taoists, we might call "personalism," the approach that found its greatest expression later in the Hebrew prophets and in the Greek dramatists. Here the stress is on developing to the fullest the capacities of the person. In the other approach—we might call it "rationalism," taught and exemplified above all by Confucius—the aim is the moulding and shaping of the individual according to pre-established ideals of rightness and perfection. I need not tell you that both these approaches still permeate our thinking about education.

Or take the military. Organized defense was a necessity for the irrigation civilization. But three different approaches emerged: a separate military class supported through tribute by the producing class, the farmers; the citizen-army drafted from the peasantry itself; and mercenaries. There is very little doubt that from the beginning it was clearly understood that each of these three approaches had very real political consequences. It is hardly coincidence, I believe, that Egypt, originally unified by overthrowing local, petty chieftains, never developed afterward a professional, permanent military class.

Even the class structure, though it characterizes all irrigation civilizations, showed great differences from culture to culture and within the same culture at different times. It was being used to create permanent castes and complete social immobility, but it was also used with great skill to create a very high degree of social mobility and a substantial measure of opportunities for the gifted and ambitious.

Or take science. We now know that no early civilization excelled China in the quality and quantity of scientific observations. And yet we also know that early Chinese culture did not point toward anything we would call science. Perhaps because of their rationalism the Chinese refrained from generalization. And though fanciful and speculative, it is the generalizations of the Near East and the mathematics of Egypt which point the way toward systematic science. The Chinese, with their superb gift for accurate observation, could obtain an enormous amount of information about nature. But their view of the universe remained totally unaffected thereby—in sharp contrast to what we know about the Middle Eastern developments out of which Europe arose.

In brief, the history of man's first technological revolution indicates the following:

1. Technological revolutions create an objective need for social and political innovations. They create a need also for identifying the areas in which new institutions are needed and old ones are becoming obsolete.

2. The new institutions have to be appropriate to specific new needs. There are right social and political responses to technology and wrong social and political responses. To the extent that only a right institutional response will do, society and government are largely circumscribed by new technology.

3. But the values these institutions attempt to realize, the human and social purposes to which they are applied, and, perhaps most important, the emphasis and stress laid on one purpose as against another, are largely within human control. The bony structure, the hard stuff of a society, is prescribed by the tasks it has to accomplish. But the ethos of the society is in man's hands and is largely a matter of the "how" rather than of the "what."

For the first time in thousands of years, we face again a situation that can be compared with what our remote ancestors faced at the time of the irrigation civilization. It is not only the speed of technological change that creates a revolution, it is its scope as well. Above all, today, as seven thousand years ago, technological developments from a great many areas are growing together to create a new human environment. This has not been true of any period between the first technological revolution and the technological revolution that got under way two hundred years ago and has still clearly not run its course.

We, therefore, face a big task of identifying the areas in which social and political innovations are needed. We face a big task in developing the institutions for the new tasks, institutions adequate to the new needs and to the new capacities which technological change is casting up. And, finally, we face the biggest task of them all, the task of insuring that the new institutions embody the values we believe in, aspire to the purposes we consider right, and serve human freedom, human dignity, and human ends.

If an educated man of those days of the first technological revolution—

an educated Sumerian perhaps or an educated ancient Chinese—looked at us today, he would certainly be totally stumped by our technology. But he would, I am sure, find our existing social and political institutions reasonably familiar—they are after all, by and large, not fundamentally different from the institutions he and his contemporaries first fashioned. And, I am quite certain, he would have nothing but a wry smile for both those among us who predict a technological heaven and those who predict a technological hell of "alienation," of "technological unemployment," and so on. He might well mutter to himself, "This is where I came in." But to us he might well say, "A time such as was mine and such as is yours, a time of true technological revolution, is not a time for exultation. It is not a time for despair either. It is a time for work and for responsibility."

T. S. ASHTON

The Industrial Revolution

Writing in 1884, the historian Arnold Toynbee popularized the term "Industrial Revolution" to characterize the massive transformation of England in the late eighteenth and nineteenth centuries from an agricultural and commercial to an industrial society, with resulting major social and political changes. Since Toynbee's time, scholars have been involved in continued controversy about the causes of the revolution, the course of its development, and even about the exact date of its beginning. The English historian T. S. Ashton shows that a number of conditions—social, economic, legal, as well as technological—were factors. Englishmen at the time noticed a quickening pace of change, but they had no idea of the momentous alterations in society that the Industrial Revolution would bring about.

In the short span of years between the accession of George III [in 1760] and that of his son, William IV [in 1830], the face of England changed. Areas that for centuries had been cultivated as open fields, or had laid untended as common pasture, were hedged or fenced; hamlets grew into populous towns; and chimney stacks rose to dwarf the ancient spires. Highroads were made—straighter, stronger, and wider than those evil communications that had corrupted the good manners of travellers in the days of Defoe. The North and Irish Seas, and the navigable reaches of the Mersey, Ouse, Trent, Severn, Thames, Forth, and Clyde were joined together by threads of still water. In

Edited selection reprinted from *The Industrial Revolution 1760–1830* by T. S. Ashton (2nd ed. 1968) by permission of Oxford University Press. © Oxford University Press 1968.

the North the first iron rails were laid down for the new locomotives, and steam packets began to ply on the estuaries and the narrow seas.

Parallel changes took place in the structure of society. The number of people increased vastly, and the proportion of children and young people probably rose. The growth of new communities shifted the balance of population from the South and East to the North and Midlands; enterprising Scots headed a procession the end of which is not yet in sight; and a flood of unskilled, but vigorous, Irish poured in, not without effect on the health and ways of life of Englishmen. Men and women born and bred on the countryside came to live crowded together, earning their bread, no longer as families or groups of neighbours, but as units in the labour force of factories; work grew to be more specialized; new forms of skill were developed, and some old forms lost. Labour became more mobile, and higher standards of comfort were offered to those able and willing to move to centres of opportunity.

At the same time fresh sources of raw material were exploited, new markets were opened, and new methods of trade devised. Capital increased in volume and fluidity; the currency was set on a gold base; a banking system came into being. Many old privileges and monopolies were swept away, and legislative impediments to enterprise removed. The State came to play a less active, the individual and the voluntary association a more active, part in affairs. Ideas of innovation and progress undermined traditional sanctions: men began to look forward, rather than backward, and their thoughts as to the nature and purpose of social life were transformed.

Whether or not such a series of changes should be spoken of as "The Industrial Revolution" might be debated at length. The changes were not merely "industrial," but also social and intellectual. The word "revolution" implies a suddenness of change that is not, in fact, characteristic of economic processes. The system of human relationships that is sometimes called capitalism had its origins long before 1760, and attained its full development long after 1830: there is a danger of overlooking the essential fact of continuity. But the phrase "Industrial Revolution" has been used by a long line of historians and has become so firmly embedded in common speech that it would be pedantic to offer a substitute.

The outstanding feature of the social history of the period—the thing that above all others distinguishes the age from its predecessors—is the rapid growth of population. Careful estimates, based on figures of burials and christenings, put the number of people in England and Wales at about five and a half millions in 1700, and six and a half millions in 1750; when the first census was taken in 1801 it was a round nine millions, and by 1831 had reached fourteen millions. In the second half of the eighteenth century population had thus increased by 40 per cent, and in the first three decades of the nineteenth century by more than 50 per cent. For Great Britain the figures are approximately eleven millions in 1801, and sixteen and a half millions in 1831.

The growth of population was not the result of any marked change in the birth rate. . . . Nor can the increase of people be attributed to an influx from other countries. . . . It was a fall of mortality that led to the increase of numbers. In the first four decades of the eighteenth century excessive indulgence in cheap gin and intermittent periods of famine took a heavy toll of lives; but between 1740 and 1820 the death rate fell almost continuously—from an estimated 35.8 for the ten years ending in 1740 to one of 21.1 for those ending in 1821. Many influences were operating to reduce the incidence of death. The introduction of root crops made it possible to feed more cattle in the winter months, and so to supply fresh meat throughout the year. The substitution of wheat for inferior cereals, and an increased consumption of vegetables, strengthened resistance to disease. Higher standards of personal cleanliness, associated with more soap and cheaper cotton underwear, lessened the dangers of infection. The use of brick in place of timber in the walls, and of slate or stone instead of thatch in the roofs of cottages reduced the number of pests; and the removal of many noxious processes of manufacture from the homes of the workers brought greater domestic comfort. The larger towns were paved, drained, and supplied with running water; knowledge of medicine and surgery developed; hospitals and dispensaries increased; and more attention was paid to such things as the disposal of refuse and the proper burial of the dead. . . .

The increase of the population in Britain occurred at a time when the output of commodities was also increasing at a rapid rate, and this coincidence has led to hasty generalizations. Some writers have drawn the inference that it was the growth of industry that led to the growth of numbers. If this were true the growth of industry must have exerted its influence, not through the birth rate (which, as we have seen, remained steady), but through the death rate. Some of the improvements in the arts of living mentioned above certainly depended on a development of industry, but it would be rash to assign to this a major part in the reduction of mortality. For population was growing rapidly, not only in Britain, but also in most other countries of western and northern Europe, where nothing in the nature of an industrial revolution occurred.

Other writers, reversing the causal sequence, have declared that the growth of population, through its effect on the demand for commodities, stimulated the expansion of industry. An increase of people, however, does not necessarily mean either a greater effective demand for manufactured goods or an increased production of these in the country concerned. (If it did we should expect to find a rapid economic development of Ireland in the eighteenth, and of Egypt, India, and China in the nineteenth century.) It may just as well lead to a lower standard of life for all. The spectre of the pressure of population on the means of subsistence which oppressed the mind of Malthus in 1798 was no chimera. It is true that the immediate pressure was less than Malthus supposed. But if, after the middle of the nineteenth century, there had been no railways in America, no opening up of the prairies, and no steamships, Britain might have learnt from bitter experience the fallacy of the view that, because with every pair of hands there is a mouth, therefore every expansion of numbers must lead to an increase of consumption and so of output. In Britain, in the eighteenth century and later, it so happened that, alongside the increase of population, there was taking place an increase of the other factors of production, and hence it was possible for the standard of life of the people — or of most of them — to rise.

There was an increase in the acreage of land under cultivation. Much attention was given to the draining of fens and marshes, to the breaking up and turning to arable of the old, rough, common pastures (which were usually spoken of as the waste), and to the hedging of land, so as to make it more productive of both crops and livestock. "In this manner," wrote an observer of

"Over London by Rail": engraving by Gustave Doré. (From Gustave Doré: Selected Engravings, by Marina Henderson; published by Academy Editions, London, and St. Martin's Press, New York.)

these developments, "was more useful territory added to the empire, at the expence of individuals, than had been gained by every war since the Revolution." Several new crops were introduced. The turnip made it possible to increase the size of the herds of cattle, and the potato, which was becoming a popular food in the North, brought substantial economies in the use of land. . . .

At the same time there was taking place a rapid increase of capital. The number of people with incomes more than sufficient to cover the primary needs of life was growing: the power to save was increasing. Stable political and social conditions, following the settlement of 1688, encouraged men to look to more distant horizons: what economists call time-preference was favourable to accumulation. The class structure also was favourable to it. It is generally recognized that more saving takes place in communities in which the distribution of wealth is uneven than in those in which it approaches more closely to modern conceptions of what is just. . . .

Accumulation does not of itself, however, lead to the creation of capital goods: it was not only a willingness to save, but also a willingness to employ savings productively, that increased at this time. In the early eighteenth century, landlords had used saved resources to improve their own estates, merchants to extend their markets, and manufacturers to engage more labour; and some of the savings of the retired and leisured classes had been lent on mortgage to local landowners, farmers or tradesmen, or invested in the shares of a turnpike trust. Gradually the market for capital widened, aided by the rise of country bankers. . . .

One thing more was necessary: the increasing supplies of labour, land, and capital had to be co-ordinated. The eighteenth and early nineteenth centuries were rich in entrepreneurs, quick to devise new combinations of productive factors, eager to find new markets, receptive to new ideas. . . . The sentiments and attitudes of mind of the period were propitious. The religious and political differences that had torn society apart in the two preceding centuries had been composed; and if the eighteenth century was not markedly an age of faith, at least it practised the Christian virtue of tolerance. The regulation of industry by gilds, municipalities, and the central government had broken down or had been allowed to sleep, and the field was open for the exercise of initiative and enterprise. It was perhaps no accident that it was in Lancashire and the West Riding, which had been exempted from some of the more restrictive provisions of the Elizabethan code of industrial legislation, that the development was most marked. It was certainly no accident that it was the villages and unincorporated towns—places like Manchester and Birmingham—that grew most rapidly, for industry and trade had long been moving away from the areas where some remnants of public control were still in operation.

During the seventeenth century the attitude of the Law had changed: from the time of Coke* judgements in the courts of Common Law had become tender indeed to the rights of property, but hostile to privilege. In 1624 the Statute of Monopolies had swept away many vested interests, and a century and a half later Adam Smith was able to say of Englishmen that they were "to their great honour of all peoples, the least subject to the wretched spirit of monopoly." Whether or not the patent system, the lines of which had been laid down by that same Statute, was stimulating to innovation in industrial practice is not easy to determine. It gave security to the inventor, but it allowed some privileged positions to be maintained for an undue length of time, and it was sometimes used to block the way to new contrivance: for nearly a quarter of a century, for example, James Watt was able to prevent other engineers from constructing new types of steam engine, even under licence from himself. Many manufacturers—not all from disinterested motives—opposed the application of the law and encouraged piracy. Associations were brought into being in Manchester and other centres of industry to contest the legality of rights claimed by patentees. The Society for the Encouragement of Arts, Manufactures and Commerce, founded in 1754, offered premiums to inventors who were willing to put their devices at the free disposal of all. And Parliament itself made awards (for example, £14,000 to Thomas Lombe when his patent for silk-throwing expired, £30,000 to Jenner for the discovery of vaccine inoculation, £10,000 to Edmund Cartwright for various contrivances, and £5,000 to Samuel Crompton for his invention of the "mule") in addition to the substantial annual grants it voted for the use of the Board of Agriculture and the Veterinary College. Without any such monetary incentive, one of the outstanding industrialists, Josiah Wedgwood, resolved "to be released from these degrading slavish chains, these mean, selfish fears of other people copying my works"; and, at a later stage, the inventors of the safety lamps, Sir Humphry Davy, Dr. Clanny, and George Stephenson, all refused, in the interest of the miners, to take out patents for their devices. It is at least possible that without the apparatus of the patent system discovery might have developed quite as rapidly as it did.

Some accounts of the technological revolution begin with the story of a dreamy boy watching the steam raise the lid of the kettle on the domestic hearth, or with that of a poor weaver gazing with stupefaction at his wife's spinning wheel, overturned on the floor but still revolving. These, needless to say, are nothing but romantic fiction. Other accounts leave the impression that the inventions were the work of obscure millwrights, carpenters, or clock-

---

*Editors' note: Sir Edward Coke (1552–1634), English jurist and chief justice under King James I. His *Institutes* and *Reports* set forth many principles of modern law.

makers, untutored in principles, who stumbled by chance on some device that was destined to bring others to fame and fortune and themselves to penury. It is true that there were inventors... who were endowed with little learning, but with much native wit. It is true that there were others... whose discoveries transformed whole industries, but left them to end their days in relative poverty. It is true that a few new products came into being as the result of accident. But such accounts have done harm by obscuring the fact that systematic thought lay behind most of the innovations in industrial practice, by making it appear that the distribution of rewards and penalties in the economic system was wholly irrational, and, above all, by over-stressing the part played by chance in technical progress. "Chance," as Pasteur said, "favours only the mind which is prepared"; most discoveries are achieved only after repeated trial and error. Many involve two or more previously independent ideas or processes, which, brought together in the mind of the inventor, issue in a more or less complex and efficient mechanism. In this way, for example, the principle of the jenny was united by Crompton with that of spinning by rollers to produce the mule; and the iron rail, which had long been in use in the coal mine, was joined to the locomotive to create the railway. In such cases of what has been called cross-mutation the part played by chance must have been very small indeed.

Yet other accounts of the industrial revolution are misleading because they present discovery as the achievement of individual genius, and not as a social process. "Invention," as a distinguished modern scientist, Michael Polanyi, has remarked, "is a drama enacted on a crowded stage." The applause tends to be given to those who happen to be on the boards in the final act, but the success of the performance depends on the close co-operation of many players, and of those behind the scenes. The men who, together, whether as rivals or as associates, created the technique of the industrial revolution were plain Englishmen or Scots. . . .

Inventors, contrivers, industrialists, and entrepreneurs—it is not easy to distinguish one from another at a period of rapid change—came from every social class and from all parts of the country. . . . Lawyers, soldiers, public servants, and men of humbler station than these found in manufacture possibilities of advancement far greater than those offered in their original callings. A barber, Richard Arkwright, became the wealthiest and most influential of the cotton-spinners; an innkeeper, Peter Stubs, built up a highly esteemed concern in the file trade; a schoolmaster, Samuel Walker, became the leading figure in the north of England iron industry. "Every man," exclaimed the ebullient William Hutton in 1780, "has his fortune in his own hands." That, it is needless to say, has never been true, or even half-true; but anyone who looks closely at English society in the mid- and late eighteenth century will understand how it was possible for it to be said, for at this time vertical mobility had reached a degree higher than that of any earlier, or perhaps any succeeding, age. . . .

The conjuncture of growing supplies of land, labour, and capital made possible the expansion of industry; coal and steam provided the fuel and power for large-scale manufacture; low rates of interest, rising prices, and high expectations of profit offered the incentive. But behind and beyond these material and economic factors lay something more. Trade with foreign parts had widened men's views of the world, and science their conception of the universe: the industrial revolution was also a revolution of ideas. If it registered an advance in understanding of, and control over, Nature, it also saw the beginning of a new attitude to the problems of human society. And here, again, it was from Scotland, and the University of Glasgow in particular, that the clearest beam of light was thrown. It is, no doubt, an academic error to overstress the part played by speculative thought in shaping the lives of ordinary men and women. . . . But, at least, there is one product of Scottish moral philosophy that cannot pass without mention in any account of the forces that produced the industrial revolution. The *Enquiry into the Nature and Causes of the Wealth of Nations* [by Adam Smith], which appeared in 1776, was to serve as a court of appeal on matters of economics and politics for generations to come. Its judgements were the material from which men not given to the study of treatises framed their maxims of conduct for business and government alike. It was under its influence that the idea of a more or less fixed volume of trade and employment, directed and regulated by the States, gave way—gradually and with many setbacks—to thoughts of unlimited progress in a free and expanding economy.

J. BRONOWSKI

Technology and Culture in Evolution

In this passage, an eminent scientist, the late Jacob Bronowski, asserts that the basic human discovery was that of the future, enabling man to store knowledge and to form plans. This finding, he argues, implies that the progress of man resulted from culture, of which the single most important component is technology. The "revolutions" described by Drucker and Ashton, then, are historical segments in the "evolution" of man.

There is no blueprint of the future: there is not even a modern bible of the future, of the kind that Karl Marx wrote—a compound of history and exhortation that might be read as a map of the promised land. Since the heyday of H. G. Wells, almost no one has written seriously about the future except the prophets of gloom, such as Aldous Huxley and George Orwell. Their books were made popular by their moralizing tone, but in fact they lack moral as well as scientific imagination: the tragic air about them derives from a complacent assurance that literary England in the first half of the century was the arbiter and expression of ethical wisdom forever. Meanwhile, the writers with a more inventive turn of mind have backed away into science fiction, where their timid and trivial adventures in whimsy do not aspire to genuine imagination and humanity, and so do not rank even as minor prophecy.

Perhaps it has become too painful to think about the future, whose melan-

J. Bronowski, "Technology and Culture in Evolution." *American Scholar* 1972. Reprinted by permission of Rita Bronowski.

choly course just ahead of us is constantly predictable, and yet which we constantly fail to steer away from patently disastrous policies. Whatever the reason, we are intellectually in the middle of a grand withdrawal from history, of which the withdrawal from the future is the less visible but the more ominous half. It is as if we were trying to close our eyes to all that has made us human, by way of biological and cultural evolution, and want instead to play at being happy foundlings in a hole in time.

The truth is, however, that the special gifts of man and his achievements are inseparable from his evolutionary history as the only substantially self-made animal. A multitude of animal species run, fly, swim, and burrow around us, shaped by and locked into their environment; and among all the species, only man has achieved enough command to have largely influenced his own biological evolution. In the past, man has molded himself for the most part unconsciously, by changing the environment so that its selective pressure on him changed. Now we are able to command at least our immediate future with a much larger understanding of the implications of what we do; and it would be ironic if we chose this new moment to bring history to a standstill. . . .

It seems to me timely to remind ourselves that man is an evolved being whose evolution is still going on. We are creatures like others in course of change, and we are unlike the others mainly in our rate and range of change. Very recent studies of the protein chemistry of primates suggest that we and the chimpanzee were one stock no longer than ten or twenty million years ago, so that our evolution has gone prodigiously fast (particularly in the growth of our brain in the last half million years). By contrast, such social insects as the ants have remained quite unchanged for at least fifty million years, locked in their rigid hierarchies of function by structure. We have to face the logic of life, which is that species reach a steady state, and stop evolving, only when the individuals fall into uniform and indeed identical types. By contrast, evolution goes on if there is a pool of viable mutations, which can express themselves in structures and in behavior different from the normal: so that it is reasonable to prophesy that the more variable the members of a species are, the more freely and unexpectedly it will evolve. If we value variety in human beings, we cannot be squeamish in admitting that, as a consequence, man will go on evolving quite strangely.

Therefore when we say aloud (and rightly) that we need to *safeguard* the environment of life, we must beware of secretly thinking that we must *stabilize* the environment—with the hidden assumption that the fullness of human life is to be equated with man as he is now. Of course, it is unwelcome and unsettling to be told that we are not the peak of nature, a museum piece for eternity; but no doubt Neanderthal man (whose line has become extinct) felt the same way before us. The quality of life is not god-given; on the contrary, since the evolutionary rise of man it has been man-made, and it must not be fixed to mean what happens to be agreeable to the kind of men that we are now—conservatives who like to pose as conservationists. It does not make sense to talk of the quality of life unless we have in mind a choice among the possible satisfactions that human life can provide, and particularly a choice among different modes of intellectual satisfaction.

Again I am deliberately invoking a long perspective in order to make the reader look hard at the Wordsworthian catchwords that are traded in the health food stores. If the basis for our disgust with the commuter city and the power state is the belief that they are unnatural to man (as surely they are) then we need to say what is natural—and we need a better ground for saying it than the authority of gut and guru. Moreover, what is natural to man must be specific to him, which is why the general accounts of animal behavior that derive from Konrad Lorenz will not do. Of course man is a poor creature if he blinds himself to the power and the satisfactions of his animal heritage; but he is even poorer, poorer than any animal, if he does not explore those satisfactions that are unique to his species. Hence the scientific search by ethologists for universals in animal behavior is distorted from its purpose, and becomes a silly piece of journalistic sensation, if it is used as a prescription for what is "natural" in human conduct. Magazine readers seem to like to be told that they share a universal beastliness with animals—perhaps because it absolves them of the responsibility to feel human; but that is *not* what has made our species man rather than any other animal.

What has made us men has been deeply documented now by the fossil finds in Africa in the last fifty years, which have traced the biological and cultural specialization of modern man back to its origins, and by the newer work in primate ethology. More than a million, perhaps two million years ago, the hominids went on from using rudimentary tools (which the chimpanzee does) to making them and keeping them for future use. That discovery, that simple lunge into technological foresight, released the brake on evolution which the environment imposes on other animals, and sent man off breakneck at a speed unmatched in the three billion years that life has existed on earth. For without that discovery, evolution is necessarily held down to the pace of biological adaptation. But from the time of the basic human *discovery of the future,* the environment ceases to set the pace, which instead is then set by the capacity to store knowledge and to form plans from it.

This is a remarkable finding, for it implies that the evolution of man has always been culture-driven, and that the driving component was technology. A culture cannot be inherited in the genes, of course; what the hominids passed from one generation to the next was greater dexterity of hand and more farsighted planning in the brain, which became able to manipulate symbols as artifacts. We assume that the choice of mates with these gifts, and the higher reproduction and survival rates of those who possessed them, produced a unique form of natural selection, namely, a self-selection for these culturally useful attributes. (The same selection is still at work: to this day, the correlation of intelligence quotients between bride and bridegroom is higher than between parents and children.) Thus human evolution owes its speed to the gift of technology by which we have shaped the environment; we have never fitted very well into any ecological niche, and instead have carved our own niches with our hands and brain. But even this metaphor is too formal: what has happened is that we have exploited a genetic accident which has made us able progressively to store and organize experience so that we can profit from almost any terrestrial environment.

On this grand scale of history, therefore, to quarrel with technology is to quarrel with the nature of man—just as if we were to quarrel with his upright gait, his symbolic imagination, his faculty for speech, or his unusual sexual posture and appetite. Of course that is no reason why those who choose should not dislike technology; now that it has helped indirectly to give them a brain two to three times larger than the chimpanzee's, they are surely free to use it to prefer the life or even the brain of the chimpanzee. But they cannot then take as their ground the claim that they want to return to nature, meaning the nature of man. For the nature of man is expressed in the same few universals in every culture, from the pygmy and the aborigine to Western man, and from the prohibition of incest to language; and one of these universals is technology.

By the same token, it is a flat denial of history to assert that cultures in which technology has flourished have stifled the development of more personal and sensitive expressions of human nature. On the contrary, the works of high culture that we admire come from the most advanced technological societies of their day: classical Greece, the Arab civilization, the Italian city states, Elizabethan and Restoration England—as soon as one looks at the monuments of art and architecture and literature that express the peaks of the human imagination for us, one sees that they match the peaks of technological sophistication in history. (We do not even take our religions from technologically backward civilizations: Buddha, Confucius, Christ and Mohammed were not the desert prophets of backward peoples, but grew up in great intellectual civilizations.) I shall not labor this historical analysis . . . because the facts here are open to everyone to inspect, and are self-evident as soon as we attend to them. The only way to get around them is to dismiss them: that is, to say that Sophocles and Michelangelo and Marlowe and Christopher Wren are fossils whose record is irrelevant to the cultural mishaps of city life in the twentieth century—the profound human problems that many citizens of America believe have been vouchsafed to them for the first time.

CARLO M. CIPOLLA

Clocks and Culture

The economic historian Carlo Cipolla describes the development of timepieces as a prime example of the evolution of complex machinery. He shows how the clock over centuries subtly and significantly influenced human behavior: as a status symbol, as eventually making punctuality an obsession, and as influencing philosophical speculation. Machines, he concludes, change people.

From the earliest antiquity, man has created various devices to solve the problem of the measurement of time. Sundials were the first solution and they continued to be widely used well into the sixteenth and seventeenth centuries, long after the appearance of mechanical clocks. Their low cost and their precision account for the great variety of types that were developed in the course of the centuries. But the sun does not always shine, and man had to invent other instruments: water clocks and fire clocks were developed in the ancient world; more recently, sand glasses came into use, and in our atomic age they are still used for timing the boiling of eggs.

Of these different devices, the water clock or *clepsydra* seems to have been the one that offered to ingenious craftsmen the greatest opportunity for developing extravagant variations on the basic principle. The primitive form of water clock was a stone vessel from which the water was allowed to escape slowly through a small hole, time being indicated by the level of the water within. In the course of time more elaborate *clepsydrae* were made, containing mechanisms that moved automata and struck the hours. It does not take an extreme flight of fancy to imagine that once *clepsydrae* with trains of wheels

Carlo M. Cipolla, *Clocks and Culture, 1300–1700*. William Collins Sons & Co., Publishers. London, 1967. Reprinted by permission.

had been constructed, some craftsman must have thought of the desirability of substituting for the flow of water some other motive power. At the same time, there were craftsmen who were bothering their heads with bells and all possible mechanisms for ringing them effectively.

Bells played a prominent role in the life of medieval towns: they ruled the life of the community and their sound lifted "all things into a sphere of order and serenity." Everybody knew their meanings, and bells rang out their messages at all times, telling the hours, announcing fire or an approaching enemy, calling the people to arms or to peaceful assemblies, telling them when to go to bed and when to get out of it, when to go to work, when to pray and when to fight, marking the opening and closing of fairs, celebrating the elections of popes, the coronations of kings, and victories in war. According to a widespread belief, the sound of bells also helped to keep away storms and epidemics. It was a matter of pride for a town, a church, a monastery to have a beautiful bell or peal of bells, and in the course of time, mechanisms were developed to ring bells with greater efficiency: it is not improbable that these contrivances, which were made of toothed wheels and oscillating levers, helped to prepare for the development of mechanical clocks. Finally, there were astronomers, astrologers and others interested in making globes and spheres and in supplying them with movements that would imitate the movements of the stars and the planets. The historian is inclined to fancy that techniques developed in the making of these devices must also have brought craftsmen closer to the making of mechanical clocks.

If a broad historical point of view invites one to stress the continuity and gradualness inherent in the process of technological change, a strictly technological point of view forces one to emphasize the fundamental difference between water clocks, bell-ringing mechanisms, and the like on the one hand and the mechanical clock on the other. Historians of technology and science have good cause to dramatize the sharp break in the history of horology brought about by the invention of the verge escapement with foliot. In the water clock, the time-keeping is governed by the rate of the flow of water through a hole, and the regulating device is simply the hole through which the water flows. In the mechanical clock, the time-keeping is governed by an oscillator or escapement which controls the unidirectional movement of the motive power and transforms it into a slow, steady and regular motion whose meaning appears on the face of the clock. The verge escapement with foliot is a most ingenious device and, as has been said, whoever invented it must have been a mechanical genius. . . . The mechanical clock was born when the verge escapement with foliot was invented. Historians have long debated the dating of this invention but the general consensus now places it in the second half of the thirteenth century. . . .

In the course of the fourteenth century mechanical clocks became progressively more numerous in Europe, and very soon they were equipped with mechanisms for striking the hours. A clock made of iron was installed at the church of St. Eustorgio in Milan in 1309. The cathedral at Beauvais had possibly a clock with a bell before 1324. In 1335, according to an Italian chronicler, the church of St. Gothard in Milan had "a wonderful clock, with a very large clapper which strikes a bell twenty-four times according to the twenty-four hours of the day and night and thus at the first hour of the night gives one sound, at the second two strokes... and so distinguishes one hour from another which is of greatest use to men of every degree."...

Thus a combination of civic pride, utilitarianism, and mechanical interest fostered the diffusion of the clock despite its relatively high cost. Since more clocks were being made, the proficiency of the makers improved, and by the end of the fourteenth century, clocks which struck the hours and the quarters had been built, but this fact should not mislead us regarding the precision of these early time-keepers. As has been said, early clocks "embodied a strange combination of brilliance in conception with a deficient technique of construction." Throughout the fourteenth and fifteenth centuries, most clocks (if and when they worked) lost or gained much time in a day, and on the other hand, contemporary requirements for precision were low: thus it was generally thought unnecessary to provide clocks with a minute hand....

The most striking occurrence in the early history of clocks is that while medieval craftsmen did not improve noticeably in precision, they soon succeeded in constructing clocks with curious and very complicated movements. It was easier to add wheels to wheels than to find better ways to regulate the escapement. On the other hand complicated movements had quite a popular appeal and most people believed that a correct knowledge of the conjunction of the heavenly bodies was essential for the success of human enterprises. One of the most remarkable pieces in this regard was the clock made about 1350 for the cathedral of Strasbourg. Enormous in size, it included a moving calendar and an astrolabe whose pointers indicated the movements of the sun, moon and planets. The upper compartment was adorned with a statue of the Virgin before whom at noon the Three Magi bowed while a carillon played a tune. On top of the whole thing stood an enormous cock which, at the end of the procession of the Magi, opened its beak, thrust forth its tongue, crowed and flapped its wings. In Bologna around the middle of the fifteenth century Master Giovanni Evangelista da Piacenza and Master Bartolomeo di Gnudolo built an impressive clock on the *Palazzo del Comune,* with a trumpeting angel, a large number of Saints and other holy dignitaries and a procession of the Magi who paid their respects to the Virgin and Child....

The clock was invented and constructed to satisfy the human need of measuring time. Soon after its appearance, however, the clock assumed the role of a status symbol. In Europe, towns competed with one another in the construction of the most lavish clocks and many of these municipal timepieces possessed elaborate movements and dials whose meaning very few could understand. Soon after the portable clock appeared, it became fashionable for kings and nobles to have the faces of these contrivances painted in their portraits. At the same time, the machine which had been devised to satisfy particular human needs created new ones. Men began timing activities that, in the absence of clocks, they had never thought of timing. People became very conscious of time, and, in the long run, punctuality became at the very same time a need, a virtue, and an obsession.* Thus a vicious circle was set into motion. As more and more people obtained clocks and watches, it became necessary for other people to possess similar contrivances, and the machine created the conditions for its own proliferation. . . .

At the same time, clocks and watches were insistently changing man's way of life and of thinking. Slowly but irresistibly hours of equal length were substituted in Europe for the unequal hours and other time divisions which were more closely connected with the seasons. A long time passed before references to "the time of the first mass" or "the time of the vespers" were completely abandoned; but the reference to the more abstract hours "of the clock" (i.e. o'clock) of equal length gained progressively more ground and eventually prevailed. In Japan the mechanical clock was first adapted to the local tradition of the "unequal" hours but eventually the mechanically more precise European solution was adopted. The impact which the clock had on philosophy and art was no less important. . . .

In the course of the sixteenth and seventeenth centuries the clock as a machine exerted deep influence on the speculations of philosophers and scientists. Kepler asserted that "The universe is not similar to a divine living being, but is similar to a clock." Robert Boyle wrote that the universe is "a great piece of clock work," and Sir Kenelm Digby wrote again that the universe was nothing but an immense clock. In the framework of this prevailing mechanistic *Weltanschauung,* God was described as an outstanding clockmaker.

Per se, these facts may seem to be of interest only to the erudite collector of historical oddities. Their meaning, however, acquires a new dimension when one notices that similar facts are quite common in the history of technology and machines. Each new machine that appears creates new needs, besides

---

*Editors' note: This was particularly true after 1700, when reasonably accurate clocks and watches came into use.

satisfying existing ones, and breeds newer machines. The new contrivances modify and shape our lives and our thoughts; they affect the arts and philosophy, and they intrude even into our spare time influencing our way of using it.

The machine is a tool. But it is not a "neutral" tool. We are deeply influenced by the machine while using it.

LEWIS MUMFORD

The All-Seeing Eye

According to Lewis Mumford, present technology is being controlled by an organization of experts which he terms the "megamachine." He views the computer as the ultimate instrument that will enable these technocrats to control the modern world. Once the bureaucracy of the "megamachine" has complete computerized dossiers on every individual, he warns, complete domination will become a reality.

In Egyptian theology, the most singular organ of the Sun God, Re, was the eye: for the Eye of Re had an independent existence and played a creative and directive part in all cosmic and human activities. The computer turns out to be the Eye of the reinstated Sun God, that is, the Eye of the Megamachine, serving as its "Private Eye" or Detective, as well as the omnipresent Executive Eye, he who exacts absolute conformity to his commands, because no secret can be hidden from him, and no disobedience can go unpunished.

The principal means needed to operate the megamachine correctly and efficiently were a concentration of power, political and economic, instantaneous communication, rapid transportation, and a system of information storage capable of keeping track of every event within the province of the Divine King: once these accessories were available, the central establishment would also have a monopoly of both energy and knowledge. No such complete assemblage had been available to the rulers of the pre-scientific ages: transportation was slow, communication over a distance remained erratic, confined to written messages carried by human messengers, while information storage, apart from tax records and books, was sporadic and subject to fire and military assault. With each successive king, essential parts would require reconstruction or replacement. Only in Heaven could there exist the all-knowing, all-seeing, all-powerful, omnipresent gods who truly commanded the system.

With nuclear energy, electric communication, and the computer, all the necessary components of a modernized megamachine at last became available: "Heaven" had at last been brought near. Theoretically, at the present moment, and actually soon in the future, God—that is, the Computer—will be able to find, to locate, and to address instantly, by voice and image, via the priesthood, any individual on the planet: exercising control over every detail of the subject's daily life by commanding a dossier which would include his parentage and birth; his complete educational record; an account of his illnesses and his mental breakdowns, if treated; his marriage; his sperm bank account; his income, loans, security payments; his taxes and pensions; and finally the disposition of such further organs as may be surgically extracted from him just prior to the moment of his official death.

In the end, no action, no conversation, and possibly in time no dream or thought would escape the wakeful and relentless eye of this deity: every manifestation of life would be processed into the computer and brought under its all-pervading system of control. This would mean, not just the invasion of privacy, but the total destruction of autonomy: indeed the dissolution of the human soul.

Half a century ago, the foregoing description would have seemed too crude and overwrought to be accepted even as satire: H. G. Wells' "Modern Utopia," which tentatively provided for a central identification system, did not dare carry the method through into every detail of life. Even twenty years ago, only the first faint outlines of this modern version of the Eye of Re could be detected by such a prescient mind as that of Norbert Wiener.* But today the grim outlines of the whole system have been laid down, with the corroborative evidence, by a legal observer, Alan F. Westin, as an incidental feature of a survey of the numerous public agencies and technological devices that are now encroaching on the domain of private freedom.

What Westin demonstrates, also in passing, is that the countless record files, compiled by individual bureaucracies for their special purposes, can already be assembled in a single central computer, thanks to the fantastic technological progress made through electro-chemical miniaturization: not merely the few I have just picked out, but civil defense records, loyalty security clearance records, land and housing records, licensing applications, trade

---

*Editors' note: Wiener (1894–1964) was instrumental in developing high-speed electronic computers.

union cards, social security records, passports, criminal records, automobile registrations, driver's licenses, telephone records, church records, job records—indeed the list becomes finally as large as life—at least of abstracted, symbolically attenuated, recordable life.

The means for such total monitoring came about through the quantum jump from macro- to micro-mechanics: so that the seemingly compact microfilm of earlier decades, to quote Westin's words, "has now given way to photochromatic microimages that make it possible to reproduce the complete bible on a thin sheet of plastic less than two inches square, or to store page by page copies of all books in the Library of Congress in six four-drawer filing cabinets." The ironic fact that this truly colossal leap was a product of research by the National Cash Register Company does not detract from the miraculous nature of this invention. . . .

If anything could testify to the magical powers of the priesthood of science and their technical acolytes, or declare unto mankind the supreme qualifications for absolute rulership held by the Divine Computer, this new invention alone should suffice. So the final purpose of life in terms of the megamachine at last becomes clear: it is to furnish and process an endless quantity of data, in order to expand the role and ensure the domination of the power system.

Here if anywhere lies the source of that invisible ultimate power capable of governing the modern world. Here is the Mysterium Tremendum, exercising unlimited power and knowledge, beside which all other forms of magic are clumsy fakes, and all other forms of control without charismic authority. Who dares to laugh at potencies of such magnitude? Who can possibly escape the relentless and unflagging supervision of this supreme ruler? What hideout so remote that it would conceal the rebellious?

HERBERT A. SIMON

What Computers Mean for Man And Society

Herbert Simon, a professor of computer science and psychology, describes the present capabilities and future potentialities of the most complex device that has thus far been invented. He heartily disagrees with Mumford's warnings about the computer. Nor does he view the computer in the same way that Cipolla does the clock, that is, as an instrument that will profoundly alter the way we think and live. Rather, Simon sees the computer as a benign device that will help us solve many of our economic and social problems, if appropriate legislative controls protect against its misuse. Far from the computer endangering us, Simon believes that it will eventually make a very important contribution to human progress in giving us knowledge of the human mind.

Energy and information are two basic currencies of organic and social systems. A new technology that alters the terms on which one or the other of these is available to a system can work on it the most profound changes. At the core of the Industrial Revolution, which began nearly three centuries ago, lay the substitution of mechanical energy for the energy of man and animal. It was this revolution that changed a rural subsistence society into an urban

---

Herbert A. Simon, "What Computers Mean for Man and Society," excerpted from *Science*, Vol. 195, pp. 1186–1191, March 18, 1977. Copyright 1977 by the American Association for the Advancement of Science.

affluent one and touched off a chain of technological innovations that transformed not only production but also transportation, communication, warfare, the size of human populations, and the natural environment.

It is easy, by hindsight, to see how inexorably these changes followed one another, how "natural" a consequence, for example, suburbia was of cheap, privately owned transportation. It is a different question whether foresight could have predicted these chains of events or have aided in averting some of their more undesirable outcomes. The problem is not that prophets were lacking—they have been in good supply at almost all times and places. Quite the contrary, almost everything that has happened, and its opposite, has been prophesied. The problem has always been to pick and choose among the embarrassing riches of alternative projected futures; and in this, human societies have not demonstrated any large foresight. Most often we have been constrained to anticipate events just a few years before their occurrence, or even while they are happening, and to try to deal with them, as best we can, as they are engulfing us.

We are now in the early stages of a revolution in processing information that shows every sign of being as fundamental as the earlier energy revolution. Perhaps we should call it the Third Information Revolution. (The first produced written language, and the second, the printed book.) This third revolution, which began more than a century ago, includes the computer but many other things as well. The technology of information comprises a vast range of processes for storing information, for copying it, for transmitting it from one place to another, for displaying it, and for transforming it.

Photography, the moving picture, and television gave us, in the course of a century, a whole new technology for storing and displaying pictorial information. Telegraphy, the telephone, the phonograph, and radio did the same for storing and transmitting auditory information. Among all of these techniques, however, the computer is unique in its capacity for manipulating and transforming information and hence in carrying out, automatically and without human intervention, functions that had previously been performable only by the human brain.

As with the energy revolution, the consequences of the information revolution spread out in many directions. First, there are the economic consequences that follow on any innovation that increases human productivity. As we shall see, these are perhaps the easiest effects of technological change to predict. Second, there are consequences for the nature of work and of leisure—for the quality of life. Third, the computer may have special consequences for privacy and individual liberty. Fourth, there are consequences for man's view of himself, for his picture of the universe and of his place and goals in it. In each of these directions, the immediate consequences are, of course, the most readily perceived. (It was not hard to foresee that New-comen's and Watt's engines would change the economics of mining in deep pits.) It is far more difficult to predict what indirect chains of effects these initial impacts will set off, for example, the chain that reaches from the steam engine through the internal-combustion engine to the automobile and the suburb.

Prediction is easier if we do not try to forecast in detail the time path of events and the exact dates on which particular developments are going to occur, but to focus, instead, upon the steady state toward which the system is tending. Of course, we are not so much interested in what is going to happen in some vague and indefinite future as we are in what the next generation or two holds for us. Hence, a generation is the time span with which I shall be concerned. . . .

Computer Capabilities

The computer is a device endowed with powers of utmost generality for processing symbols. It is remarkable not only for its capabilities but also for the simplicity of its underlying processes and organization. Of course, from a hardware standpoint it is not simple at all but is a highly sophisticated elec-tronic machine. The simplicity appears at the level of the elementary informa-tion processes that the hardware enables it to perform, the organization for execution and control of those processes, and the programming languages in terms of which the control of its behavior is expressed. A computer can read symbols from an external source, output symbols to an external destination, store symbols in one or more memories, copy symbols, rearrange symbols and structures of symbols, and react to symbols conditionally—that is, follow one course of action or another, depending on what symbols it finds in its memory or in its input devices. . . .

There is great dispute among experts as to what the generality of the com-puter implies for its ability to behave intelligently. There is also dispute as to whether the computer, when it is behaving more or less intelligently, is using processes similar to those employed by an intelligent human being, or quite different processes. The views expressed here will reflect my own experience in research with computers and my interpretation of the scientific literature. First, no limits have been discovered to the potential scope of computer intel-ligence that are not also limits on human intelligence. Second, the elementary processes underlying human thinking are essentially the same as the com-puter's elementary information processes, although modern fast computers can execute these processes more rapidly than can the human brain. In the past, computer memories, even in large computers, have probably not been as capacious as human memory, but the scale of available computer memories is increasing rapidly, to the point where memory size may not be much longer an effective limit on the capacity of computers to match human performance. Any estimate of the potential of the computer in the near or distant future depends on one's agreement or disagreement with these assumptions. . . .

Of course humans, through the processes called learning, can improve their strategies by experience and instruction. By the same token, computers can be, and to some extent have been, provided with programs (strategies) for improving their own strategies. Since a computer's programs are stored in the same memories as data, it is entirely possible for programs to modify themselves—that is, to learn. . . .

How, then, have computers actually been used to date? At present, computers typically spend most of their time in two main kinds of tasks: carrying out large-scale engineering and scientific calculations and keeping the financial, production, and sales records of business firms and other organizations. Although precise statistics are not available, it would be safe to estimate that 95 percent of all computing power is allocated to such jobs. Now these tasks belong to the horseless-carriage phase of computer development. That is to say, they consist in doing things rapidly and automatically that were being done slowly and by hand (or by desk calculator) in the pre-computer era.

Such uses of computers do not represent new functions but only new ways of performing old functions. Of course, by greatly lowering the cost of performing them, they encourage us to undertake them on a larger scale than before. The increased analytic power provided by computers has probably encouraged engineers to design more complex structures (for example, some of the very tall new office buildings that have gone up in New York and Chicago) than they would have attempted if their analytic aids were less powerful. Moreover, by permitting more sophisticated analyses to be carried out in the design process, they have also brought about significant cost reductions in the designs themselves. In the same way, the mechanization of business record-keeping processes has facilitated the introduction of improved controls over inventories and cash flows, with resulting savings in costs. Thus, the computer not only reduces the costs of the information-processing operations that it automates but also contributes to the productivity of the activities themselves.

The remaining 5 percent of computer uses are more sophisticated. Let us consider two different ways in which a computer can assist an engineer in designing electric motors. On the one hand, the engineer can design the motor using conventional procedures, then employ the computer to analyze the prospective operation of the design—the operating temperature, efficiency, and so on. On the other hand, the engineer can provide the computer with the specifications for the motor, leaving to the computer the task of synthesizing a suitable design. In the second, but not the first, case the computer, using various heuristic search procedures, actually discovers, decides upon, and evaluates a suitable design. In the same way, the role of the computer in managing inventories need not be limited to record-keeping. The computer program may itself determine (on the basis of usage) when items should be reordered and how large the orders should be. In these and many other situations, computers can provide not only the information on which decisions are made but can themselves make the decisions. Process-control computers, in automated or semiautomated manufacturing operations, play a similar role in decision-making. Their programs are decision strategies which, as the system's variables change from moment to moment, retain control over the ongoing process.

It is the capability of the computer for solving problems and making decisions that represents its real novelty and that poses the greatest difficulties in predicting its impact upon society. An enormous amount of research and developmental activity will have to be carried out before the full practical implications of this capability will be understood and available for use. In the single generation that modern computers have been in existence, enough basic research has been done to reveal some of the fundamental mechanisms. Although one can point to a number of applications of the computer as decision-maker that are already 20 or 25 years old, development and application on a substantial scale have barely begun.

Economic Effects of Computers

... Now the rate of technological change depends both upon the rate of discovery of new innovations and upon the availability of capital to turn them into bricks and steel (or wire and glass). In this process, computers compete with other forms of technology for the available capital. Hence, the process of computerization is simply a part, currently an important part, of the general process of technological change. . . .

In taking this very global and bird's-eye view of the economics of mechanization, we should not ignore the plight of the worker who is displaced by the computer. His plight is often genuine and serious, particularly if the economy as a whole is not operating near full employment, but even if it is. Society as a whole benefits from increased productivity, but often at the expense of imposing transient costs on a few people. But the sensible response to this problem is not to eschew the benefits of change; it is rather to take institutional steps to shift the burdens of the transition from the individual to society. Fortunately, our attitudes on these questions appear to be maturing somewhat, and our institutional practices improving, so that the widespread introduction of the computer into clerical operations over the past generation has not called forth any large-scale Ludditism. In fact, during the depression that we are currently experiencing, in contrast to some earlier ones, technology has not been accused as the villain. . . .

Control and Privacy

The potential of computers for increasing the control of organizations or society over their members and for invading the privacy of those members has caused considerable concern. The issues are important but are too complex to be discussed in detail here. I shall therefore restrict myself to a few comments which will serve rather to illustrate this complexity than to provide definitive answers.

A first observation is that our concern here is for competitive aspects of society, the power of one individual or group relative to others. Technologies tend to be double-edged in competitive situations, particularly when they are available to both competitors. For example, the computerization of credit information about individuals facilitates the assembly of such information from many sources, and its indefinite retention and accessibility. On the other hand, it also facilitates auditing such information to determine its sources and reliability. With appropriate legal rules of the game, an automated system can provide more reliable information than a more primitive one and can be surrounded by more effective safeguards against abuse. Some of us might prefer, for good reasons or bad, not to have our credit checked at all. But if credit checking is a function that must be performed, a strong case can be made for making it more responsible by automating it, with appropriate provision for auditing its operation.

Similarly, much has been said of the potential for embezzlement in computerized accounting systems, and cases have occurred. Embezzlement, however, was known before computers, and the computer gives auditors as well as embezzlers powerful new weapons. It is not at all clear which way the balance has been tilted.

The privacy issue has been raised most insistently with respect to the creation and maintenance of longitudinal data files that assemble information about persons from a multitude of sources. Files of this kind would be highly valuable for many kinds of economic and social research, but they are bought at too high a price if they endanger human freedom or seriously enhance the opportunities of blackmailers. While such dangers should not be ignored, it should be noted that the lack of comprehensive data files has never been the limiting barrier to the suppression of human freedom. The Watergate criminals made extensive, if unskillful, use of electronics, but no computer played a role in their conspiracy. The Nazis operated with horrifying effectiveness and thoroughness without the benefits of any kind of mechanized data processing.

Making the computer the villain in the invasion of privacy or encroachment on civil liberties simply diverts attention from the real dangers. Computer data banks can and must be given the highest degree of protection from abuse. But we must be careful, also, that we do not employ such crude methods of protection as to deprive our society of important data it needs to understand its own social processes and to analyze its problems.

Man's View of Man

Perhaps the most important question of all about the computer is what it has done and will do to man's view of himself and his place in the universe. The most heated attacks on the computer are not focused on its possible economic effects, its presumed destruction of job satisfactions, or its threats to privacy and liberty, but upon the claim that it causes people to be viewed, and to view themselves, as "machines."

To get at the real issues, we must first put aside one verbal confusion. All of us are familiar with a wide variety of machines, most of which predated the computer. Consequently, the word "machine" carries with it many connotations: of ridigity, of simplicity, of repetitive behavior, and so on. If we call anything a machine, we implicitly attribute these characteristics to it. Hence, if a computer is a machine, it must behave rigidly, simply, and repetitively. It follows that computers cannot be programmed to behave like human beings.

The fallacy in the argument, of course, lies in supposing that, because we have applied the term "machine" to computers, computers must behave like older forms of machines. But the central significance of the computer derives from the fact that it falsifies these earlier connotations. It can, in fact, be programmed to behave flexibly, in complex ways, and not repetitively at all. We must either get rid of the connotations of the term, or stop calling computers "machines."

There is a more fundamental question behind the verbal one. It is essentially the question that was raised by Darwinism, and by the Copernican revolution centuries earlier. The question is whether the dignity of man, his sense of worth and self-respect depends upon his being something special and unique in the universe. As I have said elsewhere:

> The definition of man's uniqueness has always formed the kernel of his cosmological and ethical systems. With Copernicus and Galileo, he ceased to be the species located at the center of the universe, attended by sun and stars. With Darwin, he ceased to be the species created and specially endowed by God

with soul and reason. With Freud, he ceased to be the species whose behavior was—potentially—governable by rational mind. As we begin to produce mechanisms that think and learn, he has ceased to be the species uniquely capable of complex, intelligent manipulation of his environment.

What the computer and the progress in artificial intelligence challenge is an ethic that rests on man's apartness from the rest of nature. An alternative ethic, of course, views man as a part of nature, governed by natural law, subject to the forces of gravity and the demands of his body. The debate about artificial intelligence and the simulation of man's thinking is, in considerable part, a confrontation of these two views of man's place in the universe. It is a new chapter in the vitalism-mechanism controversy.

Issues that are logically distinct sometimes become stuck together with the glue of emotion. Several such issues arise here:

To what extent can human behavior be simulated by computer?

In what areas of work and life should the computer be programmed to augment or replace human activities?

How far should we proceed to explore the human mind by psychological research that makes use of computer simulation?

The first of these three issues will only be settled, over the years, by the success or failure of research efforts in artificial intelligence and computer simulation. Whatever our beliefs about the ultimate limits of simulation, it is clear that the current state of the art has nowhere approached those limits.

The second question will be settled anew each year by a host of individual and public decisions based on the changing computer technology, the changing economics of computer applications, and our attention to the social consequences of those applications.

The answer to the third question depends upon our attitudes toward the myths of Pandora and Prometheus. One viewpoint is that knowledge can be dangerous—there are enough historical examples—and that the attempt to arrive at a full explanation of man's ability to think might be especially dangerous. A different point of view, closer to my own, is that knowledge is power to produce new outcomes, outcomes that were not previously attainable. To what extent these outcomes will be good or bad depends on the purposes they serve, and it is not easy, in advance, to predict the good and bad uses to which any particular technology will be put. Instead, we must look back over human history and try to assess whether, on balance, man's gradual emergence from a state of ignorance about the world and about himself has been something we should celebrate or regret. To believe that knowledge is to be preferred to ignorance is to believe that the human species is capable of progress and, on balance, has progressed over the centuries. Knowledge about the human mind can make an important contribution to that progress. It is a belief of this kind that persuades researchers in artificial intelligence that their endeavor is an important and exciting chapter in man's great intellectual adventure.

Summary

From an economic standpoint, the modern computer is simply the most recent of a long line of new technologies that increase productivity and cause a gradual shift from manufacturing to service employment. The empirical evidence provides no support for the claim sometimes made that the computer "mechanizes" and "dehumanizes" work. Perhaps the greatest significance of the computer lies in its impact on Man's view of himself. No longer accepting the geocentric view of the universe, he now begins to learn that mind, too, is a phenomenon of nature, explainable in terms of simple mechanisms. Thus the computer aids him to obey, for the first time, the ancient injunction, "Know thyself."

SIMON RAMO

The Systems Approach

Along with the growth of complexity in our technological society, there is also an increase in the number of specialists or experts in particular fields. Dr. Simon Ramo, a founder of TRW—one of the world's largest communications companies—describes how experts can be formed into teams and profitably employed in what is known as a "systems approach" to solve not only engineering but also complex societal problems. For Ramo, the systems approach will provide a cure for the chaos that threatens our society.

In the systems approach, concentration is on the analysis and design of the whole, as distinct from the analysis and design of the components or the parts. It is an approach that insists upon looking at a problem in its entirety, taking into account all the facets, all the intertwined parameters. It is a process for understanding how they interact with one another and how these factors can be brought into proper relationship for the optimum solution of the problem. The systems approach relates the technology to the need, the social to the technological aspects; indeed, it starts by insisting on a clear understanding of exactly what the problem is and of the goals that should dominate the solution and lead to the criteria for evaluating alternative avenues. As the end result, the approach seeks to work out a detailed description of a specified combination of men and machines—with such concomitant assignment of function, designated use of matériel, and pattern of information flow that the whole system represents a compatible, optimum, interconnected ensemble for achieving the performance desired.

The systems approach is the application of logic and common sense on a sophisticated technological basis. It is quantitative and objective. It makes possible the consideration of vast amounts of data, requirements, and (often conflicting) considerations that usually constitute the heart of a complex, real-life problem. It recognizes the need for carefully worked out compromises, for tradeoffs among the competing factors (such as time versus cost). It provides for simulation and modeling so as to make possible the predicting of performance before the entire system is brought into being. And it makes feasible the selection of the best approach from the many alternatives. . . .

Of course, you don't have to be a "professional" to use a "systems" approach. When any one of us has a problem of any kind—preparing a budget, choosing where to live or what job to seek, designing a chair, producing nails, building a house, or selecting a route to take on a trip—in every instance, it is well to be logical, to use common sense, to consider objectively all the factors involved in the problem. It should be realized that there are an infinite number of alternatives to any objective, and a "best" way, if you can be clear enough about goals or criteria, or at least a contest for first place among alternative solutions. In this sense, then, the systems approach is old indeed. But if the problem is simple to understand and the candidate solutions are easy to identify, optimize and compare, then the concept of the systems approach as applicable is supported only by leaning heavily on the extreme definition that such an approach means merely the use of logic and common sense. It is not supported if we understand the systems approach to mean the assembly of a large team of interdisciplinary experts and the formal execution of a powerful, quantitative listing and analysis of every little detail and interaction in sight.

What makes it now appear new? What makes it justified and significant to talk about it now as a "mobilizing" technology that is ready for application to the big civil-systems problems of our times? It is partly because of the great acceleration of the development of the tools of systems engineering in recent years. Some of this in turn has resulted from the need for a deep application of this kind of methodology to the highly complex and costly defense and space programs. It is partly also the consequence of the general expansion of the technological aspects of our society, dealing with which has justified large-scale advances in the techniques of the systems approach.

There are today, as compared with a decade or two ago, a substantially greater number of professionals who are well seasoned in interdisciplinary problems, who know how to relate the many facets of one technology to another and to relate these in turn to all the nontechnological factors that characterize practical problems. The teams of highly specialized and trained individuals who are expert at applying the systems approach could and probably should be called "systems engineers." This is appropriate if the word "engineer" is used in what probably should be its true meaning but which, when applied to the work most engineers do, is too broad. Engineering is generally defined by members of the profession as the application of science to the problems of society, the efficient employment of resources to the solution of problems. Regardless of how the definition is usually stated, it pretty much amounts to this, and it is usually an overstatement in this form. This is because most engineers, even those who have had a broad formal education in college liberally including humanities and the social sciences, are specialists exclusively in a particular branch of science or technology and its application to society's needs. If a group of people are really going to be professional in the overall application of science to society, then some of the group must know society well. Such qualifications are essential for the competent systems-engineering team.

This is an age of specialization, and a good systems-engineering team includes many individual specialists who have learned how to work their areas into sensible interfaces with the contributions of the other specialists. It is the total team, or if you prefer, the multiheaded "engineer," who must include in "his" head the total intelligence, background, experience, wisdom, and creative ability in all aspects of the problem of applying science, and particularly, who must integrate this total intelligence—... who must *mobilize* it all—to get real-life solutions to real-life problems.

In this sense, then, a good systems-engineering team—and we have begun in the United States to develop such groups—combines individuals who have specialized variously in mathematics, physics, chemistry, other branches of physical science, the many branches of engineering, economics, political science, psychology, sociology, business finance, government, and so on. The systems engineer knows how to attack the interaction problems amongst these specialties that characterize any practical problem. This is worth stressing because it is quite often assumed that the systems approach means that narrowly conditioned engineers, skillful in the details of technology, but with no knowledge of people and the workings of our social systems, are brought in to revolutionize these systems. They do this by automating everything or by putting all of the facts on a computer so that the computer can come up with a perfect answer, or by some other "technology-pure" approach that disregards the human elements. Such a concept of the systems approach is completely erroneous. If it were correct, the systems approach would be quite unsuitable for problems that involve people. And if, indeed, the systems approach meant only the application of technology to the technological aspects of a human problem, we would then need a new name—something bigger and broader—to cover a "look-at-the-problem-as-a-whole" with the technology and the sociology factors properly put together into the right balance....

The public is ready for the systems approach and... the professionals are beginning to be available so that it can be put to use in practical problems. We are not yet at the point where this approach is being broadly used. We are not set up for it, as applied to the big social-engineering problems of our society. What is true is largely that we now know we have problems and we want to see them attacked. We also appreciate that there is a technological ingredient in most of these problems, and we are recognizing that if technology can be properly synthesized with considerations of a nontechnological nature, with the factors of economics, sociology, and political forces, then we might have a new and superior tool for going after these problems.

How do we break away from the present pattern of fragmentary, embryo efforts, spottily applied here and there, and rise to a heightened activity somewhere near what is needed in total attack to meet the problems in a timely way? One answer is to note that the systems approach itself contains within it the elements for furthering its own use. The systems approach is a bottleneck-breaker. The more it is used, the easier it is for it to get used.

Thus, the systems approach is often a first step in answering the question of how much money is needed. It helps to articulate the goals that might have been only crudely understood before. If systems work is done competently, it is inherent in it that it is logical and quantitative as much as is truly possible, and it provides comparisons. You know what you will get for what you pay. When the systems results are in, you also know how much it costs to do the systems studies. It is probably then also apparent that the cost savings, if an optimum design uncovered by the analyses is chosen over an ordinary design, greatly exceed the expenditures on the analyses.

Starting with the acquisition of all the data and facts, a systems effort describes performance, cost, equipment, matériel, and information flow patterns, and the people required to work at prescribed, defined tasks. It shows how the proposed system integrates with the existing operations of real-life society. Accordingly, the systems approach, when applied properly, answers a good many questions everyone has who is involved in decision-making, whether he be a government official, a businessman, a head of a hospital, a voter, or a professional participant. One of these must decide on a public or personal stand to take; another one must make a commitment on risking capital to start the development of the equipment that might be marketable to go into the system; still another has to conceive of how the system will fit into the present society.

The pace of the application of science and technology to the big third area of society, civilian systems, oftentimes is bogged down because the problems are so difficult, complex, little understood, and controversial, and involve so many semiautonomous groups with selfish interests. Many times nothing can conceivably be done unless a new level of objectivity somehow is caused to be attained. The possibilities must be laid out from a platform of adequate breadth in consideration of all the factors and adequate detail as to the criteria for judging alternative approaches. Solid specifics as to what the choices are, and what will be the consequence of taking these various choices or doing nothing at all, must be brought out.

So, perhaps in the end, the systems approach is most essential as a tool to be developed and applied because it encourages and makes possible action. We badly need that encouragement in a society of people who must, in majority, move along together in their thinking, approval, interest, and appreciation before such action is possible. In fact, we are today controlled too much by crisis action; nothing gets done until a problem has reached crisis proportions. Then we are likely to go off in a frenzy. The habit of the use of the systems approach, if we can acquire it, will provide a steady flow of clues to predict and forestall cataclysmic effects of inaction.

There is still one more important point concerning the way in which the systems approach should militate for social advances, for decisive implementation to solve our problems. The systems approach suggests organizational innovation, and such innovation is usually required in our social structure if we are to get on with handling our unsolved problems. The systems approach, in showing the interconnections between various aspects of a problem and in bringing these together into appropriate tradeoffs, compromises, and optimizations, automatically lays the foundation for something else, namely the system's implementation. A practical systems approach to any truly existent problem contemplates implementation to solve the problem after the analysis and synthesis have taken place. It also points out how the interacting factors must be kept under control by proper reporting and decision procedures.

Thus, it might be that a systems approach would show how to unpollute a major river into which many cities pour refuse and whose waters are used in various ways by the industries and population along its banks for a considerable distance. The systems approach would show what can be done, what it will cost, why it is beneficial, and would consider all the negatives, such as the need for moving certain industrial operations. But is also would include the relocation expenses, and in so doing it would contrast the bad effects, such as the dislocations with their cost and impact on human lives, with benefits to those same human lives. Now, if all of these things have been considered on a thorough and objective basis, and if the people of the area in considerable majority wish to go ahead and implement the steps that are called out, then they obviously need an organization that has the power to do so. They are led by this previously unavailable understanding to the idea of modifying existing organizations or creating new ones. They are led, moreover, by the systems approach to see what kind of organization, with what powers and responsibilities, over what aspects of their society, controlled in what way, must now be created if they really want to get on with the solutions. In a sound beginning way, this is happening already with some water basins being defined and regional commissions being created.

It is not very helpful to make a systems analysis showing how much smog is produced by automobiles, and how this could be changed, unless, since the problem is that of an area covering many cities, there is going to be some kind of legislation, binding on the whole area and not just on one component city of the area, to implement the rules, regulations, and practices that are required. A systems study of the smog problem in an area is not competently carried out unless its results make evident what rules and control organizations are needed. Again, we see beginnings in the United States; some one hundred air quality regions have already been defined. . . .

Once most people are wedded to creative logic and objectivity to get solutions to society's problems, the world is going to be a lot better. Then maybe we can say an important thing, namely, that science and technology are then being used to the fullest on behalf of mankind.

THEODORE ROSZAK

The Citadel
Of Expertise

Theodore Roszak, one of the most persuasive voices among the anti-technologists, is particularly critical of the experts, in whom Ramo sees society's salvation. To Roszak, the experts constitute an elite—a technocracy that exists for its own sake, does not serve any human purposes, and destroys human freedom.

The great paradox of the technological mystique [is] its remarkable ability to grow strong by virtue of chronic failure. While the treachery of our technology may provide many occasions for disenchantment, the sum total of failures has the effect of increasing our dependence on technical expertise. After all, who is there better suited to repair the technology than the technicians? We may, indeed, begin to value them and defer to them more as repairmen than inventors—just as we are apt to appreciate an airplane pilot's skill more when we are riding out a patch of severe turbulence than when we are smoothly under way. Thus, when the technology freezes up, our society resorts to the only cure that seems available: the hair of the dog. Where one technique has failed, another is called to its rescue; where one engineer has goofed, another—or several more—are summoned to pick up the pieces. What other choice have we? If modern society originally embraced industrialism with hope and pride, we seem to have little alternative at this advanced stage but to cling on with desperation. So, by imperceptible degrees, we license technical intelligence, in its pursuit of all the factors it must control, to move well beyond the sphere of "hardware" engineering—until it begins to orchestrate the entire surrounding social context. The result is a proliferation of what Jacques Ellul has called "human techniques": behavioral and management sciences, simulation and gaming processes, information control, personnel administration, market and motivational research, etc.,—the highest stage of technological integration.

Once social policy becomes so determined to make us dance to the rhythm of technology, it is inevitable that the entire intellectual and moral context of our lives should be transformed. As our collective concern for the stability of urban-industrialism mounts, science—or shall we say the scientized temperament?—begins its militant march through the whole of culture. For who are the wizards of the artificial environment? Who, but the scientists and their entourage of technicians? Modern technology is, after all, the scientist's conception of nature harnessed and put to work for us. It is the practical social embodiment of the scientific worldview; and through the clutter of our technology, we gain little more than an occasional and distorted view of any other world. People may still nostalgically honor prescientific faiths, but no one—no priest or prophet—any longer speaks with authority to us about the nature of things except the scientists. . . .

Given the extent of this empire of expertise, it might seem at first glance incomprehensible that urban-industrial society has any serious problems left at all. With such a growing army of trained specialists at work accumulating data and generating theory about every aspect of our lives, we should surely have long since entered the New Jerusalem. Obviously we have not. But why? The answer most commonly forthcoming in recent years is that our expertise has simply not been properly co-ordinated; it has been practiced in too narrow, disorganized, and myopic a way. So we begin to hear of a new panacea: the "systems approach"—and in years to come, we shall undoubtedly be hearing a great deal more about it.

Systems analysis, derived from World War II and Cold War military research, is the attempt to solve social problems by ganging up more and more experts of more and more kinds until every last "parameter" of the situation has been blanketed with technical competence and nothing has been left to amateurish improvisation. The method is frequently mentioned nowadays as the most valuable civilian spin-off of our aerospace programs. We are told that the same broad-gauged planning, management, and decision making that have succeeded in putting a man on the moon can now be used to redesign cities and reform education. In brief, systems analysis is the perfection of Ellul's "human technique." To quote the military-industrialist Simon Ramo, one of the great boosters of the approach, it is the effort to create a "multi-headed engineer," a "techno-political-econo-socio" expert. . . .

Notably, the good systems team does not include poets, painters, holy men, or social revolutionaries, who, presumably, have nothing to contribute to "real-life solutions." And how should they, when, as Ramo tells us, the relevant experts are those who understand people "as members of a system of people, machines, matériel, and information flow, with specific, well-described and often measurable performance requirements"? Above all, the experts are those who can place "quantitative measures on everything—very often, cost and time measures." . . .

The message is clear. The ills that plague urban-industrial society are not techno-genetic in essence; they are not the result of a radically distorted relationship between human beings and their environment. Rather, they result from an as yet incomplete or poorly co-ordinated application of scientific expertise. The province of expertise must, therefore, be broadened and more carefully administered. Urban-industrial society, having hopelessly lost touch with life lived on a simpler, more "primitive" level, and convinced beyond question of the omnipotence of technical intelligence, can do no other than trust to ever greater numbers of experts to salvage the promise of industrialism. Accordingly, the principal business of education becomes the training of experts, for whom progressively more room is made available in government and the economy. From our desperate conviction that the endangered artificial environment is the *only* livable environment, there stems the mature social form of industrial society: the technocracy.

VICTOR C. FERKISS

Bureaucracy

Victor Ferkiss, a political scientist, provides a historical interpretation of the growth of bureaucracy and the rule of the experts. He describes how bureaucracy originated as a liberating force in western European society, and emphasizes the role of technology in the consolidation of bureaucracy. Ferkiss concludes that sophisticated information technology, as provided by the computer, can be used to decentralize bureaucracy, returning decision-making to the people and diminishing the power of the experts.

What I have to say may seem startling and far out simply because it is so commonplace. But it is sometimes useful to know where we are and how we got here before we try to ascertain where we are going.

I may be wrong, but I think that for most people in America today... bureaucracy is a dirty word. This is likely to be true regardless of whether one is pessimistic or optimistic about technology. For many persons pessimism about technology includes the fear that it will make this terrible thing called bureaucracy stronger or more pervasive, while optimism entails the belief that technology will somehow make possible liberation from the thralldom of bureaucracy. If this is true, then before we talk about the relationship between technology and bureaucracy we must first gain a realistic understanding of the nature of bureaucracy and its impact upon society.

It is hard for many people today to recognize that bureaucracy came into the world—the Western world at least—as a liberating force, as did its

Reprinted by permission of the publisher, from Victor C. Ferkiss, "Symposium on Bureaucracy, Centralization, and Decentralization," in *Technology, Power, and Social Change,* edited by Charles A. Thrall and Jerold M. Starr (Lexington, Mass.: Lexington Books, D. C. Heath and Company, copyright 1972 D. C. Heath and Company).

concomitant, administrative centralization. Everyone is now so into the idea of community, face to face contact, and total relationships that it is hard to accept the fact that much of the Western past has been devoted to the attempt to break away from these very things.

The King's peace came to localities in England and Europe as an alternative to the often arbitrary rule of the lord of the manor, an alternative eagerly welcomed by the common man. Centralization, written law, and fixed rules were originally regarded as liberating. This is equally true with regard to the segmented kind of personal utilitarian relationships which so many people deplore. Just a few years ago . . . theologian Harvey Cox wrote a popular and influential book called *The Secular City,* celebrating among other things the impersonality of relationships in the large modern technopolitan metropolis. Since then he has changed his mind at least once, but many persons would still argue that there is much to be said for centralized bureaucratic power. If you are a black in Mississippi you are a lot better off dealing with a federal court operating under formal and universalized rules than with a local sheriff operating on the basis of personalized local criteria. But most of us today—and this is not purely a matter of intellectual fashion—fear impersonal power, segmented relationships, and all the other alleged evils we associate with bureaucracy.

Bureaucracy and centralization have historically gone hand in hand, both made possible by technology. Technology provided the kinds of communications devices which made the governing of large empires not only possible but easy. It made possible all the apparatus of reporting, inspection, conferences, and coordination which are part of any modern bureaucratic state. Technology allowed for the meticulous record keeping required by taxation, conscription, and public health and education. When Max Weber* describes the essence of modern bureaucracy—in a tone overlain with a pessimism which has had much to do with influencing the attitudes of subsequent generations of intellectuals toward bureaucracy—he is describing a system for which modern technology is the necessary if not the sufficient condition.

With technologically-conditioned modern bureaucracy came centralized and hierarchically organized power. The dominant rule of capital cities over political life and of major industrial and commercial centers over economic life was greatly strengthened. And, as Weber noted, hierarchical relationships—power at the top and obedience at the bottom—were a basic feature of the bureaucratic systems themselves and of the political and economic systems of which they were the instrument and underpinning.

---

*Editors' note: Weber (1864–1920) was a German sociologist and economist who argued that bureaucracy was the most important feature of modern society.

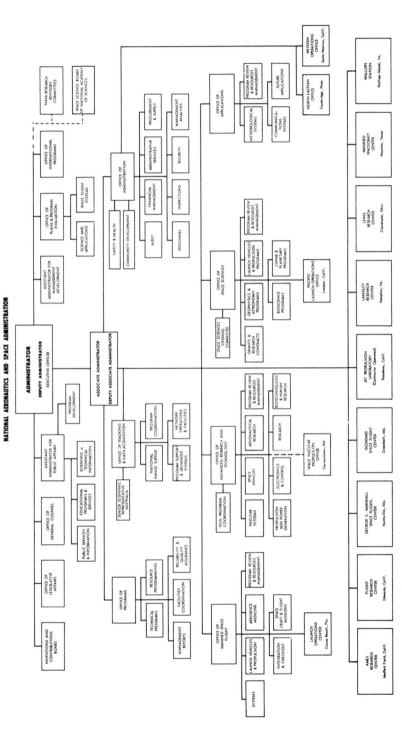

NATIONAL AERONAUTICS AND SPACE ADMINISTRATION, AUGUST 17, 1962

NATIONAL AERONAUTICS AND SPACE ADMINISTRATION

Organization of NASA: This chart exemplifies the complexity of modern bureaucracy.

But does the fact that in the past technology has been the source of hierarchical, centralized, bureaucratic political and economic systems mean that this must always be the case? Might not new forms of technology change these relationships and make possible a greater degree of decentralization, local autonomy, and individual freedom? New forms of physical power— especially electric motors—have made possible a high degree of decentralization of productive processes as compared with the early days of the industrial revolution. It can be argued that most of the economic centralization in industry today is the result of financial rather than technological considerations, that it is not physical but market conditions that give the giant corporation its competitive advantages. In the same vein, new technologies have led to a greater decentralization of population within existing metropolitan areas. Some would argue that Mr. Mumford's dream of garden cities is, ironically, already with us in the nightmare of suburbia, made possible by the automobile's decentralizing impact upon patterns of population settlement.

Beyond this, modern technology has made possible dispersion of control centers as well as of population. Some urbanologists would argue that this is an oversimplification, that birds of a feather continue to flock together, that even if Gulf Oil can have its technical headquarters in Houston and service your credit card from there, its executives want to have their real headquarters on the same blocks in New York where the big banks are. Many industrial and especially financial and intellectual activities still remain concentrated. But it is now technologically possible to run a large modern state from anywhere, from a San Clemente or Florida "White House" as well as from Washington. And certainly it is not necessary to have every activity run from the same location, even if individual activities are run from central headquarters. The National Institutes of Health are not in Washington, but in suburban Maryland; the Pentagon is in suburban Virginia; the Space Flight Center in Houston; the Strategic Air Command in Colorado; and so on. Communications today make it possible for individuals to operate from almost anywhere and for societies to be controlled from almost anywhere. There is a tremendous amount of choice possible.

But wherever it is located, centralized direction has its problems; problems which, paradoxically, are created by the very improved technological means which make a high degree of communication possible. The more information coming into a central place, the more decisions which can be referred to a central headquarters, the greater the information and decision overload. This is not a wholly new problem of course. It used to be said that the Emperor Franz Joseph of Austria-Hungary used to try to read and sign every important government document himself and as a result was five years behind when he died. However apocryphal, this story illustrates how bad things could get just using ordinary pen and paper supplemented by the telegraph wire. Paradoxically, the vast technological advances of the mere half century since Franz Joseph's time have so confounded the problem of administration that it is harder to get needed information than ever before. . . .

Nor is information overload the only block to centralized direction of large scale activities created by improvements in communications technology. The larger an organization, the more efficient its internal communications, the more things to coordinate as far as decisions are concerned and the harder it is to get agreement and consistency. There is so much business to be transacted that in order to avoid information and work overload officials must rigidly and arbitrarily narrow their fields of attention, as a result of which decisions are made by subcenters of power which are mal-integrated or even directly at cross-purposes with each other. This kind of counterproductive activity will also occur in large organizations as a result of political power considerations, independently of technological factors, as when the Agriculture Department helps export tobacco and cancer to our friends abroad at the same time the Surgeon General's office is trying to curtail tobacco-caused cancer at home. But communications overload makes the situation much worse.

Suppose one wanted to overcome the difficulties caused by information overload in order to make more efficient social control possible? Does technology offer any options for solving some of the problems it has helped to create or exacerbate? Recent concern with the interrelated problems of our cities and the ecological situation of our planet as a whole has led to a burst of interest in the possible use of the computer as a tool for mapping, or modeling, or managing large systems. A computer is essential to handle the vast number of variables involved in systems analyses of the kind proposed by such theorists as Jay Forrester,* which involve operations too complex for the unaided human mind.

Would the perfection of such computer technology, enabling us to better understand what is going on in the complex systems of which we are a part, lead to structures which are more centralized, more bureaucratic, and more hierarchical than those which so many deplore at present? Not necessarily. Organic unity and efficient communication—which may be necessary to planetary survival—may not require as much hierarchy and centralization as sometimes assumed.

Our oldest organic theories of government go back to Plato, who compared the body politic to a human body, its organs under the control of man's mental faculties. But while it is true that a good deal of the time our voluntary

---

*Editors' note: Forrester, a professor at Massachusetts Institute of Technology, has designed complex computer simulations of world conditions, in which such factors as food, energy, population, etc., can be studied.

muscles are controlled by deliberate intellectual decisions—and it may be possible that many more aspects of our physiology are subject to mental control than we have heretofore believed—Plato's analogy between the human body and society is based on a false premise. From the biologist's point of view, the body is a complicated system of information feedback where much of what is thought in the brain depends on the signals it gets from nerves, glands, and other sources. The body is far more democratic than Plato realized.

Thus, if by bureaucracy one means a necessarily hierarchically controlled organization, it is possible to conceive of a system possessing a high degree of coordination which is not really bureaucratic in the usual sense of the word. Information technology could be devised which could make everyone a participant in social decision making even at the national or world level. People could participate in coordinated large scale communal activity without the feeling that they were mere cogs in an engine operated from outside, or simply pawns being pushed about by unreachable, unmovable leaders.

If freedom means absolute individual or local autonomy, independent of the desires and needs of the larger community, then freedom cannot exist in the modern world. But if freedom is conceived of as effective participation in coordinated decision making, involving a high degree of understanding of what is actually going on, it may be possible. Kenneth Boulding talks about the need to move from a cowboy economy to a spaceship economy if our densely populated, highly interdependent world is to survive. But it is equally important that we move from a cowboy mentality about politics to a spaceship mentality, which recognizes that understanding and participation rather than isolation and complete independence is the essence of freedom.

In summation, then, bureaucracy as it has existed in the past is going to change radically—or at least can be radically changed—as a result of new technological options. New possibilities exist which can help us cope with the major political task of our age: the creation of mechanisms adequate for dealing with the problems of an interdependent world while at the same time permitting the individual a measure of freedom, power, and control of his own identity and destiny. Technology, however, can never do more than provide us with possibilities; the task of choosing is our own.

LYNN WHITE, JR.

Dynamo and Virgin Reconsidered

Written by a descendant of American presidents and statesmen, The Education of Henry Adams *is a classic autobiography. In it, Henry Adams sharply contrasted the symbolic power of the Virgin Mary as evidencing the spirituality of the Middle Ages and the awesome power of the electric dynamo as expressing the materiality of the modern era. In this selection, Lynn White argues that Adams' analysis was incomplete, because a profound change occurred in the medieval period, which escaped him. Labor-saving power machinery first appeared in profusion in the Middle Ages, and free men, not slaves, used this machinery to build the great cathedrals honoring the Virgin. Power machinery relieved monotonous human drudgery and contributed to spirituality by enhancing human freedom. Thus, the Virgin and the dynamo are not symbolic opposites, but allies.*

The *Education of Henry Adams* is a classic in the proper sense that, while it is read by few, it has helped to shape the notions and emotions of millions of Americans who have never even heard of it. Adams, both a grandson and a great-grandson of Presidents of the United States, was thwarted in his own political ambitions and so turned his devious and subtle mind to the problem of the dynamics of human destiny. In his old age he became less a historian than a philosopher of history. "I don't give a damn what happened," he exclaimed to a friend. "What I want to know is why it happened."

Lynn White, Jr., "Dynamo and Virgin Reconsidered," *The American Scholar*, Spring 1958, pp. 183–194 as excerpted. Reprinted by permission of the author.

But he was a historian-philosopher with the soul of an artist even though, by his own wry diagnosis, being "a quintessence of Boston . . . [his] instinct was blighted from babyhood." As an artist he groped for symbols—not only words but images.

His first great image came to him in the summer of 1895 when, with Senator and Mrs. Henry Cabot Lodge, he toured Normandy and the neighboring regions: Amiens, Bayeux, Coutances, Mont-Saint-Michel, Vitré, Le Mans, Chartres. He had, of course, known of the things he saw; but now, suddenly, he grasped them. Or rather, these incredible churches of the twelfth and thirteenth centuries seized him and shook him until his intellectual teeth rattled. What force had impelled men to infuse stone with such striving and restlessness, and to fling it high into heaven? Why had they filled acres of windows with throbbing colors such as a Yankee had never dreamed of? Why had they enriched every portal and pinnacle with figures from earth, hell, and paradise?

All of these buildings were shrines of St. Mary the Virgin, the Mother of God. It was she, the Queen of Heaven, who had commanded and inspired this titanic outburst of cultural vitality. To explain to himself and to America "why it happened," Henry Adams wrote *Mont-Saint-Michel and Chartres,* a work of genius but very personal in its interpretation. For Adams' Virgin is not what medieval Catholicism had thought its Virgin was: she is, rather, the life force, the wellspring of fecundity, "reproduction—the greatest and most mysterious of all energies." She had been the Oriental fertility goddesses, Diana of the Ephesians, Venus. In her Christian garb, however, whether "symbol or energy, the Virgin had acted as the greatest force the Western world had ever felt, and had drawn man's activities to herself more strongly than any other power, natural or supernatural, had ever done."

But Henry Adams was puzzled as to why this force, formerly so irresistible, had declined. "The Woman had once been supreme; in France she still seemed potent, not merely as a sentiment, but as a force. Why was she unknown in America? . . . American art, like American language and education, was as far as possible sexless. Society regarded this victory over sex as its greatest triumph." Clearly, America was bowing down before some new power. What was it?

In 1900, at the Trocadero Exposition in Paris, Adams found his second image, the dynamo. "Among the thousand symbols of ultimate energy, the dynamo . . . was the most expressive." To him it became "a symbol of infinity." He felt it to be "a moral force, much as the early Christians felt the Cross." A profound change had taken place in the nature of civilization: "The new American, like the new European, was the servant of the powerhouse, as the European of the twelfth century was the servant of the Church." . . .

And yet these two great symbols, designed to achieve intelligibility for the human condition, only served to underscore human defeat. *The Education of Henry Adams* is one of the most reticent of autobiographies: emotion is many layers below the surface. But at his death, in a wallet of very special papers, a lyrical and agonized "Prayer to the Virgin of Chartres" was found which in-corporates a prayer to the dynamo clearly prophesying—in 1906 at the latest—the destruction of humanity by atomic energy. The Virgin represented all that was distinctively human, all that the Cartesians regarded as "subjective"; the dynamo pointed to the annihilation of all human values, first by the achieve-ment of an antlike society, and then by the victory of impersonal cosmic force over all life.

This is the dread of the twentieth century. Admitting that we have many totalitarianisms and may have more, admitting that the atom is a peril, never-theless are we justified in believing with Henry Adams that human values are engaged in a last and hopeless duel not only with the death wish in our own natures but with the entropy of the universe as well? Symbols are not merely means of expressing thought about experience; often they impel us to sche-matize our understanding of experience in their own terms. Are the dynamo and the Virgin in fact opposing and mutually exclusive energies?

Even Adams saw the joke when, in the summer of 1904, he bought a French automobile in which to go on pilgrimage to the shrines of Our Lady. However dissimilar spiritual and mechanical forces might seem, the student of history was able to fuse them happily for professional purposes. Was this just a convenient accident, or was it a pragmatic symbol with a validity as great as Adams' dualism?

And when one goes to Beauvais or Laon, what does one see? Structures which are the greatest engineering feats in human history up to the time of their building. The technicians of the twelfth and thirteenth centuries, far from being traditionalists, were creating an entirely new concept of architec-ture, dynamic rather than static. In their cathedrals we see a sublime fusion of high spirituality and advanced technology.

Moreover, these are the first vast monuments in all history to be built by free—nay, unionized!—labor. Almost the only branch of technology to make little progress during the Middle Ages was weight-lifting. The freemasons of those centuries, unlike the voiceless slaves and conscript peasants of past ages, would not often be bothered with handling big stones. Both Romanesque and Gothic engineers achieved enormous scale with small blocks. That these churches are filled not only with gifts and monuments of the feudal aristocracy but perhaps to an even greater extent with chapels, windows, and the like offered by the guilds of craftsmen and merchants who dominated the walled cities above which the Virgin's temples towered points to a social revolution closely connected with the technological revolution. And these changes had their roots not merely in economic developments but in religion, which, if vital, can never be divorced from its context.

St. Benedict of Nursia, the founder of the Benedictine Order, is probably the pivotal figure in the history of labor. Greco-Roman society had rested on the backs of slaves. Work was the lot of slaves, and any free man who dirtied his hands with it, even in the most casual way, demeaned himself. Plato once sharply rebuked two friends who had constructed an apparatus to help solve a geometrical problem: they were contaminating thought. Plutarch tells us that Archimedes was ashamed of the machines he had built. Seneca remarks that the inventions of his time, such as stenography, were naturally the work of slaves, since slaves alone were concerned with such things. In the classical tradition there is scarcely a hint of the dignity of labor. The provision of Benedict, himself an aristocrat, that his monks should work in fields and shops therefore marks a revolutionary reversal of the traditional attitude toward labor; it is a high peak along the watershed separating the modern from the ancient world. For the Benedictine monks regarded manual labor not as a mere regrettable necessity of their corporate life but rather as an integral and spiritually valuable part of their discipline. During the Middle Ages the general reverence for the laboring monks did much to increase the prestige of labor and the self-respect of the laborer. Moreover, since the days of St. Benedict every major form of Western asceticism has held that "to labor is to pray," until in its final development under the Puritans, labor in one's "calling" became not only the prime moral necessity but also the chief means of serving and praising God. The importance of frugal living and consecrated labor in building up fluid investment capital and in fostering the rapid expansion of capitalist economy in the regions of Europe and America most deeply affected by the puritan spirit is a commonplace of the economic history of early modern times. The Benedictine ancestry of the puritan attitude toward work is less often emphasized.

Moreover, although St. Benedict had not intended that his monks should be scholars, a great tradition of learning developed in the abbeys following his *Rule:* for the first time the practical and the theoretical were embodied in the same individuals. In Antiquity learned men did not work, and workers were not learned. Consequently, ancient science consisted mostly of observation and abstract thought; experimental methods were rarely used. The craftsmen had accumulated a vast fund of factual knowledge about natural forces and substances, but the social cleavage prevented classical scientists from feeling that stimulus from technology which has been so conspicuous an element in the development of modern experimental science. The monk was the first intellectual to get dirt under his fingernails. He did not immediately launch into scientific investigation, but in his very person he destroyed the old artificial barrier between the empirical and the speculative, the manual and the liberal arts, and thus helped create a social atmosphere favorable to scientific and technological develoment. It is no accident, therefore, that his ascetic successors, the friar and the Puritan, were eminent and ardent in experiment.

The first power machine, the water wheel, appeared in the Roman Empire in the last century before Christ; but it may have been a barbarian invention, since we know that by the time of Christ it was already used in northern Denmark and China. Not until after the disintegration of the Western Roman Empire did it become common. By 1086, William the Conqueror's *Domesday Book* shows us that in England there were over 5600 water mills for some 3000 communities. Since there is no reason to believe that Britain was technologically in advance of the Continent, this means that by the end of the eleventh century every European was living daily in the presence of a major engine motored by nonhuman power.

This is the sort of situation that breeds new ideas. From the late tenth century onward, water power gradually began to be applied not only to grinding grain but to all sorts of industrial processes: forging, tanning, making the mash for beer, sawing wood, doing the laundry, polishing armor, and sharpening knives. In cloth-making one of the most tedious and labor-consuming processes was fulling: for ages men had endlessly tramped the raw cloth in troughs filled with water and fuller's earth until it was properly felted. Now hammers powered by water wheels did the job so effectively that in the thirteenth century the English cloth industry shifted from the flat southeast to the more rugged northwest part of the country, where the streams ran a bit faster and good millsites were more available.

The creak of water wheels, then, was a sound as typical of Henry Adams' age of cathedral building as was the *Ave stella maris* chanted in praise of the Virgin. Was this pure coincidence, or perhaps even the embryonic form of the modern schizophrenia which separates spiritual values from material concerns? The Middle Ages did not seem to think so. St. Bernard's Cistercian monks were so devoted to the Virgin that every one of their hundreds of monasteries was dedicated to her; yet these White Benedictines seem often to have led the way in the use of power. Some of their abbeys had four or five water wheels, each powering a different workshop. . . . For two hundred years and more there had already been a rapid replacement of human by nonhuman energy wherever industry demanded large amounts of power or where the required motion was so simple and repetitive that a man could be replaced by a mechanism. The chief glory of the Middle Ages was not, as Henry Adams thought, its cathedrals, its epics, its vast structures of scholastic philosophy, or even its superb music, which Adams' nieces learned to sing for his delectation; it was the building for the first time in history of a complex civilization which was upheld not on the sinews of sweating slaves and coolies but primarily by nonhuman power. The century which achieved the highest expression of the cult of the Virgin Mary likewise first envisaged the concept of a labor-saving power technology which has played so large a part in the formation of the modern world. . . .

The Virgin and the dynamo are not opposing principles permeating the universe; they are allies. The growth of medieval power technology, which escaped Adams' attention, is a chapter in the conquest of freedom.

More than that, to those who search out "why it happened," it is part of the history of religion. The humanitarian technology which in later centuries has grown from medieval seeds was not rooted in economic necessity; for such "necessity" is inherent in every society, yet has found expression only in the Occident, nourished in the tradition of Western theology. It is ideas which make necessities conscious. The labor-saving power machines of the later Middle Ages were harmonious with the religious assumption of the infinite worth of even the most seemingly degraded human personality, and with an instinctive repugnance toward subjecting any man to a monotonous drudgery which seems less than human in that it requires the exercise neither of intelligence nor of choice. The Middle Ages, believing that the Heavenly Jerusalem contains no temple, began to explore the practical implications of this profoundly Christian paradox. Although to labor is to pray, the goal of labor is to end labor. . . .

ROBERT H. GUEST

Scientific Management and The Assembly Line

In this selection, Robert Guest, a professor of business administration, describes the origin and development of the Scientific Management movement, whose guiding spirit was the engineer Frederick W. Taylor. Taylorism, as the movement was properly termed, was used by Henry Ford in setting up assembly line manufacture, and by the mid-1930s, scientific management was in full bloom. It became highly controversial and significantly aided the rise of industrial unions. Despite its many limitations, Guest believes that scientific management, through the mediation of industrial psychology, will lead to a better understanding of the nature of human work.

Few men in the history of American technology have had greater impact on the organization of work than Frederick W. Taylor (1856–1915), the father of Scientific Management. What Eli Whitney and others did to lay the groundwork of mass production in the 1800s, later perfected in the continuous-flow technology of Henry Ford in the twentieth century, Frederick Taylor applied to the motions of men at work. Today's large fraternity of industrial engineers, systems and methods experts, work-standards specialists,

From *Technology in Western Civilization*, Volume II: *Technology in the Twentieth Century*, edited by Melvin Kranzberg and Carroll W. Pursell, Jr. Copyright © 1967 by The Regents of the University of Wisconsin. Reprinted by permission of the Oxford University Press, Inc.

and a whole host of management experts in a very real sense owe their jobs and allegiance to Taylor. While many would credit America's great industrial leap forward in the twentieth century in large measure to the work of this man, others—especially in the trade union movement—would condemn Taylor for "making man just another machine."

The Rise of Taylorism

In the closing decades of the nineteenth century Taylor became concerned with vast inefficiencies he found in the burgeoning industries of the period. In many industrial occupations the need for craft skills was disappearing. Basic industries that were formerly isolated and small-scale grew in size and complexity. In steel, for example, there were large concentrations of production facilities operated by unskilled and semi-skilled manpower in the Bessemers, open hearths, rolling mills, and in the fabricating segments of the industry. The technical processes determined the character of the jobs, the pace at which men worked, and the degree of control workers could exercise over their jobs. The work itself was becoming hardly more than the use of muscle power to feed the furnaces and operate the machines. Taylor watched at close range how the workers performed and became convinced that most were working at a low degree of efficiency.

As a disciplined and methodical thinker well versed in engineering logic, Taylor, in effect, asked the simple question, "Why can we not apply the same principles of efficiency to the hand and muscle of man that we apply to the design of machines?" In one of his first experiments in the steel industry late in the 1890s, he tackled the case of a hardworking but unskilled pig-iron handler. Taylor's own words show vividly his own thinking as he described the incident later.

> "Now, Schmidt, you are a first-class pig-iron handler and know your business well. You have been handling at a rate of 12½ tons per day. I have given considerable study to handling pig iron, and feel sure that you could do a much larger day's work than you have been doing. Now don't you think that if you really tried you could handle 47 tons of pig iron per day, instead of 12½ tons?"
>
> What do you think Schmidt's answer would be to this?
>
> Schmidt started to work, and all day long, and at regular intervals, was told by the men who stood over him with a watch, "Now pick up a pig and walk. Now sit down and rest. Now walk—now rest," etc. He worked when he was told to work, and rested when he was told to rest, and at half past five in the afternoon had his 47½ tons loaded on the car. And he practically never failed to work at this pace and do the task that was set him during the three years he was at Bethlehem; and throughout this time he averaged a little more than $1.85 per day, which was the ruling rate of wages at that time in Bethlehem. That is, he received 60 per cent higher wages than were paid to other men who were not working on task work. One man after another was picked out and

trained to handle pig iron at the rate of 47½ tons per day until all of the pig iron was handled at this rate, and the men were receiving 60 per cent more wages than other workmen around them.

In this example one can see the raw data from a single experiment that were to become the rudiments of the great Scientific Management movement. Taylor had observed and timed all of Schmidt's movements to determine which motions were necessary to perform the tasks and which were not. By instructing the worker in the precise motions required, with adequate time for rest, the operation could be performed more efficiently in a machine-like manner. Taylor proved here, and in many applications that followed, that it was possible to manipulate human activity and to control it by logical procedures in much the same way that physical objects could be measured and controlled.

In deciding to make work "scientific," Taylor had been influenced not only by the tendency of workers to avoid hard work, but also by the current practices of trying to force men to work. As a plant supervisor himself, Taylor had become dissatisfied with the usual methods of threatening workers with discharge or with other types of persuasion. And until he attempted the experiment described above he had been frustrated by the seeming ability of groups of workers to "peg" the pace at which they worked. Contrary to what critics have since said about Taylor, he placed the blame not on the worker as a person, but rather on the procedures that were expected of him in performing the task. Once the proper procedures were established, it would be easy to perform the work, Taylor thought. He also felt that the rate at which work ought to be done could be determined scientifically.

The hallmarks of the Taylor method were job analysis and time study. Job analysis depended upon one's being able to break down a series of operations into elements made up of simple constituent motions. The elements of the operation would be analyzed to determine which were superfluous and which were essential for the job. These elements could then be arranged, and rearranged, so long as the sum total added up to the total operation for the worker. In the total complex of a worker's operation, Taylor discovered that much time was wasted by improper tools and equipment. In mining operations, for example, if a worker had to take time to sharpen his cutting tool, this delay constituted "down time" which was a waste. Taylor thus decided that such operations were more properly jobs for someone else. Careful examination showed many other elements of the job that could be taken out and delegated elsewhere. In short, Taylor was applying at the immediate job level the same principle of division of labor that had been applied to the factory or industry as a whole.

The other ingredient of Taylor's system was time study. Once the job had been broken down into its constituent parts and extraneous motions eliminated, the remaining operations could be timed. The timing was done by a stopwatch, a device which, in addition to serving its utilitarian function, later became to workers a symbol of suspicion and distrust. Taylor was able to take each element of the job, time it, and then calculate how much time was involved in the work cycle. In machining operations, for example, some of the elements he timed were functions of the machine tool itself, while others were physical motions of the operator's own functions. Thus, when the worker was merely supervising and adjusting the machine while it operated, the machine time was recorded. But when the operator was setting up the work and removing it, these elements were considered part of the handling time.

Taylor was realistic in his recognition that certain allowances had to be made for unanticipated problems, both in the machine and in the worker. Although Taylor claimed that these allowances themselves were subject to time and motion study, they were, in fact, usually arbitrary and based upon general experience. Moreover, individual differences between workers were not considered. When Taylor was questioned about who should be timed he gave different responses: "Select a good, fast man," or "Give me an average, steady man." Furthermore, when asked who was capable of making the time study, his usual reply was, "Give me a man with experience." He did insist that all of the technical aspects of work should be standardized and put at the highest level of efficiency before the stopwatch was applied. Taylor made simplistic assumptions with respect to the psychological motivations of the individual and his needs as a member of a human work group. Indeed, Taylor insisted on doing everything to exclude those factors that make up what we now regard as the total environment of a job. It should be emphasized, however, that some of Taylor's greatest contributions resulted from his insistence that technical operations themselves, apart from the human element, be standardized, synchronized, and operated efficiently.

The Worker on the Assembly Line

The principal criticism of Taylor and the work rationalization concept is that it ignored the workers' feelings and motivations. Taylor himself vigorously denied this, believing that men would respond favorably to the obvious logic of its benefits. Taylor expected the worker to appreciate the elimination of wasteful and unproductive motions; to be happy to have tasks simplified so as not to have to make complicated decisions; and to welcome guidance in the "best way" of performing the task. The worker would be given the right tools to do the job, and the machines would be kept in proper adjustment. Furthermore, the worker would be paid fairly for his effort. Wage incentives could be established on a piecework basis to give extra compensation for extra effort. The worker, like any normal human being, would respond to man's natural desire to benefit himself economically—at least Taylor thought so.

One way to illustrate how some of Taylor's assumptions might be questioned is to look at the job which represents an extreme form of work rationalization: assembly-line operations, and more specifically, those in the automobile industry. As has been pointed out, "The extraordinary ingenuity that has gone into the construction of automobile assembly lines, their perfected synchronization, the 'all but human' or 'more than human' character of the machines, the miracle of a car rolling off the conveyor belt each minute under its own power—all this has caught and held the world's imagination. . . . On the other hand, the extreme subdivision of labor (the man who puts a nut on a bolt is the symbol) conjoined with the 'endlessly moving belt' has made the [automobile] assembly line the classic symbol of the subjection of man to the machine in our industrial age."

Utilizing the two basic principles of standardization and interchangeability, Henry Ford was able to work out and apply three additional "principles" of progressive manufacture in pioneering the automobile assembly line: (1) the orderly progression of the product through the shop in a series of planned operations so arranged that the right part always arrives at the right place at

the right time; (2) the mechanical delivery of these parts to the operators and the mechanical delivery of the product from the operators, as they are assembled; and (3) a break-down of operations into their simple constituent motions.

These principles are purely mechanical. Extended to the human component of the total work-flow system, and when combined with the Taylor principle, they mean the following for the worker: (1) a mechanically controlled work pace; (2) repetition of simple motions; (3) minimum skill requirements; (4) predetermined operating procedures; (5) a small fraction of the total product worked on; and (6) superficial mental attention....

The Underlying Assumptions of Scientific Management

The basic assumption Taylor made was that men were motivated by the desire to maximize economic gain. Anything done to make their work more efficient would make workers produce more to get more money. Therefore, the worker would be happy and satisfied on his job. A second assumption was one that seemed quite consistent with the long tradition of "individualism" in American industry. It held that the worker's world was focused on the individual worker and that the group of people with whom he works was not important to him. Hence the worker's relationship to his fellow workers, to his boss, and to the total organization, were virtually ignored. Because of this the further assumption was made that man, being the logical extension of the machine, could have not only his physical motions but his thinking processes standardized. It assumed that there was only one correct way to do a job, and that that way was determined by the requirements of the machinery. All one had to do to have a "standard" worker was to standardize his motions, his hours, his basic wages, and the entire routines of his work life. The result was that the worker became hardly more than a passive agent of the machine process.

The inadequacy of these assumptions was reflected in one form of worker response, namely, the growth of unionism. The union movement was an organized recognition not only that men had to be protected from arbitrary action by management, but that workers had, indeed, a need for expression as members of social groups. It recognized, too, that the motives or norms of the groups are not always directed toward the maximization of work and efficiency. In fact, one of the most common phenomena in the history of work is the powerful force that workers can generate in work restriction or "doing as little as one can get away with."

The Human Relations Movement

The first serious research to expose the limited assumptions of Scientific Management was that inspired by Elton Mayo (who has been called America's first industrial sociologist) during the 1920s. These studies, frequently hailed as the beginnings of a counter-movement against Scientific Management, opened up the era of what is commonly called the human relations movement.

Mayo's study took place at the Hawthorne plant of the Western Electric Company, just outside of Chicago. The researchers began their studies in much the same way that Taylor began his, except that they were interested in problems of physical fatigue among the workers. During World War I the British Medical Research Council had done considerable work on fatigue among factory workers. Now, a few years later, the Hawthorne group sought to discover first, the relationship between the workers' efficiency (as measured in parts produced) and the amount of illumination that would affect the performance of their tasks. They chose two groups of employees with similar backgrounds working under similar conditions. The plan was to have one experimental group and vary the intensity of the light under which they worked in order to determine whether it had any effect on their output. Conditions for the "control" group were to be held constant.

The first shocking revelation was that there did not seem to be any relationship between the rate of production and changes in the amount of lighting. The researchers were curious about this fact; they thought perhaps there was something wrong with their experimental design, or that there might be factors other than illumination.

Mayo, who had earlier performed research in textile mills in Philadelphia, was convinced that the problem was not simply physiological, or the amount of light and its effect upon the workers' vision. Rather, he believed that there were also psychological factors at work. The researchers therefore switched the experiment slightly in order to get at these; they told the workers that the light bulbs had been changed (although they had not been), and implied that the illumination would be brighter. The workers responded by saying that they liked the "increased" illumination. This in turn was proof to the researchers that something other than physiological conditions was needed to explain the paradox.

The experiment continued in hopes that the researchers would still investigate the problem of the illumination as such and isolate the psychological factors. To do so, the researchers decided to make some other changes related to rest pauses and the hours of work. They further reasoned that if they could isolate the workers they could identify other physical conditions of work which might explain output behavior. A new group was then organized and placed in a separate room. The six girls chosen were average workers assembling a not-too-complicated telephone relay component. The job cycle was approximately one relay per minute. Thus, any changes in output would be clearly measurable. By this time, and from past experience, the investigators expected that they would probably not be able to identify any single factor, such as illumination. However, they were curious about other kinds of conditions and attitudes related to the work experience. In carrying out the experiment, the researchers kept careful records of the time of day when defects might occur. Weather conditions, temperature, and humidity in the test room were also controlled; and medical examinations were held periodically. Activities outside of working hours and types and amount of food eaten were also considered. Through direct observation the investigators watched carefully every aspect of worker behavior hour by hour. They also took notes of conversations. These later proved one of the keys to the "great discovery" that pointed to the importance of social factors at work.

After the experiment had been under way for some time, it was decided to call the girls in when changes were contemplated. More than that, their own comments were encouraged, and they were even allowed veto power over any change that was made or proposed. They were also encouraged to work at a natural pace and not to push themselves too hard just to satisfy the experiment.

The researchers felt that over an extended period of time it would be possible to find out what made for the most satisfactory output both in terms of quantity and quality. In the first experimental period no change was made in the working conditions and the hours stayed the way they always had been. This period had been preceded by a two-week period when production records were kept without the girls' knowledge. In the third series of weeks the method of payment was changed so that each girl was paid more closely in relation to her own effort. Several weeks later, efforts were made to alter the rest pauses, varying them in frequency and length. In some phases of the experiment, the rest periods lasted as much as half an hour each day.

These experiments went on for a year, with the experimenters becoming frustrated over the fact that the results were not what they had expected. Instead of changes that could be related directly to various types of changes in rest periods, they discovered that there had been a steady increase in output throughout the year. Even when they reverted to the original conditions of work with no rest pauses, including no special lunch period, the daily and weekly production rose to a point higher than it had ever been before. The physiological experimental work thus went down the drain because of this mysterious and unexplained continued rise in output.

The girls' own comments eventually suggested the key to the puzzle: they looked upon the experiments as fun. They also enjoyed not being told by their boss what to do, and being able to share ideas with the experimenters. More important, the girls had the feeling that they were part of an experiment that could lead to improvements in working conditions for all the workers. It was, in other words, the girls' *involvement* which improved their output, not the specific conditions of their work environment. The Hawthorne studies paved the way for literally hundreds of other experiments since that time by many researchers. All of them led to the conclusion that the measurement of work itself and the application of Scientific Management principles ignored the importance of the human group and the motivation associated with group behavior.

Contributions of Scientific Management

Despite the subsequent modification of Scientific Management, Taylor's work stands as an inevitable though unique contribution to modern industrial technology. When all other resources were being rationalized, labor could not hope to escape. Paradoxically, improvement over the years has come in two opposite directions. First, it is now recognized that many of Taylor's "scientific" findings were, in fact, mere rationalizations of the class interests of management rather than necessary conditions implicit in the work itself. At the same time, the abandoning of the more extreme claims of Scientific Management and the recognition of psychological factors, through the introduction of human relations studies, have actually made the approach more "scientific" over the years.

What will be required in the future will be a combination of the advantages of Scientific Management with the contributions of behavioral science to the understanding of the nature of human work. In an automated society where we will be less concerned with the efficient use of hand and brawn our efforts will need to be directed toward releasing the great potential of the mind. This is the challenge for tomorrow's enlightened management.

GARY BRYNNER AND DAN CLARK

Worker Alienation

Two automobile workers, Gary Brynner and Dan Clark, testifying before a Senate subcommittee on worker alienation in July, 1972, describe vividly the experience of working on an assembly line and the effects of alienation on workers. Note that Clark states that the longest period that anyone has worked at the plant is about six and a half years.

Statement of Gary Brynner, President, UAW Local 1112, Lordstown, Ohio, Accompanied by Dan Clark, Member, UAW Local 1112, Lordstown, Ohio

Mr. Brynner: Thank you. I might say it is our pleasure to offer testimony to your committee. We feel it is an avenue that should have been traveled, and we are glad that it has taken place now. There are problems with the workers.

"Alienation" is, I guess, a good term. We offer some suggestions and some criticisms as they be.

There are symptoms of the alienated worker in our plant where we specialize, where I am President, and Dan is an assembler. Absentee rate, as you said, has gone continually higher. Turnover rate is enormous. The use and turning to alcohol and drugs is becoming a bigger and bigger problem, and apathy—apathy within our union movement toward union leaders and to the Government.

I think those lead from the alienation of the worker. In our plant we make 101.6 Vegas an hour, the fastest line speed in the country. A guy has about thirty-six seconds to do an operation. The jobs are so fragmented that he is

From *Worker Alienation,* Hearings Before the Subcommittee on Employment, Manpower and Poverty of the Committee on Labor and Public Welfare, U.S. Senate, 92nd Congress, 2nd session, July 25–26, 1972, pp. 8–13.

offered very little as far as input to that product. He cannot associate with it or he does not realize what he is doing to it.

Conveyor lines in our plant, the heights, and every movement of the conveyor line is determined to make the guy a little more efficient, to take movement of the bending and stretching, to make him more efficient.

They believe in discipline. When they do not have the discipline rule to cover a violation, they create one. Drugs has been on the increase in plants as you well know. So our corporation drafts up another shop rule to cover misuse or selling of drugs on company premises, another form or way to discipline the workers. They correct and they deal with the problem, but they never look for the cause. They further alienate the worker by more discipline. You cannot control people with fear. They should have learned that in our fight in just this past year, where they disciplined some one thousand workers who refused to speed up on their lines, and out of that one thousand—and I do not want you to feel sorry for us, because it cost them tens of thousands of dollars in back-pay, because the union was able to establish they were wrong, the workers were not suppressed to that kind of action without regress or redress.

I brought with me a guy who is an assembler in the plant where we make the 101.6 Vegas an hour. His name is Dan Clark. Dan is faced with those problems. I am not because as president I am full-time, and he can tell you how an assembler feels and how he suffers and the inequities that go on.

At this time I would like to have Dan present the assembler's view.

SENATOR KENNEDY: Very good.

MR. CLARK: Senator Kennedy, members of the committee: I have briefly a few problems that face the workers. I work in the body shop, and that is where your car first begins. You start putting certain parts together—

SENATOR KENNEDY: How long have you worked there?

MR. CLARK: I have been there now about five years, which is a year and a half short of the highest seniority there.

The problems that face the workers are monotony of the job, repetition, and boredom. We are constantly doing the same job over and over again. Where you have problems of hours, like right now—I am on sick leave, just had an operation, but before I left, four days ago, we were working ten and one half to eleven hours, which you have no excuse to leave.

Eight hours, a working day, which it should be eight hours, you cannot leave in eight hours unless you have an excuse to go to the doctor's, hospital, or emergency call of one of your kids are sick. That is the only way you are going to leave.

You stay in that plant for ten and one half to eleven hours of their choosing, and it may be maybe 95 degrees outside, but inside you can almost bet in the body shop it is a good 110. But that is not the warmest place in the whole plant. The warmest place in the whole plant is the paint shop. That is on the second level.

When it is 95 outside, it is 120 or so in that paint shop, and your ventilation system, the air you are getting blown in on you, is supposed to be so cooled down—that is that 95-degree temperature outside that is coming on. You have no ventilation really at all. They are situated in such a place, they are not on the job anyhow. Your job may be off to the lefthand side and the ventilation is on the righthand side.

They say there is nothing we can do about it, that is the way it was designed, and that is the way it is. There is nothing we can do.

There is the noise level in the plant. The body shop is worse. One man says, he is a supervisor there, has taken under his control, he says, noise level. I know there are problems there, it is above the noise level that it is supposed to be. He says I will take care of it. That is a year and a half now which has gone by and nothing has been done yet. There has been nothing provided unless you want to provide it yourself.

I know I put cotton in my ears, because I cannot take it too much longer. There is pollution in the plant.

In the body shop that consists of, where you are welding, you are assembling the car together, and you are welding. You will have fumes like smoke or something that come on over, and dust and fumes and smoke coming out, and they do not have anything for that either. You put on your safety glasses and grin and bear it. That is about it.

You are going to find men today, who are younger—most of the men there are my age, I am twenty-five years old, and most of us agree that we do not want to spend all of our life in this plant working under these conditions.

In the 1930s our fathers or forefathers, whatever you want to say, they revolted. They wanted the rights for a union.

In 1970 we revolted and all we want to do is improve on things. That is all we want. Why should we be criticized for something like that? All we want is improvement in working conditions.

MARSHALL McLUHAN

Automation

The introduction of the computer into some work processes permitted what is known as automation, a situation in which human labor is replaced by automatic devices and systems. While some social scientists believe that automation is merely a refinement of the assembly line, others claim that it marks a radical break with the past and has introduced a "post-industrial" society. Marshall McLuhan, in this selection, elicits a number of the implications of the automation of work. Because it releases humans from the necessity of labor, it frees them to make learning the principal commodity of production and consumption. People will thus become more educated and participate more in the varied facets of life. Instead of uniformity and a faceless standardization, McLuhan sees automation as making possible a future world of self-realization and personal diversity.

A newspaper headline recently read, "Little Red Schoolhouse Dies When Good Road Built." One-room schools, with all subjects being taught to all grades at the same time, simply dissolve when better transportation permits specialized spaces and specialized teaching. At the extreme of speeded-up movement, however, specialism of space and subject disappears once more. With automation, it is not only jobs that disappear, and complex roles that reappear. Centuries of specialist stress in pedagogy and in the arrangement of data now end with the instantaneous retrieval of information made possible by electricity. Automation is information and it not only ends jobs in the world of work, it ends subjects in the world of learning. It does not end the world of learning. The future of work consists of learning a living in the automation age. This is a familiar pattern in electric technology in general. It ends the old dichotomies between culture and technology, between art and commerce, and between work and leisure. Whereas in the mechanical age of fragmentation leisure had been the absence of work, or mere idleness, the reverse is true in the electric age. As the age of information demands the simultaneous use of all our faculties, we discover that we are most at leisure when we are most intensely involved, very much as with the artists in all ages....

Automation is not an extension of the mechanical principles of fragmentation and separation of operations. It is rather the invasion of the mechanical world by the instantaneous character of electricity. That is why those involved in automation insist that it is a way of thinking, as much as it is a way of doing. Instant synchronization of numerous operations has ended the old mechanical pattern of setting up operations in lineal sequence. The assembly line has gone the way of the stag line. Nor is it just the lineal and sequential aspect of mechanical analysis that has been erased by the electric speed-up and exact synchronizing of information that is automation.

Automation or cybernation deals with all the units and components of the industrial and marketing process exactly as radio or TV combine the individuals in the audience into new interprocess. The new kind of interrelation in both industry and entertainment is the result of the electric instant speed. Our new electric technology now extends the instant processing of knowledge by interrelation that has long occurred within our central nervous system. It is that same speed that constitutes "organic unity" and ends the mechanical age that had gone into high gear with Gutenberg. Automation brings in real "mass production," not in terms of size, but of an instant inclusive embrace. Such is also the character of "mass media." They are an indication, not of the size of their audiences, but of the fact that everybody becomes involved in them at the same time. Thus commodity industries under automation share the same structural character of the entertainment industries in the degree that both approximate the condition of instant information. Automation affects not just production, but every phase of consumption and marketing; for the consumer becomes producer in the automation circuit, quite as much as the reader of the mosaic telegraph press makes his own news, or just *is* his own news....

The very same process of automation that causes a withdrawal of the present work force from industry causes learning itself to become the principal kind of production and consumption. Hence the folly of alarm about unemployment. Paid learning is already becoming both the dominant employment and the source of new wealth in our society. This is the new *role* for men in society, whereas the older mechanistic idea of "jobs," or fragmented tasks and specialist slots for "workers," becomes meaningless under automation. . . .

Such is also the harsh logic of industrial automation. All that we had previously achieved mechanically by great exertion and coordination can now be done electrically without effort. Hence the specter of joblessness and property-lessness in the electric age. Wealth and work become information factors, and totally new structures are needed to run a business or relate it to social needs and markets. With the electric technology, the new kinds of instant inter-dependence and interprocess that take over production also enter the market and social organizations. For this reason, markets and education designed to cope with the products of servile toil and mechanical production are no longer adequate. Our education has long ago acquired the fragmentary and piece-meal character of mechanism. It is now under increasing pressure to acquire the depth and interrelation that are indispensable in the all-at-once world of electric organization.

Paradoxically, automation makes liberal education mandatory. The electric age of servomechanisms suddenly releases men from the mechanical and specialist servitude of the preceding machine age. As the machine and the motor-car released the horse and projected it onto the plane of entertainment, so does automation with men. We are suddenly threatened with a liberation that taxes our inner resources of self-employment and imaginative participation in society. This would seem to be a fate that calls men to the role of artist in society. It has the effect of making most people realize how much they had come to depend on the fragmentalized and repetitive routines of the mechanical era. Thousands of years ago man, the nomadic food-gatherer, had taken up positional, or relatively sedentary, tasks. He began to specialize. The development of writing and printing were major stages of that process. They were supremely specialist in separating the roles of knowledge from the roles of action, even though at times it could appear that "the pen is mightier than the sword." But with electricity and automation, the technology of fragmented processes suddenly fused with the human dialogue and the need for over-all consideration of human unity. Men are suddenly nomadic gatherers of knowledge, nomadic as never before, informed as never before, free from fragmentary specialism as never before—but also involved in the total social process as never before; since with electricity we extend our central nervous system globally, instantly interrelating every human experience. Long accustomed to such a state in stock-market news or front-page sensations, we can grasp the meaning of this new dimension more readily when it is pointed out that it is possible to "fly" unbuilt airplanes on computers. The specifications of a plane can be programmed and the plane tested under a variety of extreme conditions before it has left the drafting board. So with new products and new organizations of many kinds. We can now, by computer, deal with complex social needs with the same architectural certainty that we previously attempted in private housing. Industry as a whole has become the unit of reckoning, and so with society, politics, and education as wholes.

Electric means of storing and moving information with speed and precision make the largest units quite as manageable as small ones. Thus the automation of a plant or of an entire industry offers a small model of the changes that must occur in society from the same electric technology. Total interdependence is the starting fact. Nevertheless, the range of choice in design, stress, and goal within that total field of electromagnetic interprocess is very much greater than it ever could have been under mechanization.

Since electric energy is independent of the place or kind of work-operation, it creates patterns of decentralism and diversity in the work to be done. This is a logic that appears plainly enough in the difference between firelight and electric light, for example. Persons grouped around a fire or candle for warmth or light are less able to pursue independent thoughts, or even tasks, than people supplied with electric light. In the same way, the social and educational patterns latent in automation are those of self-employment and artistic autonomy. Panic about automation as a threat of uniformity on a world scale is the projection into the future of mechanical standardization and specialism, which are now past.

PART THREE

Conditions Of Technological Development

Introduction

Industrialization had its start in England in the late eighteenth century. Most other European countries, the United States, Canada, and Japan became industrialized during the course of the nineteenth century, and the Soviet Union emerged as a great industrial power in the early decades of the twentieth century. Intensive agriculture through the use of machines and commercial fertilizers, the exploitation of mineral and energy resources, the mass production of goods, and the growing concentration of population in large urban centers were the chief features of this process of industrialization. Modern and rapidly changing technology, of which we are so aware today, is thus primarily a product of Western European and North American civilization.

The Third World countries are those in Asia, Africa, and Central and South America that have not become highly industrialized. In some of these nations, people farm the land as their ancestors did centuries ago, and artisans produce goods by hand as in the past. In many parts of the world, also, there still exist primitive societies whose people are for the most part unaware of and unaffected by modern technology. Further, within such highly industrialized nations as the United States, there are communities that have not fully accepted modern technology and still follow their ancestral traditions.

The development of technology in various societies and in different areas of the world, then, has not been uniform. This is because there are preconditions underlying technological change. Among the most important are the physical environment, the cultural values of societies, and the size and distribution of the population. This section will explore these three preconditions and will indicate how at times they are intimately related with the advance of technology.

The physical environment and the available natural resources of Eskimo peoples living in the Arctic areas, for example, are strikingly different from those of the island peoples of the South Pacific. As a result, the ways in which these societies provide themselves with food, shelter, and clothing—that is, their technologies—are very unlike. But also, the physical environment of the Arctic and South Pacific areas has been almost completely unaffected by the technology of the inhabitants, a complete contrast to the environmental

117

damage that has occurred in the highly industrialized parts of the world. One suspects, then, that either the sparse population in these areas, or their less complex technologies, or the cultural values of their peoples, or a combination of these factors is responsible for the lack of environmental deterioration.

The first five reading selections in this section describe the ideas of Western European cultures concerning the environment and the effects produced by these attitudes. Two central ideas that have been held since antiquity, Clarence Glacken notes, are that the environment has influenced human culture and that man is an agent of environmental change. It has been only relatively recently, however, he points out, that we have begun to realize fully to what a great extent modern technology and population growth can affect and alter our physical environment.

As Nathan Rosenberg shows, the availability of natural resources and the geography of an area have an important influence on the types of technologies that are developed. The level of technological knowledge in a culture, in turn, is a key factor in resource exploitation. People, he stresses, must know the uses to which resources may be put, and also have the capability of extracting them economically. According to Walter Hibbard, however, industrialized nations are currently rapidly exhausting their high-grade mineral resources. Hibbard does not question the desire for affluence. Instead he calls for the development of new technologies to exploit low-grade ores and hitherto inaccessible resources, the substitution of plentiful for scarce materials, and the recycling of scrap.

New technologies, however, often have serious consequences for the environment. Nor is this problem a new one. Jean Gimpel describes how technology in the Middle Ages caused deforestation and air and water pollution. Echoing Glacken's statement, René Dubos writes that only within the last few decades has environmental damage been fully comprehended and efforts made to protect environmental quality. Socially objectionable technological methods, he says, are now being abandoned, and he believes that this reorientation of attitudes is a result of aroused public opinion.

Dubos' belief that humans can reverse dangerous trends brings to the fore the question of the impact of cultural values on technological development. In what ways, if any, does—or can—society influence or direct the course of technological advance? The next four selections explore this area. Robert Heilbroner argues that technology develops sequentially, and that it has followed a predictable course in Western civilization for at least the last two centuries. Social policies, he thinks, may influence technology to some degree, and any new technology must fit in with existing social and economic institutions. Despite these reservations, however, Heilbroner concludes that technology is the principal force shaping modern industrial society, and that cultural values have little effect on the course of technology.

William Ogburn also believes that technology is the major determining factor in social change. Advancing technology, he thinks, creates a continually changing technological environment, over which people have little control. They therefore must adapt to change. George Daniels, on the other hand, completely repudiates the ideas of Heilbroner and Ogburn. It is the culture, the values of society, Daniels contends, that determines the course of technology and decides which technologies will be accepted and which rejected.

A brief look at how the Amish communities of Pennsylvania confront the technological innovations occurring in the larger American culture sheds some light on this dispute. Because of their religious beliefs, John Hostetler writes, the Amish refuse to work on Sundays, even though they would profit economically. On the other hand, the realization that farm tractors save hours of labor and that autos offer convenient transportation finally convinced many Amish communities to accept these innovations. The traditional agricultural life and technology of the Amish, then, is affected by innovations, but change is very gradual and occurs only after much deliberation.

The Amish communities, like the primitive Eskimo and South Pacific societies, are sparsely populated. This fact leads to the puzzling question as to the relationship of population size and density and technological change. T. S. Ashton's article on the Industrial Revolution, included in Part Two, pointed out that population growth occurred during that period, but that it could not be considered as the sole cause of contemporary technological advance, nor could advancing technology in the Industrial Revolution be the cause of the growth in population at the time. However, it is an established fact that progress in medical and public health technologies in the last half of the nineteenth century was largely responsible for the world population explosion that began at that time. And, under certain conditions, a population increase has stimulated technological advance. Thus, Kingsley Davis describes how the migration of peoples into sparsely populated countries resulted in the exploitation of resources and technological development in many parts of the world.

A most important contemporary problem is the rapidly growing population of many Third World nations. Most authorities agree that technological development in these countries must occur to alleviate the situation. Producing enough food to prevent starvation is one facet of the problem, and Neal Jensen is not optimistic that large-scale famine can be prevented in the long run, given the rate of population growth. Another facet of the problem is attempting to make these countries self-sufficient by introducing new technologies. S. Husain Zaheer describes vividly how the problems of population growth and agricultural productivity are intertwined in his country, India, and argues that only advanced technology can provide a solution. A beginning in establishing scientifically-based technology in India has been made, he declares, but the industrialized nations must be generous in sending technicians and providing money, if India's effort is to succeed.

Wilson Clark, however, thinks that the transfer of modern industrial technologies is a mistake, and serves primarily to increase the long-term debt of the developing nations. Instead, he thinks that the establishment of "intermediate" technologies in the developing countries would be more beneficial. He also believes that in many instances the adoption of such technologies would be advantageous to the developed nations. What both Zaheer and Clark stress, however, is that the developed nations have a moral duty to carry out the developmental task.

After studying the selections in this section, the reader will profit by thinking about the technological enterprises in his or her own town or city. Did the geography, climate, or natural resources of the area contribute to the industries or businesses that developed there? Did the traditions or customs of the local inhabitants influence the kind of technology that developed? Has there been any relation between the local technologies and the size or density of the population? In attempting to answer such questions, the reader will begin to comprehend the more complex problems of world-wide technological development.

CLARENCE J. GLACKEN

Nature and Culture In Western Thought

In this selection Professor Clarence Glacken describes three ideas that have influenced Western thought and culture from ancient times to the present: the idea of the earth having been designed by God chiefly as an abode for man; the belief that the environment influences man; and the idea of man as a modifier of nature.

In the history of Western thought, men have persistently asked three questions concerning the habitable earth and their relationships to it. Is the earth, which is obviously a fit environment for man and other organic life, a purposefully made creation? Have its climates, its relief, the configuration of its continents influenced the moral and social nature of individuals, and have they had an influence in molding the character and nature of human culture? In his long tenure of the earth, in what manner has man changed it from its hypothetical pristine condition?

From the time of the Greeks to our own, answers to these questions have been and are being given so frequently and so continually that we may restate them in the form of general ideas: the idea of a designed earth; the idea of environmental influence; and the idea of man as a geographic agent. These ideas have come from the general thought and experience of men, but the first owed much to mythology, theology, and philosophy; the second, to pharmaceutical lore, medicine, and weather observation; the third, to the plans,

Clarence J. Glacken, *Traces on the Rhodian Shore,* University of California Press, 1967. Reprinted by permission of the University of California Press.

activities, and skills of everyday life such as cultivation, carpentry, and weaving. The first two ideas were expressed frequently in antiquity, the third less so, although it was implicit in many discussions which recognized the obvious fact that men through their arts, sciences, and techniques had changed the physical environment about them.

In the first idea, it is assumed that the planet is designed for man alone, as the highest being of the creation, or for the hierarchy of life with man at the apex. The conception presupposes the earth or certain known parts of it to be a fit environment not only for life but for high civilization.

The second idea originated in medical theory. In essence, conclusions were drawn by comparing various environmental factors such as atmospheric conditions (most often temperature), waters, and geographical situation with the different individuals and peoples characteristic of these environments, the comparisons taking the form of correlations between environments and individual and cultural characteristics.... Although environmentalistic ideas arose independently of the argument of divine design, they have been used frequently as part of the design argument in the sense that all life is seen as adapting itself to the purposefully created harmonious conditions.

The third idea was less well formulated in antiquity than were the other two; in fact, its full implications... were not explored in detail until Marsh* published *Man and Nature* in 1864. Like the environmental theory, it could be accommodated within the design argument, for man through his arts and inventions was seen as a partner of God, improving upon and cultivating an earth created for him. Although the idea of environmental influences and that of man as a geographic agent may not be contradictory—many geographers in modern times have tried to work out theories of reciprocal influences—the adoption by thinkers of one of these ideas to the exclusion of the other has been characteristic of both ancient and modern times; it was not perceived, however, until the nineteenth century that the adoption of one in preference to the other led to entirely different emphases. One finds therefore in ancient writers, and in modern ones as well, ideas both of geographic influence and of man's agency in widely scattered parts of their work without any attempt at reconciling them; since Greek times the two ideas have had a curious history, someties meeting, sometimes being far apart....

The idea of a designed earth, the doctrine of final causes applied to the natural processes on earth, is an important segment—but only that—of a much broader and deeper body of thought suffused throughout all types of writings: science, philosophy, theology, literature. This is the idea of teleology

*Editors' note: George Perkins Marsh (1801–1882) was a statesman, linguist, and conservationist whose *Man and Nature* was highly influential in stimulating a conservationist movement in the late nineteenth century.

in general. One cannot deny, however, its immense historical force in the field of nature and earth study, nor the reinforcements to the broader area of teleological explanation coming from its use here. The idea of a designed earth, whether created for man or for all life with man at the apex of a chain of being, has been one of the great attempts in Western civilization, before the theory of evolution and modern ecological theories emerging from it, to create a holistic concept of nature, to bring within its scope as many phenomena as possible in order to demonstrate a unity which was the achievement of an artisan-creator. It is a doctrine at home with the religious interpretation of nature, with pre-evolutionary thought which was congenial to the belief in special creation and the fixity of species. (Evolution, admittedly, could be and has been interpreted as part of a design.) The combination of special creation and fixity of species meant the existence of harmonies in nature from the beginning. . . .

Much has been learned since the end of the eighteenth century in the study of nature based on evolutionary theory, genetics, ecological theory; but it is no accident that ecological theory—which is the basis of so much research in the study of plant and animal populations, conservation, preservation of nature, wildlife and land use management, and which has become the basic concept for a holistic view of nature—has behind it the long preoccupation in Western civilization with interpreting the nature of earthly environments, trying to see them as wholes, as manifestations of order. . . .

Since the design argument applied to the earth was an all-encompassing attempt to bring a unity into the observed phenomena—nonliving matter, plants, animals, and man—it is only natural that it involved the other two ideas, but they . . . also enjoyed an independent existence and history of their own.

The idea of environmental influences on culture is as important historically for the questions it suggested as for its own intellectual and philosophical content. It is part of that broad and ancient contrast between *physis* and *nomos,* between nature and law or custom. It is an idea deeply involved with interpreting the endlessly fascinating array of human differences, rich new materials for which were furnished in the ancient world during the Hellenistic period, in the modern by the age of discovery. It probably grew out of medicine; travel and voyaging have both helped it along, for men apparently lived everywhere—in deserts, on hot sandy coasts, near swamps, in mountains— and brought it into disfavor as examples appeared which contradicted it. . . . The important point in its impingement on ethical and religious theory was the implication that people living under a certain environment could be expected to act as they did, the environment rather than human frailty being responsible for shortcomings. Thus climate was a favorite explanation for inebriety or sobriety of whole peoples. Nomos, however, was never completely forgotten. Men's customs, their governments, their religions, were great cultural molding forces. . . . Theories of environmental influence are compatible with design arguments because adaptation of life to environment is assumed in both cases; both provided answers to the question why men were living where they were, how they prospered, why and how they lived in inhospitable and bleak environments. In the modern period, . . . theories of environmental influence have had strong affiliations with writings on national character; more often than not they have encouraged monolithic summation: the Germans, the French, the Arabs, could be characterized in a few sentences. On the other hand, they were safeguards—albeit negative ones—against a purely cultural determinism. Surely it is a mistake to think the history of civilization can be written purely as cultural, social, or economic history. . . .

In the eighteenth century, the writings of Montesquieu, Wallace, Hume, and particularly Malthus bring to maturity and influence a difference kind of environmental theory, emphasizing not the elements of climate or the physical differences in environment but the limitations which the environment as a whole imposes on all life. This idea in varying forms has produced some of the most polemical writing of the last century and a half.

The idea of design in nature really focused attention on God as artisan, man and nature being in the subordinate position as the created. The idea of environmental influence centered on nature; if it were expressed within a religious context, God was there as creator and man was largely plastic in the molds of nature. The idea of man as a modifier of nature, however, centered on man; if the idea is expressed in a religious context, God often becomes an artisan purposefully leaving the creation unfinished; nature is there to be improved by human skill. In many ways it is the most interesting of the three ideas because it assumes choices; different results come from different skills. Its sources lie, I think, in ideas of artisanry and order-bringing, and when man's activity is seen within a religious framework, he becomes a finisher of nature, set on earth as its guardian and custodian. One of the most distinctive aspects of this idea is the fundamental distinction, often implicit, between the nature of human and animal art. It was not that animals or lowly insects lacked skills—we need only read the homilies on the ants and bees and the eager moralizing in them. In the ancient world, the distinction was implied in the combined power of hand and mind. In the modern, the quarrel between the ancients and the moderns, and the development of the idea of progress throw light on the nature of the human endowment: it is greater because what is known in one generation may be communicated to the next, and an accumulation of skills, knowledge, artisanry, may take place, and over a period of time broaden the gulf between human and animal skill. In secular thought, the idea in general has optimistic overtones. . . . The pessimism comes later in the nineteenth century with more frequent communication of ideas, knowledge of the historical depth of the changes, and observation of the unprecedented pace of the changes consequent upon the increase in technological ability and the growth in world population. The historical march of this idea has been from local to more general observation and generalization. At the present day it is all-pervasive, a natural and expected outcome of the tremendous force of human agency.

NATHAN ROSENBERG

Technology And Resource Endowment

From prehistoric times until the present, humans have used technology to solve the problems presented by the natural environment. At the same time, they have used the natural resources of their environment—from wood for fire to silicon for transistors—in their effort to control nature for their own needs. Thus the environment has provided both the challenges for man's technology and the raw material with which to meet those challenges. In the following selection Nathan Rosenberg, an economic historian, stresses that the kinds of technology that emerge in any area depend upon the availability of natural resources and its geography. However, the level of technological knowledge in a society is also crucial. People must not only know how their resources can be used to advantage; they must also have the ability to extract them economically.

Technological knowledge ought to be understood as the sort of information which improves man's capacity to control and to manipulate the natural environment in the fulfillment of human goals, and to make that environment more responsive to human needs. The intimate relationship between technology and environment becomes apparent as soon as one asks the question: What constitutes a natural resource? The answer is not a simple

Nathan Rosenberg, *Technology and American Economic Growth*. Harper & Row, 1972. Reprinted by permission of the author.

Black Mesa Coal Mine

one, but the safest way to begin such an answer is by saying: "It all depends." If we define resources in terms of mineral deposits or acres of potentially arable land, qualifications spring to mind. The Plains Indian did not cultivate the soil; neither coal, oil, nor bauxite constituted a resource to the Indian population or, for that matter, to the earliest European settlers in North America. It was only when technological knowledge had advanced to a certain point that such mineral deposits became potentially usable for human purposes. Even then, the further economic question turns, in part, upon accessibility and cost of extraction. Improvements in oil drilling technology (as well as changing demand conditions) make it feasible to extract oil today from depths which would have been technically impossible fifty years ago and prohibitively expensive twenty years ago. Similarly, low grade taconite iron ores are being routinely exploited today which would have been ignored earlier in the century when the higher-quality ores of the Mesabi range were available in abundance. Oil shale, known to exist in vast quantities—for example, in the Green River formation in Colorado, Utah, and Wyoming— is not yet worth exploiting but might well be brought into production if petroleum product prices rise very much above their present levels. The rich and abundant agricultural resources of the Midwest were of limited economic importance until the development of a canal network beginning in the 1820s with the completion of the Erie Canal, and later a railroad system which made possible the transportation of bulky farm products to eastern urban centers at low cost. Natural resources, in other words, cannot be catalogued in geographic or geological terms alone. The economic usefulness of such resources is subject to continual redefinition as a result of both economic changes and alterations in the stock of technological knowledge. Whether a particular mineral deposit is worth exploiting will depend upon all of the forces influencing the demand for the mineral, on the one hand, and the cost of extracting it, on the other. . . .

From the perspective of the economic historian surveying the historical experience of the wealth—and poverty—of nations, the production and use of technological knowledge must be seen against the backdrop of specific societies with different cultural heritages and values, different human capital and intellectual equipment, and confronting an environment with a very specific collection of resources. The emphasis on the specificness of resources is important, because resources establish the particular framework of problems, of constraints, and opportunities, to which technological change is the (occasional) human response. . . . Technological change . . . does not *occur* in the abstract but rather in very specific historical contexts. It occurs, that is, as a successful solution to a particular problem thrown up in a particular resource context. For example, the cutting off of an accustomed source of supply during wartime has often been an important stimulus for the development of new techniques. Thus France's early commercial leadership in the production of synthetic alkalis (utilizing the Leblanc process) was, in large measure, a result of her loss of access to her traditional supplies of Spanish barilla during the Napoleonic wars. The Haber nitrogen fixation process was developed by the Germans during World War I when the British blockade deprived them of their imports of Chilean nitrates. The loss of Malayan natural rubber as a result of Japanese occupation in World War II played a critical role in the rapid emergence of the American synthetic rubber industry. On the other hand, the fact that the British led the world in the development of a coal-using technology was hardly surprising in view of the abundance and easy accessibility of her coal deposits and the growing scarcity of her wood-fuel supplies which increasingly constrained the expansion of her industries in the seventeenth and eighteenth centuries. Indeed, the steam engine itself originated as a pump for solving the problem of rising water levels which impeded extractive activity in British mines—coal as well as other minerals. It seems equally fitting and proper that the British are currently performing the pioneering work in the development of techniques for the instrument-landing of airplanes in dense fog; and conditions of the natural environment make it appropriate for the Israelis to be devoting much effort to cheap desalination techniques, the Dutch to the development of salt-resistant crop varieties, and the students at California Institute of Technology to the attempt to perfect an electric motor for use in automobiles. In all these cases, technological exploration is intimately linked up with patterns of resource availability or conditions of the natural environment in particular locational contexts. But, it is important to add, although the demand or need for a particular kind of technique is established in relation to some aspect of the natural environment, the capacity to respond creatively to this need is an altogether different matter. Past history and the contemporary world abound, to put it mildly, in unsolved problems. Moreover, the *kind* of solution which any society can produce to the problems presented by its natural environment will turn on the level of knowledge and expertise which is available to it. Thus, the Israeli kibbutz with its sophisticated irrigation techniques represents a very different response to the barrenness of the Negev from that of the Arab bedouins; and the Ifugao tribesmen of northern Luzon increased rice output for over a thousand years by expanding their extraordinary labor-intensive system of mountain rice terraces, whereas not far from these terraces today farmers are expanding output by exploiting the new, high-yielding rice varieties which have been made possible by modern genetics and botany. Although the "slash and burn" system of agriculture, widely practiced in the humid tropics, permits a limited population to eke out a precarious existence (always provided that natural fertility is restored by sufficiently long fallow periods), a sophisticated knowledge of soil chemistry may pinpoint the soil nutrients which might permit continuous cultivation and far higher levels of output per acre.

WALTER R. HIBBARD, JR.

Mineral Resources: Challenge or Threat?

The abundance of natural resources—and the development of technologies to exploit them—has long made Americans, in the words of historian David Potter, a "people of plenty." But, warns Walter Hibbard, Jr., former director of the U.S. Bureau of Mines, our mineral resources are limited, and we cannot continue to exploit them as we have in the past. In this selection, he spells out what tasks lie ahead with respect to mineral resources for the United States to maintain its affluence. Although the United States is a major mineral producer, companies prefer to exploit foreign resources because of their readier availability, lower costs, and less regulation. In the future, Hibbard warns, we must recycle more mineral-containing products, reclaim more metal wastes, and above all perfect technologies for mining the oceans and deeper mineral deposits.

A requisite for affluence, now or in the future, is an adequate supply of minerals—fuels to energize our power and transportation; nonmetals, such as sulfur and phosphates to fertilize farms; and metals, steel, copper, lead, aluminum, and so forth, to build our machinery, cars, buildings, and bridges. These are the materials basic to our economy, the multipliers in our gross national product. But the needed materials which can be recovered by known methods at reasonable cost from the earth's crust are limited, whereas their rates of exploitation and use obviously are not. This situation cannot continue. . . .

Walter R. Hibbard, Jr., "Mineral Resources: Challenge or Threat?" *Science*, Vol. 160, April 12, 1968. Copyright 1968 by the American Association for the Advancement of Science.

The Time Problem

Winning from the earth the minerals needed for prosperity and well-being depends not only on the capital available for investment, but also on the technology that can be applied.... Our resources are limited less by the amounts of raw materials than by the technology of treatment and extraction and by the capacity to produce at a reasonable cost.... [According to one authority,] "Unlike land, which becomes more valuable as population increases and good prospects are snapped up, technology can be improved, and the supply of capital can be stretched."

Although technology may stretch capital by less-expensive production facilities, permitting utilization of lower grade ores, only long-range planning can remedy the time problem....

Technology, willingness to risk capital, and planning have made the United States a major producer of minerals. The U.S. Bureau of Mines shows complete world production data on 54 minerals commodities for 1966; U.S. led in the production of 27. In the case of 11 additional commodities for which complete figures were not available, it is believed the United States led in the production of six.

However, as long as mineral deposits in other parts of the world can be profitably developed, the incentives for radical innovation in technology are slight and investment capital is attracted abroad. It is axiomatic that investors seek out ventures that are the most profitable. Hence, capital will flow to those countries with lower labor costs, greater government incentives (such as tax benefits and subsidies), and minimum costs for pollution control (relative to the United States) as well as high-grade reserves which can be readily exploited by well-established procedures and available equipment. Already, American investment in mining is going abroad at an increasing rate—to Australia, Canada, Spain, South Africa, and South America. If this trend continues, by 1985 we may be importing a major portion of our large-tonnage metals such as iron, copper, lead, and zinc, thus adding commodities for which the United States is already primarily dependent on overseas sources. Advanced technology at home, economically applied to domestic reserves, can reverse this trend.

Technology Can Expand Resource Base

Most of our mineral industries are mature; they have been operating for a long time, and the cream has been skimmed from the richest and most easily recovered ores. Yet technological innovation is continually injecting new life into these mature industries. I believe that technology can help increase our mineral resource base in the following four ways:

1. Exploration and discovery. The minerals so far used by man have come from very near the surface. Most were discovered from outcrops. We must learn how to explore at depths, and we must develop methods to find and extract minerals in the deeper layers of the earth's crust and from under the sea.

2. Improved mining, beneficiation, and processing. More efficient methods for mining ores and for upgrading them before smelting and refining can make the use of leaner ores technically and economically feasible.

3. Recycling of scrap and waste. There is tremendous opportunity in "mining" our scrap heaps and junkyards.... Such salvage programs could be greatly extended through research and improved collection and processing techniques.

4. Substitution. Using abundant materials in place of those in short supply is the challenge of the physical metallurgist—and of the polymer chemist. Not only is there strong economic pressure to find substitutes (because they are usually cheaper), but there is technical pressure also. The materials engineer, redesigning from basic principles, often finds that the traditional materials are not the best technically.

Potentially one of the most rewarding opportunities for dramatic expansion of our mineral resource base—and one of the greatest challenges to our ingenuity—is the exploration and exploitation of the almost untapped three-quarters of the earth's crust beneath the oceans. The deepwater sections are beyond our reach at present, but very encouraging progress has been made on the continental shelves, defined as offshore sea bottom to a depth of 200 meters. These shelves are geologically similar to the adjacent dry land; and we can assume with some confidence that they contain ore bodies of similar types and distributions.

Invisible Gold

Two reforms are urgently needed to extend the use and to expand the reuse of valuable materials, even though they may seem to run counter to the affluent status our society seems to be trying to maintain. (1) We should design our durable, mineral-containing products to last longer before they go out of style or wear out, and (2) we should design such products to make it easier to collect and separate their mineral content for recycling after they are discarded.

An American automobile, for instance, lasts about seven years, or 160,000 kilometers. From a technical standpoint, doubling those figures should not be difficult, and there would be a tremendous saving in metals and other materials.

Moreover, most automobiles seem to be designed on the assumption that no one, not even a mechanic, will ever want to take one apart. Workers at the Bureau of Mines have dismantled several dozen cars of different ages and makes in the course of current work on solid-waste disposal problems. The manufacturers were unable to tell them the composition and distribution of materials in their cars, since many components were supplied by vendors. The placement of these components and the overall design of the car are subject to many restrictions: conservation of space, esthetic appeal, ease of manufacture, safety, and others. The result is that not only the exterior design but also the materials which are used in automobiles change from year to year. Wiring becomes more complex in order to take care of additional electrical equipment. Increasing use is made of stainless, aluminized, and galvanized steel and of aluminum and zinc castings and other materials which make salvage difficult.

I propose that in designing automobiles—and refrigerators, ranges, and other metallic consumer products—manufacturers should provide greater durability, retard obsolescence, and anticipate the need for recycling. If engineering design were to include this concept, valuable materials could then be readily saved when the product is obsolete or worn out. This is a stiff requirement but a necessary one. The annual addition to the scrap market of millions

UPI PHOTO

of tons of metal is such a valuable potential resource that we cannot afford to overlook any means of making it easier to salvage.

The continuing failure, on the one hand, to retard the flow of usable materials into scrap piles, and, on the other hand, to utilize this above-ground bonanza more fully to satisfy our proliferating requirements is shortsighted, in fact, criminal.

There is no reason why, with skill in design and materials application, we cannot make products more durable while we salvage every bit we can from our unusable and discarded products, and thereby extend the mineral resource base of the nation.

JEAN GIMPEL

Environmental Pollution in the Middle Ages

Social historian Jean Gimpel illustrates with ample detail that environmental destruction and pollution caused by the growth of technology are old problems. Gimpel describes the various technological innovations of the Middle Ages, such as the shift from wood to coal and the tanning process, which increased industrial waste and created significant pollution. At that time, also, governments faced the problem of protecting their peoples and the environment from the results of technological growth.

The industrialization of the Middle Ages played havoc with the environment of western Europe. Millions of acres of forests were destroyed to increase the area of arable and grazing land and to satisfy the ever greater demand for timber, the main raw material of the time. Not only was timber used as fuel for the hearths of private homes and for ovens, it was also in one way or another essential to practically every medieval industry. In the building industry wood was used to build timber-framed houses, water mills and windmills, bridges, and military installations such as fortresses and palisades. In the wine industry wood was used for making casks and vats. Ships were made of wood, as was all medieval machinery such as weavers' looms. Tanners

From *The Medieval Machine: The Industrial Revolution of the Middle Ages* by Jean Gimpel. Copyright © 1976 by Jean Gimpel. Reprinted by permission of Holt, Rinehart, and Winston.

needed the bark of the trees and so did the rope makers. The glass industry demolished the woods for fuel for its furnaces, and the iron industry needed charcoal for its forges. By 1300, forests in France covered only about 32 million acres—2 million acres less than they do today. . . . Medieval man brought about the destruction of Europe's natural environment. He wasted its natural resources, and very soon felt the consequences of his destructive activities, the first of which was the considerable rise in the price of timber as a result of its increasing scarcity. At Douai, in northern France, in the thirteenth century wood had already become so scarce and expensive that families from the lower income groups could not afford to buy a wooden coffin for their dead. They had to rent one, and when the ceremony at the cemetery was over, the undertaker would open the coffin, throw the corpse into the earth, and bring back the coffin to use again. . . .

A few figures from building accounts serve to show how quickly medieval man could destroy his environment. An average house built of wood needed some twelve oaks. In the middle of the fourteenth century, for the building operation at Windsor Castle, a whole wood was bought and all the trees felled—3,004 oaks. This was still not sufficient, for some ten years later 820 oaks were cut in Combe Park and 120 in Pamber Forest, bringing the total for this one castle up to 3,944 oaks. The *Times* of London reported on August 24, 1971, that only 300 to 400 oaks were still standing in Robin Hood's famous Sherwood Forest. The article was headed: "Hearts of Oak grow Faint in Sherwood as Age, Thirst and Pollution take Toll." But the greatest toll was taken in the Middle Ages.

The building of thousands of furnaces in hundreds of medieval forests to satisfy the extensive demand for iron was a major cause of deforestation. Iron ore, unlike gold ore, is practically never found in its natural state except in meteorites, and it requires a special fuel to smelt and reduce it. From the very beginning, the fuel used was charcoal, the black porous residue of burned wood. This absolute reliance on charcoal made it essential for iron smelters up to the late eighteenth century to build their furnaces in the forests, where wood for the making of charcoal was directly at hand.

The extent of the damage caused by iron smelters to forests can be appreciated when one realized that to obtain 50 kilograms of iron it was necessary at that time to reduce approximately 200 kilograms of iron ore with as much as 25 steres (25 cubic meters) of wood. It has been estimated that in forty days, one furnace could level the forest for a radius of 1 kilometer.

It is not surprising to hear that certain authorities took measures to halt or at least slow down the massacre of the forests. It was in their financial interest. In the forest of Dean by the opening decades of the thirteenth century, the Crown was restricting the right of working solely to the royal forges in the

"I just invented fire and pollution."

DRAWING BY JOSEPH FARRIS; © 1970 THE NEW YORKER MAGAZINE, INC.

Forest, and in 1282 a report was made by the regarders of the Forest on the wastage of timber caused by the sixty or so forges located there. In the Dauphiné in 1315 the representatives of the Dauphin were greatly alarmed at the widespread destruction of the woods of that region. They formally accused the iron-producing factories of being directly responsible for this disaster and recommended that forcible measures should be taken to arrest the situation. . . .

The decreasing availability of timber and the progressive rise in the price of wood led England to import timber from Scandinavia. The first fleet of ships loaded with Norwegian fir trees sailed into Grimsby harbor, on the east coast of England, in 1230. And in 1274, the master carpenter of Norwich Cathedral went to Hamburg to buy timber and boards. During this same period a substitute fuel for wood was found—coal.

Some of the great European coalfields of the nineteenth and twentieth centuries were first mined in the thirteenth century: in Belgium those of Liège, Mons, and Charleroi, and in France those of the Lyonnais, Anjou, and Forez. A charter of the Priory of Saint-Sauveur-en-Rue in the Forez mentions coal as early as the year 1095.

In England the vast coalfields that were to play such a decisive role in the Industrial Revolution of the eighteenth and nineteenth centuries appear in innumerable documents of the thirteenth and fourteenth centuries: those of Newcastle-upon-Tyne, and also those in the Midlands, in Derbyshire, and Nottinghamshire, as well as the Shropshire coalfields which later enabled Abraham Darby in 1709 to launch the second Industrial Revolution by using coke to reduce iron ore. The Scottish and Welsh coalfields were also already fully active.

Coal in France was called *terre houille* or *charbon de roche,* and in England the terms "pit coal" and "sea coal" were common, the latter from the fact that coal was originally worked and picked up on the beaches of maritime counties such as Durham or Northumberland.

As early as 1226, we find in London a Sea Coal Lane, also known as Lime Burners Lane. The lime-burning industry was one of the first to convert to the use of coal, along with the iron industry. Brewers, dyers, and others followed. In 1243 the first recorded victim of coal mining, Ralph, son of Roger Ulger, drowned in an open pit. At first coal was mined in shallow pits, usually 6 to 15 meters (20 to 50 feet) deep, but sometimes, as in the French coal mines of Boussagues in the Languedoc, there were already underground galleries. In Newcastle there were such extensive diggings around the city that it was dangerous to approach it by night, lest one fall and break one's neck in the open trenches. Here and in many other places, the medieval environment was already an industrial environment. . . .

With the burning of coal, western Europe began to face atmospheric pollution. The first person recorded to have suffered from medieval pollution was a Queen of England, Eleanor, who was driven from Nottingham Castle in 1257 by the unpleasant fumes of the sea coal burned in the industrial city below. Coal smoke was considered to be very detrimental to one's health, and up to the sixteenth century coal was generally used as a domestic fuel only by the poorer members of society, who could not afford to buy wood. Medieval coal extracted from the surface was of inferior quality, with more bitumen in it than the coal mined today. As it burned, it gave off a continuous cloud of choking, foul-smelling, noxious smoke. The only good domestic coal was that extracted from the coalfields bordering the Firth of Forth, which was burned by the Scottish kings, and the coal extracted at Aachen in Germany, which was used to make fires in the town hall and in the mayor's chambers.

By the last decades of the thirteenth century, London had the sad privilege of becoming the first city in the world to suffer man-made atmospheric pollution. In 1285 and 1288 complaints were recorded concerning the infection and corruption of the city's air by coal fumes from the limekilns. Commissioners of Inquiry were appointed, and in 1307 a royal proclamation was made in Southwark, Wapping, and East Smithfield forbidding the use of sea coal in kilns under pain of heavy forfeiture. . . . The proclamation does not seem to have been very successful. Complaints continued and a commission . . . was appointed, with instructions "to inquire of all such who burnt sea-coal in the city, or parts adjoining and to punish them for the first offence with great fines and ransoms, and upon the second offence to demolish their furnaces." The commission was no more successful than the proclamation, and London remained a polluted town. . . .

In the towns people suffered also from industrial water pollution. Two industries in particular were held responsible in the Middle Ages for polluting the rivers: the slaughtering and the tanning industries, especially tanning. Municipalities were always trying to move the butchers and the tanners downstream, outside the precincts of the town.

The slaughtering and quartering of livestock in the Middle Ages was generally done on the butcher's premises. A French parliamentary decree of September 7, 1366, compelled Paris butchers to do their slaughtering and cleaning alongside a running stream beyond the city. This decree was certainly necessary, as some 250,000 head of livestock were slaughtered each year in Paris. The author of the *Menagier de Paris* worked out that 269,256 animals had been slaughtered in 1293: 188,522 sheep, 30,346 oxen, 19,604 calves, and 30,784 pigs. Quite enough to pollute the Seine.

The Paris authorities tried to limit the degree of this pollution not only by restricting the slaughtering of animals within the precincts of the city but also by imposing restrictions on the tanners, who dressed ox, cow, and calf hides, and the tawers, who dressed the skins of deer, sheep, and horses. "In 1395 the king's representative at the *Châtelet* wanted to compel the tawers who were dressing their leather on the banks of the Seine, between the Grand-Pont and the Hôtel du duc de Bourbon to move downstream, because industry corrupted the water of the riverside dwellers, both those lodging in the Louvre and those lodging in the Hôtel du duc de Bourbon."

Tanning polluted the river because it subjected the hides to a whole series of chemical operations requiring tannic acids or lime. Tawing used alum and oil. Dried blood, fat, surplus tissues, flesh impurities, and hair were continually washed away with the acids and the lime into the streams running through the cities. The water flowing from the tanneries was certainly unpalatable, and there were tanneries in every medieval city. . . .

RENÉ DUBOS

The New Environmental Attitude

Although the problem of environmental damage through savage exploitation is an old one, as Gimpel pointed out, concern for the problem is relatively recent. Until the 1860s, only a few isolated voices were heard calling for conservation and protection of the environment. Still another hundred years passed before the general public, in the 1960s, expressed concern over some of the environmental consequences of our rapidly expanding technological growth. In this selection, biologist René Dubos discusses the recent change in public attitudes which has led to measures to ensure that socially undesirable technologies were modified or eliminated.

Until our time, disasters caught the world by surprise, and it was therefore difficult, if not impossible, to control their manifestations. In contrast, modern societies are beginning to develop methods for anticipating future upheavals, whether of natural or man-made origin. The long-range effects to be expected from social or technological innovations are now discussed long before the event, especially if there is any likelihood that these effects might be dangerous. One of the beneficial results of the campaign for

René Dubos, "The Despairing Optimist." Reprinted from *The American Scholar,* Volume 45, Number 2, Spring 1976. Copyright © 1976 by the United Chapters of Phi Beta Kappa. By permission of the publishers.

a better environment has been to foster studies that will help to prevent, or at least to minimize and rapidly correct, the environmental damage caused by industry. Clumsy and difficult to administer as it is, the legislation that makes it necessary to file a statement of "environmental impact" before undertaking any large-scale project is a first step in a direction that will certainly be followed in the future.

It is admittedly difficult, perhaps impossible, to determine completely in advance the "environmental impact" of human interventions into natural and social systems; but it is certain nevertheless that much of the damage that was done in the past could have been prevented if environmental awareness had been as widespread as it is now becoming. A few examples will suffice to illustrate the wide range of changes that have taken place in technological development as a result of environmental awareness.

Industrial growth has long been considered desirable, because of its contribution to health and happiness, for the creation of wealth, or simply for its own sake. Until recent times, progress was indeed identified with quantitative growth. In contrast, modern societies have begun to question the desirability of certain innovations that are technologically feasible and economically profitable, but that have undesirable social aspects. The shelving of the SST is a case in point, and so is the delay in the development of various nuclear technologies for the production of energy—such as the breeder reactor. The evaluation of potential long-range dangers for human beings and for the environment is becoming one of the crucial factors in the formulation of technological policies.

The partial banning of pesticides exemplifies a situation in which a technology that had first been accepted with enthusiasm was brought under strict control once its dangers had been recognized. . . .

These examples, and many others that could be cited, are evidence of a rather new attitude in the modern world: the willingness to abandon policies and practices that are technologically possible and economically profitable but socially objectionable. The symbolic importance of such reversals is that they are caused, not by political upheavals or scientific breakthroughs, but by public awareness of environmental consequences.

The social attitude toward innovations can undergo rapid changes even while these innovations are still in the stage of scientific or cultural development. Witness the heated controversies and progressive evolution of views about genetic engineering and behavioral control, despite the paucity of evidence that effective techniques will soon—if ever—be developed for the manipulation of genes or of behavior on a population scale.

Perhaps the most interesting in the long run are the indications of reversals of attitudes concerning the role of human beings in various technologies. Up to now, one of the explicit ideals of technological civilization has been to substitute machines for human beings whenever possible. At first, this policy had the advantage of increasing industrial productivity; then, progressively, machines proved more efficient than human beings in most industrial operations. It is now becoming apparent, however, that the unreasonable use of Western technology often creates conditions that are inimical to the expression of human resourcefulness and creativity. Gandhi's saying that the future of India is not in mass production but in production by the masses may have meaning for the Western world as well. The human tragedies resulting from unemployment may lead industrial societies to rediscover that, except for the dullest, most repetitive, and painful tasks, human beings are better than machines— and certainly more creative.

One of the characteristics of our time evident in practically all fields of human endeavor is the rapidity with which steps can be taken to impose a new orientation on social and technological trends, once public opinion has been alerted to their potential consequences. It is worthy of note, in passing, that the direction of trends at any given time is more influenced by grass-roots movements than by official directives. They are less a result of conventional education than of the widespread awareness generated by concerned citizen groups and by the news media.

Phrases such as "limits to growth" and "anti-technological movement" lose much of their disturbing and somewhat frightening connotations when read in the light of the changes that are now occuring in social attitudes. It is obvious of course that all forms of growth have limits from the *quantitative* point of view; it is certain also that protests will continue against *certain* types of technology. These limits and protests, however, correspond to a desire for *qualitative* change rather than to a rejection of development. The forms in which growth takes place are continuously undergoing evolution, just as do all the manifestations of life.

One of the most remarkable achievements of the campaigns conducted by those who have been castigated as eco-freaks or anti-technologists is that the opportunity to live in a good environment is coming to be regarded as one of our "unalienable" rights, along with the rights to freedom, to education, and to some form of medical care. A good environment, furthermore, means not only conditions that are favorable to the maintenance of physical health, but also certain emotional and aesthetic qualities of the surroundings. These criteria, which are very different from those that governed quantitative industrial growth in the past, are likely to generate new forms of growth in which the quality of life will take precedence over the quantity of goods produced.

Such a change of attitude will certainly lead to the downgrading of certain technologies and resources, and will foster the development of others that are of little importance today. Belching smokestacks were regarded as a sign of prosperity during the first phase of the industrial revolution; they are now read as evidence of poor technological management and of social irresponsibility. There are no visible limits to the kind of growth associated with qualitative changes, because social evolution is endlessly inventive of new resources and new technologies that can be developed to achieve the goals imagined by the new mentality. . . .

In my opinion, the difficulties of our time, or of any time, are not reasons for discouragement. History shows that crises usually foster renewal and herald new phases of creativeness, different from the past. The most interesting characteristic of human beings is that they can transcend social as well as biological determinism. Animals are almost completely prisoners of biological evolution, but human beings are blessed with the freedom and inventiveness of social evolution. They can retrace their steps and start on a new course whenever they see danger ahead; they can integrate the raw materials of the earth, with the knowledge derived from past experience and from new learning, in a continuous evolutionary process of creation.

ROBERT L. HEILBRONER

Do Machines Make History?

In this selection economist Robert L. Heilbroner asks the question: "Do machines make history?" and answers in the affirmative. He believes that technology has determined the general course of social change, at least since the beginnings of capitalism. Until man gains more control over technology, he concludes, it will continue to determine the direction of social change. His analysis is extremely important, since if the course of technological change is at most only partially responsive to social values and policies, then establishing control over technology may well be impossible.

> *The hand-mill gives you society with the feudal lord; the steam-mill, society with the industrial capitalist.*
>
> MARX, *The Poverty of Philosophy*

That machines make history in some sense—that the level of technology has a direct bearing on the human drama—is of course obvious. That they do not make all of history, however that word be defined, is equally clear. The challenge, then, is to see if one can say something systematic about the matter, to see whether one can order the problem so that it becomes intellectually manageable.

To do so calls at the very beginning for a careful specification of our task. There are a number of important ways in which machines make history that

Robert L. Heilbroner, "Do Machines Make History?" from *Technology and Culture* 8 (1967). Published by the University of Chicago Press, Chicago 60637. © 1967 by the Society for the History of Technology. All rights reserved. Reprinted by permission.

will not concern us here. For example, one can study the impact of technology on the *political* course of history, evidenced most strikingly by the central role played by the technology of war. Or one can study the effect of machines on the *social* attitudes that underlie historical evolution: one thinks of the effect of radio or television on political behavior. Or one can study technology as one of the factors shaping the changeful content of life from one epoch to another: when we speak of "life" in the Middle Ages or today we define an existence much of whose texture and substance is intimately connected with the prevailing technological order.

None of these problems will form the focus of this essay. Instead, I propose to examine the impact of technology on history in another area—an area defined by the famous quotation from Marx that stands beneath our title. The question we are interested in, then, concerns the effect of technology in determining the nature of the *socioeconomic order.* In its simplest terms the question is: did medieval technology bring about feudalism? Is industrial technology the necessary and sufficient condition for capitalism? Or, by extension, will the technology of the computer and the atom constitute the ineluctable cause of a new social order?

Even in this restricted sense, our inquiry promises to be broad and sprawling. Hence, I shall not try to attack it head-on, but to examine it in two stages:

1. If we make the assumption that the hand-mill does "give" us feudalism and the steam-mill capitalism, this places technological change in the position of a prime mover of social history. Can we then explain the "laws of motion" of technology itself? Or to put the question less grandly, can we explain why technology evolves in the sequence it does?

2. Again, taking the Marxian paradigm at face value, exactly what do we mean when we assert that the hand-mill "gives us" society with the feudal lord? Precisely how does the mode of production affect the superstructure of social relationships?

These questions will enable us to test the empirical content—or at least to see if there *is* an empirical content—in the idea of technological determinism. I do not think it will come as a surprise if I announce now that we will find *some* content, and a great deal of missing evidence, in our investigation. What will remain then will be to see if we can place the salvageable elements of the theory in historical perspective—to see, in a word, if we can explain technological determinism historically as well as explain history by technological determinism.

*   *   *

We begin with a very difficult question hardly rendered easier by the fact that there exist, to the best of my knowledge, no empirical studies on which to base our speculations. It is the question of whether there is a fixed sequence to technological development and therefore a necessitous path over which technologically developing societies must travel.

I believe there is such a sequence—that the steam-mill follows the hand-mill not by chance but because it is the next "stage" in a technical conquest of nature that follows one and only one grand avenue of advance. . . .

What evidence do we have for such a view? I would put forward three suggestive pieces of evidence:

1. *The Simultaneity of Invention*

The phenomenon of simultaneous discovery is well known. From our view, it argues that the process of discovery takes place along a well-defined frontier of knowledge rather than in grab-bag fashion. Admittedly, the concept of "simultaneity" is impressionistic, but the related phenomenon of technological "clustering" again suggests that technical evolution follows a sequential and determinate rather than random course.

2. *The Absence of Technological Leaps*

All inventions and innovations, by definition, represent an advance of the art beyond existing base lines. Yet, most advances, particularly in retrospect, appear essentially incremental, evolutionary. If nature makes no sudden leaps, neither, it would appear, does technology. To make my point by exaggeration, we do not find experiments in electricity in the year 1500, or attempts to extract power from the atom in the year 1700. On the whole, the development of the technology of production presents a fairly smooth and continuous profile rather than one of jagged peaks and discontinuities.

3. *The Predictability of Technology*

There is a long history of technological prediction, some of it ludicrous and some not. What is interesting is that the development of technical progress has always seemed *intrinsically* predictable. This does not mean that we can lay down future timetables of technical discovery, nor does it rule out the possibility of surprises. Yet I venture to state that many scientists would be willing to make *general* predictions as to the nature of technological capability twenty-five or even fifty years ahead. This too suggests that technology follows a developmental sequence rather than arriving in a mere chancy fashion.

I am aware, needless to say, that these bits of evidence do not constitute anything like a "proof" of my hypothesis. At best they establish the grounds on which a prima facie case of plausibility may be rested. But I should like now to strengthen these grounds by suggesting two deeper-seated reasons why technology *should* display a "structured" history.

The first of these is that a major constraint always operates on the technological capacity of an age, the constraint of its accumulated stock of available knowledge. . . .

The gradual expansion of knowledge is not, however, the only order-bestowing constraint on the development of technology. A second controlling factor is the material competence of the age, its level of technical expertise. To make a steam engine, for example, requires not only some knowledge of the elastic properties of steam but the ability to cast iron cylinders of considerable dimensions with tolerable accuracy. . . .

Yet until a metal-working technology was established—indeed, until an embryonic machine-tool industry had taken root—an industrial technology was impossible to create. Furthermore, the competence required to create such a technology does not reside alone in the ability or inability to make a particular machine . . . , but in the ability of many industries to change their products or processes to "fit" a change in one key product or process.

This necessary requirement of technological congruence gives us an additional cause of sequencing. For the ability of many industries to co-operate in producing the equipment needed for a "higher" stage of technology depends not alone on knowledge or sheer skill but on the division of labor and the specialization of industry. And this in turn hinges to a considerable degree on the sheer size of the stock of capital itself. Thus the slow and painful accumulation of capital, from which springs the gradual diversification of industrial function, becomes an independent regulator of the reach of technical capability. . . .

In the future as in the past, the development of the technology of production seems bounded by the constraints of knowledge and capability and thus, in principle at least, open to prediction as a determinable force of the historic process.

*   *   *

The second proposition to be investigated is no less difficult than the first. It relates, we will recall, to the explicit statement that a given technology imposes certain social and political characteristics upon the society in which it is found. Is it true that, as Marx wrote in *The German Ideology,* "A certain mode of production, or industrial stage, is always combined with a certain mode of cooperation, or social stage," or as he put in the sentence immediately preceding our hand-mill, steam-mill paradigm, "In acquiring new productive forces men change their mode of production, and in changing their mode of production they change their way of living—they change all their social relations"?

As before, we must set aside for the moment certain "cultural" aspects of the question. But if we restrict ourselves to the functional relationships directly connected with the process of production itself, I think we can indeed state that the technology of a society imposes a determinate pattern of social relations on that society.

We can, as a matter of fact, distinguish at least two such modes of influence:

1. The Composition of the Labor Force

In order to function, a given technology must be attended by a labor force of a particular kind. Thus, the hand-mill (if we may take this as referring to late medieval technology in general) required a work force composed of skilled or semiskilled craftsmen, who were free to practice their occupations at home or in a small atelier, at times and seasons that varied considerably. By way of contrast, the steam-mill—that is, the technology of the nineteenth century— required a work force composed of semiskilled or unskilled operatives who could work only at the factory site and only at the strict time schedule enforced by turning the machinery on or off. Again, the technology of the electronic age has steadily required a higher proportion of skilled attendants; and the coming technology of automation will still further change the needed mix of skills and the locale of work, and may as well drastically lessen the requirements of labor time itself.

2. The Hierarchical Organization of Work

Different technological apparatuses not only require different labor forces but different orders of supervision and co-ordination. The internal organization of the eighteenth-century handicraft unit, with its typical man-master relationship, presents a social configuration of a wholly different kind from that of the nineteenth-century factory with its men-manager confrontation, and this in turn differs from the internal social structure of the continuous-flow, semi-automated plant of the present. As the intricacy of the production process increases, a much more complex system of internal controls is required to maintain the system in working order....

There is a danger, in discussing the structure of the labor force or the nature of intrafirm organization, of assigning the sole causal efficacy to the visible presence of machinery and of overlooking the invisible influence of other factors at work. Gilfillan,* for instance, writes, "engineers have committed such blunders as saying the typewriter brought women to work in offices, and

---

*Editors' note: S. F. Gilfillan, a sociologist and author of *The Sociology of Invention.*

with the typesetting machine made possible the great modern newspaper, forgetting that in Japan there are women office workers and great modern newspapers getting practically no help from typewriters and typesetting machines." In addition, even where technology seems unquestionably to play the critical role, an independent "social" element unavoidably enters the scene in the *design* of technology, which must take into account such facts as the level of education of the work force or its relative price. In this way the machine will reflect, as much as mould, the social relationships of work.

These caveats urge us to practice what William James called a "soft determinism" with regard to the influence of the machine on social relations. Nevertheless, I would say that our cautions qualify rather than invalidate the thesis that the prevailing level of technology imposes itself powerfully on the structural organization of the productive side of society....

We cannot say whether the society of the computer will give us the latter-day capitalist or the commissar, but it seems beyond question that it will give us the technician and the bureaucrat.

*   *   *

Frequently, during our efforts thus far to demonstrate what is valid and useful in the concept of technological determinism, we have been forced to defer certain aspects of the problem until later. It is time now to turn up the rug and to examine what has been swept under it. Let us try to systematize our qualifications and objections to the basic Marxian paradigm:

1. *Technological Progress Is Itself a Social Activity*

A theory of technological determinism must contend with the fact that the very activity of invention and innovation is an attribute of some societies and not of others. The Kalahari bushmen or the tribesmen of New Guinea, for instance, have persisted in a neolithic technology to the present day; the Arabs reached a high degree of technical proficiency in the past and have since suffered a decline; the classical Chinese developed technical expertise in some fields while unaccountably neglecting it in the area of production. What factors serve to encourage or discourage this technical thrust is a problem about which we know extremely little at the present moment.

2. *The Course of Technological Advance Is Responsive to Social Direction*

Whether technology advances in the area of war, the arts, agriculture, or industry depends in part on the rewards, inducements, and incentives offered by society. In this way the direction of technological advance is partially the result of social policy. For example, the system of interchangeable parts, first introduced into France and then independently into England failed to take root in either country for lack of government interest or market stimulus. Its success in America is attributable mainly to government support and to its appeal in a society without guild traditions and with high labor costs. The general *level* of technology may follow an independently determined sequential path, but its areas of application certainly reflect social influences.

3. *Technological Change Must Be Compatible with Existing Social Conditions*

An advance in technology not only must be congruent with the surrounding technology but must also be compatible with the existing economic and other institutions of society. For example, labor-saving machinery will not find ready acceptance in a society where labor is abundant and cheap as a factor of production. Nor would a mass production technique recommend itself to a society that did not have a mass market. Indeed, the presence of slave labor seems generally to inhibit the use of machinery and the presence of expensive labor to accelerate it.

These reflections on the social forces bearing on technical progress tempt us to throw aside the whole notion of technological determinism as false or misleading. Yet, to relegate technology from an undeserved position of *primum mobile** in history to that of a mediating factor, both acted upon by and acting on the body of society, is not to write off its influence but only to specify its mode of operation with greater precision. Similarly, to admit we understand very little of the cultural factors that give rise to technology does not depreciate its role but focuses our attention on that period of history when technology is clearly a major historic force, namely Western society since 1700.

* * *

What is the mediating role played by technology within modern Western society? When we ask this much more modest question, the interaction of society and technology begins to clarify itself for us:

1. *The Rise of Capitalism Provided a Major Stimulus for the Development of a Technology of Production*

Not until the emergence of a market system organized around the principle of private property did there also emerge an institution capable of systematically guiding the inventive and innovative abilities of society to the problem of facilitating production. Hence the environment of the eighteenth and nineteenth centuries provided both a novel and an extremely effective encouragement for the development of an *industrial* technology. In addition, the slowly opening political and social framework of late mercantilist society gave rise to

---

*Editors' note: Prime mover.

social aspirations for which the new technology offered the best chance of realization. It was not only the steam-mill that gave us the industrial capitalist but the rising inventor-manufacturer who gave us the steam-mill.

2. The Expansion of Technology within the Market System Took on a New "Automatic" Aspect

Under the burgeoning market system not alone the initiation of technical improvement but its subsequent adoption and repercussion through the economy was largely governed by market considerations. As a result, both the rise and the proliferation of technology assumed the attributes of an impersonal diffuse "force" bearing on social and economic life. This was all the more pronounced because the political control needed to buffer its disruptive consequences was seriously inhibited by the prevailing laissez-faire ideology.

3. The Rise of Science Gave a New Impetus to Technology

The period of early capitalism roughly coincided with and provided a congenial setting for the development of an independent source of technological encouragement—the rise of the self-conscious activity of science. The steady expansion of scientific research, dedicated to the exploration of nature's secrets and to their harnessing for social use, provided an increasingly important stimulus for technological advance from the middle of the nineteenth century. Indeed, as the twentieth century has progressed, science has become a major historical force in its own right and is now the indispensable precondition for an effective technology.

*   *   *

It is for these reasons that technology takes on a special significance in the context of capitalism—or, for that matter, of a socialism based on maximizing production or minimizing costs. For in these societies, both the continuous appearance of technical advance and its diffusion throughout the society assume the attributes of autonomous process, "mysteriously" generated by society and thrust upon its members in a manner as indifferent as it is imperious. This is why, I think, the problem of technological determinism—of how machines make history—comes to us with such insistence despite the ease with which we can disprove its more extreme contentions.

Technological determinism is thus peculiarly a problem of a certain historic epoch—specifically that of high capitalism and low socialism—*in which the forces of technical change have been unleashed, but when the agencies for the control or guidance of technology are still rudimentary.*

The point has relevance for the future. The surrender of society to the free play of market forces is now on the wane, but its subservience to the impetus of the scientific ethos is on the rise. The prospect before us is assuredly that of an undiminished and very likely accelerated pace of technical change. From what we can foretell about the direction of this technological advance and the structural alterations it implies, the pressures in the future will be toward a society marked by a much greater degree of organization and deliberate control. What other political, social, and existential changes the age of the computer will also bring we do not know. What seems certain, however, is that the problem of technological determinism—that is, of the impact of machines on history—will remain germane until there is forged a degree of public control over technology far greater than anything that now exists.

WILLIAM F. OGBURN

Technology
As Environment

In this selection, the late sociologist William Fielding Ogburn draws attention to the different types of environments in which man functions. In addition to the natural environment and the social enviroment, there is the technological environment, which is created entirely by man. While changes in the natural environment occur slowly and undramatically, the technological environment is rapidly changing and being reshaped. Consequently, Ogburn asserts, man does not adapt to the natural world so much, but instead must continually adjust to the changes in his technological environment. In short, Ogburn, like Heilbroner, pinpoints technology as the determining factor in social change.

The environment to which plants and animals are adjusted we call natural environment, and we think of it in terms of temperature, altitude, precipitation, atmosphere, soil, water, light, darkness, other animals, and vegetation. But there are other environments. For instance, there is a social environment. Many insects and higher animals live in groups, as do bees, termites, wolves, cattle, apes, and, notably, human animals. Thus men must adjust to their community as well as to nature. The social environment is in addition to the natural environment.

I am now to introduce to you still another environment for man—technology—that is, the material products of technology, which is the implication of the word "technology" in the title. The word is used loosely to comprise the applications of scientific discovery and the material products of technology. In short, it includes the objects of material culture. Thus a technological environment consists of such fabricated objects as buildings, vehicles, processed foods, clothing, machines, ships, laboratories. As an illustration, an urban employee working with the machines in a factory would be working in a technological environment. A technological environment is not exclusive. Such an employee is also working in a social environment, for he interacts with his fellow employees and employers. He is also working in a natural environment, since he is working in nature's air and light and moisture and pressure.

Environment, as thus thought of, is seen as a sort of envelopment, a near-totality in which a man is immersed. Thus a man in a factory is surrounded by technologically produced objects. A wild animal is surrounded by nature. With this enveloping environment a living animal or plant must be in a relationship that is more or less harmonious to its environment. This harmonious relationship relates to the whole environment.

However, an environment consists of parts; that is, it is made up of elements, such as trees, water, houses, foods. Man's adjustment to environment is more appropriately viewed as an adjustment to the various elements that compose the environment. Particularly is his maladjustment seen in reference to some particular element in the environment. Thus an animal from the tropics is maladjusted to the cold of the Arctic. People have died for lack of Vitamin C or because of so small an object as the proboscis of the *Anopheles,* carrying even smaller objects, protozoans, that produce malaria. Animals adjust, as well as maladjust, to small elements of environment. Thus tens of thousands of sheep have been prevented from dying by adding one part of cobalt to two million parts of water and have thus become adjusted to those grazing lands where there was no trace of cobalt.

So, in thinking of our technological environment, it is well to think of it in terms of the individual elements to which we adjust. Thus the tin can is an invention of a century ago to which we adjust by processing food in factories instead of in the family kitchen and so letting the housewife spend more of her time elsewhere than in the kitchen.

The technological elements are many, counted in the millions, comparable to, though not so numerous as, the elements in our natural environment. They also vary greatly in size, from, say, a needle to a skyscraper office building. A large object like a skyscraper, however, may not bring about as much adjustment on our part as does the tiny needle. The extent of adjusting we do to a technological element is not closely related to its physical size, nor is the complexity of an element of technology an indication of the amount of adjusting we may make. The electronic digital computers that perform the seemingly magical functions of an electrical brain are an extraordinarily complex invention; yet they may call forth less adjusting than the simple invention of the wheel.

We have been using the expression "the amount of adjusting." That there are degrees of adjustment may appear strange to those who derive their concept of adjustment from biology, where the measure of adjustment is living and the lack of adjustment is death. But the extent of adjustment varies, especially in the human individual. Being sick is not as good an adjustment as being well. A neurotic has a less satisfactory adjustment than a normal person. The word "adaptation" implies variation more than the word "adjustment." Thus varying degrees of adaptation are suggested by such terms as "strain," "tension," "nervousness," "vitality," "energy," "illness," "strength."

Another extension of the meaning of adjustment is necessary when such a term is taken from biology and applied to humans living in communities. In biology we think of an individual or an aggregation of individuals living or dying because of adjustment or lack of it. But with humans, aggregations are societies, and their group life is characterized by various institutions, such as schools, families, churches, states, clubs, economic and political organizations. When humans adjust to environment by groups, as well as by individuals, their group adjustment implies changes in social activities, such as those of religion, education, marriage, political and productive occupations. The lower animals have no schools and no churches, no parliaments, and no factories. Though the lower animals live in groups, their group adaptation is not greatly different from the adaptation of a group of plants. In either case the group adaptation to environment is something like the arithmetic sum of the adaptation of a collection of individuals.

But with humans, the adjustment of a collection of individuals is an adjustment of their group life and may mean an adjustment of their schools, factories, parliaments, and churches. In other words, with mankind, adjustment to environment means more than life and death of an aggregation of individuals; it means degrees of adaptation of social institutions and customs.

Thus when we added the steam engine to our technological environment and applied mechanical power to our tools instead of muscle, we worked in large buildings called factories, instead of in the family dwelling. Hence our adjustment to this technological element, the steam engine, meant an adjustment of the institution of the family and of our economic institutions.

One of the earliest technological changes in our environment concerned producing fire by friction. Before the acquisition of fire, the habitat of early man was Africa and southern Asia. He could not go outside this area because of the cold and the shortage of fruits. But when fire was used in cooking, the hard indigestible fibers of many plants and leaves were more edible. By migration, his food supplies could be increased greatly, and he could live outside the semitropics. Thus the adaptation to fire was migration, and men thus were spread more widely over the earth than any other animal.

For hundreds of thousands of years men were wanderers within a generally large though limited area. The little band of humans would eat a locality out of its supply of animals and wild plants and then move on to fresh food supplies. Then was added to this technological environment a most important implement, a digging stick, with which he would dig holes in the ground, drop in a seed, cover it, and then dig away weeds from the growing plant. This digging stick was a simple tool, a hard stick with a point or a flattened blade-like end, as in a small spade, or a stick with a joint at the end, which suggests a hoe. Yet the adjustment to this simple digging stick changed the wandering band of a dozen or more individuals to a more or less settled community of several scores of inhabitants. This was quite a change in the social life of man, and sociologists should recognize the influence of technology in this transformation of his society.

Various sociologists in the past have written of the influence of natural environment on our social life, but, strangely, few have studied the influence of the technological environment.

As the hoe evolved into the plough, food was raised from seeds of grasses—notably barley, oats, wheat, and rice—all of which could be preserved longer than fruits; and animal food, particularly milk, was produced from tamed animals. This increased food supply, based upon technology, made possible communities much larger than were possible in the hoe, or digging-stick, culture. Villages of several thousand inhabitants were possible where the climate and soil were suitable. In general, early agricultural villages were smaller than these. These early agriculturalists lived compactly in villages and went out to cultivate their fields and tend their flocks. But as the thickly cultivated plants annually took out of the soil chemical elements important for the growth of plant food, the soil became less fertile, and the villagers with their ploughs, domesticated animals, and seeds moved on to seek new lands. To find these lands they often cut and burned the forests. As they came in contact with the peoples living by hunting and gathering wild food, they killed them, conquered and married them, or enforced their culture on them. So the world became peopled by agriculturalists rather than by hunters and gatherers of wild foods. The adjustment to the technology of agriculture led to the replacement of the hunting people and to larger communities.

These adaptations I have just recounted—namely, migrations, increases in population, and stability of residence—are only the immediate adjustments that come directly from the uses of these technological elements. But the group adjustment to technological environment is more complex than the adjustment of the lower animals. Group adjustment to a technological element is made only in a few customs or institutions, not in the totality of them. The first adaptations are those coming from direct uses. But to these changed customs and institutions coming directly from their use, secondary, indirect, or derivative adjustments are in turn made.

Thus the first direct adjustment to the technology that increases the food supply and makes it more assured from season to season and from year to year is a larger population. But the adjustment in turn to a larger population may be a greater division of labor, a specialization of occupation, different religious ceremonies, a differentiation of age societies, or the creation of social classes. These are derivative adaptations to the original or direct adaptation to the technological innovation. The original or direct adaptation is a change in some element or part of the society which we may call A. A, then, has adjusted to T, the technological innovation. But in a society there are other parts or elements than A, as for instance, B, C, D, etc., where other elements, B, C, D, etc., may be interconnected with A. Hence B, C, D, etc., adapt to the new adaptation of A, which has resulted from an adaptation to T.

Men adjust to the steam engine by letting it drive their tools for them. Consequently, they work away from home in factories. Then the family, a social institution, adjusts to the absence of workers and to the new production and to the additional source of income. The adjustments in the family are the decline in the authority of the husband and father, the removal of economic production from the home, the separation of husband and wife, and

the different type of education for the children. These are not the direct adaptations to the steam engine but are adaptations to the uses of steam-driven tools away from the homestead.

These derivative or indirect adaptations to the technological elements in our environment are not usually recognized or appreciated, for many sociologists are interested more in descriptions than in causes, and when they search for causes they look only to the direct cause, not to the derivational causes. Causes are like links in a chain and occur in a succession. A sequence may begin with an adjustment A to a technological element T. B, another adjustment in another element of society, is seen as an adjustment to A but not to T. The decline in the authority of the husband and father in the family is not interpreted as an adjustment to the steam engine, but only as an adjustment to the transfer of production away from the home, which was in turn an adjustment to the steam engine. The most numerous adjustments to a technological environment are the derivative ones; for any one direct adaptation to a technological element creates a change in a custom or an institution to which several other customs or institutions will adjust. But commonly these derivative adjustments are not seen as adjustments to the technological element in the first instance.

However, there are some reports of both direct and derivative adjustments to a technological element. Ralph Linton has studied the Tanala adjustments to the technology of a wet rice cultivation. Formerly the people had cultivated dry rice, which required a large or joint family. Under the wet cultivation, a

© NIKOLAY ZUREK/JEROBOAM, INC.

single family, instead of a joint family, did the work, and the village became permanent. The displaced families moved off into the jungle to seek new fields, but the kinship ties of the joint family held, and a tribal organization developed through intermarriage. With the increased wealth and property came kings, slaves, and warfare.

Similarly, in several different parts of the world the adjustment to cattle-raising has led to increased stealing, to war, to slavery, and to the creation of a nobility.

The most extensive adaptations to a technological environment are not to a single element but to a cluster of elements. Thus cities were a dramatic community adjustment to three basic technological elements: (1) an agricultural technology which enabled a farm family to feed more than its members, (2) a transportation technology which would bring food into the city and goods exchanged out of the city, and (3) tools of manufacture. With cities, as all sociologists know, came radical changes in many customs and institutions.

So also the modern family in the United States and western Europe is an adaptation to a cluster of technological and scientific elements, namely, the steam engine, contraceptives, and scientific discoveries affecting religion. Religious beliefs have made extensive adaptations to scientific discoveries which affected the forms of belief in miracles, healing, life after death, the location of heaven and hell, and creation.

The technological environment in modern times differs from natural environment in that it changes more rapidly. Natural environment has changed: four times northern Europe and America were covered with glaciers, but these glaciers came and went only a few feet a year; whereas in modern times there have come, within a couple of centuries, the steam engine, the internal-combustion engine, the dynamo, and now the atomic reactor. The railroad, the automobile, the airplane, and now the guided missile have come in an equally short time. Quite as rapid has been the advent of the telephone, radio, motion picture, television, microfilm, the tape recorder, and now the putting of vision on magnetic tape.

Unlike the natural environment, the technological environment is a huge mass in rapid motion. It is no wonder then that our society with its numerous institutions and organizations has an almost impossible task in adjusting to this whirling technological environment. It should be no surprise to sociologists that the various forms and shapes which our social institutions take and the many shifts in their function are the results of adjustments—not to a changing natural environment, not to a changing biological heritage—but adaptations to a changing technology.

GEORGE H. DANIELS

Technological Change and Social Change

Historian George H. Daniels disputes the idea of technological determinism as advanced by Heilbroner as well as William F. Ogburn's theory of "social lag," that society must continually adapt to technological change. Daniels, in contrast, stresses the degree to which technological innovation reflects and responds to felt social needs.

Roger Burlingame was the first American historian who tried to deal with these big questions [regarding the role of technology] against the broad expanse of American history, and in the essentials, at least, most historians of technology have followed his lead. The picture of technology as a motive force in American civilization which Burlingame sketched in two books written in the late thirties—that of technology as the major force leading to our unity as a nation—is a dramatic one which assigns to technology a central— one might even say a determining—role in American history.

The key methodological concept in the Burlingame analysis is that of the "social lag," a concept formulated by the sociologist Ogburn in 1923 and generally adopted by historians of technology ever since. Technology, so the

George H. Daniels, "The Big Questions in the History of Technology," from *Technology and Culture* 11 (1970). Published by the University of Chicago Press, Chicago 60637. © 1970 by the Society for the History of Technology. All rights reserved. Reprinted by permission.

notion goes, changes society by changing our environment, to which we, in turn, adapt. Between the change and the adaptation, however, there is always a lapse of time, the social lag. Technology, in this view, is the primary active force; sooner or later other institutions come into conformity with it. The social history of technology, then, is the story of institutions trying to catch up with technological realities. Thus the failure of politics to catch up with technology was a cause of the Civil War; the individualism of the 1870s was another example of social lag; the persistence of privately owned utilities is still another.

There is probably something to the concept of a social lag, at some times and in some places, but I would like to suggest that uncritical adoption of it is one of the great difficulties besetting us in the history of American technology. For the question of whether, on the whole, technology causes social change or social change causes technological change is one of the "big," and still unresolved, questions before us. Treating it as a fully resolved matter has been a cause of much confusion and, I think, misdirected effort, and it has obscured the nature of some of the other big questions.

I do not wish to be understood as arguing that technology does not have social consequences—of course it makes a difference in the life styles of every human being in a society whether production is by machine or by hand. In American history, it is perfectly true that the drift to the cities which accompanied growing industrialization altered the whole social hierarchy and political control of the nation. Nor would I be inclined to argue that the atomic bomb did not have important implications. What I do believe is that no single technological innovation—and no group of them taken together in isolation from nontechnological elements—ever changed the direction in which a society was going before the innovation. Even when the innovation is imposed from without, anthropologists have demonstrated that societies display a remarkable ability to adapt it to their own life styles. Urbanization in American history, for example, was a phenomenon in which technology was involved—but the process itself was a broad social movement, beginning before the technological innovations that are often cited as its determinants; and its explanation involves immigration, population growth, finance, and other matters as well as technology. And I further believe that the direction in which the society is going determines the nature of its technological innovations. Of course, these are at present unproved assumptions, although I think that recent studies lend them a certain amount of credence.

At any rate, the biggest question of all has to do with the nature and direction of causation, and at present we know very little about it. There is a pressing need for studies that will shed light on this question; the only advice I have is that we realize that it *is* a question.

Despite the lack of evidence, most scholars who have concerned themselves with the relations between technology and society have implicitly adopted the view of Ogburn and of Burlingame by framing their research questions in such terms as: What was the effect of the automobile, the railroad, the typewriter, or the radio, on society? They have then observed the uses of such single innovations and assumed that the innovation was the direct cause of the uses. This was the "impact" of the innovation. In this, both those who deplore technology and those who believe that it can solve all of man's problems agree. Thus Ogburn and Nimkoff studied certain changes that have taken place in the American family and attributed them all to recent technology, with no consideration of trends antedating the technology. In the same manner, Ogburn counted well over a hundred "impacts" of the airplane.

General American historians, on those rare occasions when they deal with technology at all, also adopt this framework. If he knows anything about technology, the general American historian is likely to have at least two facts available which he places somewhere in his book to demonstrate his virtuosity: (1) The cotton gin fastened slavery on the South and thereby was a major cause of the Civil War, and (2) the typewriter brought women to work in offices and thereby "liberated" them. An unscientific survey, which consisted of going to the library and looking through a shelfful of high school and college texts, indicated that almost all of them contained these two "facts." Gilfillan, in commenting upon the second of these claims, made the obvious point that in Japan there were women office workers, neither liberated nor getting much help from the typewriter. More to the point than Gilfillan's objection is the fact that American women had been in the process of being "liberated" for a full generation before the appearance of a typewriter. Women had already been working in American offices, and they were beginning to do so in increasing numbers. Women had, in fact, been working previously at a great variety of jobs in America; their moving into offices, I suspect, can be directly correlated with the increase in the total number of workers at this kind of job. One may as well credit the invention of the tin can, the contraceptive, or any of a hundred other things with the emancipation of women, a purely *social* process which took advantage of appropriate inventions when they appeared and perhaps, in fact, stimulated the appearance of those inventions. The fact that the typewriter *in America* brought more women into the offices is not altogether irrelevant. It fit neatly into a preexisting social process and facilitated that process, as did the tin can and the contraceptive. Would the contraceptive have been accepted in a society that firmly believed the lot of women was to be forever bearing children? Contemporary experience suggests that it would not. What would have happened had the typewriter been invented in a society where the very idea of women in offices was unthinkable? This is probably an unanswerable question, although there is a related one that could be answered; namely, what happens when typewriters are introduced into societies where they have not been used before and where women do not hold jobs outside the home? Do they bring women into offices? I doubt it, but the answer would help us to understand the role of our own technology better.

The same case could be made about the other familiar example. It is perfectly true that technology contributed to the profitability of slave labor, as it did to free labor, not simply in the form of the cotton gin, but with spinning machinery, power looms, and other equipment which left only the production of raw materials relatively unmechanized. But slavery was already fastened on the South, and it was on the rise before the introduction of the gin because of the opening of new lands. The allegation that slavery would have disappeared and there would have been no Civil War without the cotton gin is therefore pure romanticism. The problem in the 1790s was that of *using* all the cotton that could be produced with the available labor supply; thus, invention was aimed at remedying the balance. It was a case of the same technological disequilibrium that Rosenberg found to be such a potent stimulus for innovation in the late nineteenth-century machine-tool industry. A high productive capacity at one level of the process stimulated invention at another. Far more significant than what the gin did to slavery is the fact that the gin was invented in a society where the labor system was based on slavery. I remain relatively certain that had there been no Eli Whitney and no cotton gin, Southerners would have found other uses for their slaves. I also remain convinced that had there been no slaves, but there had still been a flying shuttle, invention would have been aimed primarily at the production stage of cotton manufacture.

In both these cases and in many others that could be mentioned, the real effect of technical innovation was to help Americans do better what they had already shown a marked inclination to do. This, I suggest, and not a lack of talent or imagination on the part of historians, is the reason that after so many efforts have been made, we are still unable to point to any broad social consequences of an innovation. Historians who deal in such matters, it seems to me, have simply been asking unanswerable questions. The use of an invention does modify the user's habits, but there is no evidence that it has markedly changed the direction of the total complex of habits. Habits seem to grow out of other habits far more directly than they do out of gadgets.

* * *

Recent trends in economic history, I think, may offer us a more realistic and more satisfying framework than that of the social lag—provided we make sufficient allowances for economists' biases. For a very long time, economists and economic historians tried resolutely to ignore technology. They looked upon it as an alien force which occasionally disturbed the equilibrium of the economy they were studying. "Assuming that the state of the arts remains constant" was a frequently encountered phrase in their work, and this simplifying assumption helped them to understand other elements of the economy. When they did discover technology, their characteristic response was to try to deny its equilibrium-disturbing character. The economist concerned with the question of economic growth generally draws smooth curves showing a steady rise in productivity. His curves can contain the introduction of the telegraph, the telephone, the railroad, the automobile, or the airplane and yet show no trace of a revolution. It is possible, of course, that the sheer weight and complexity of the economic structure, especially one so weighty and complex as that of the United States, can disguise revolutionary changes. The shifting of resources and effort as some industries become obsolescent, the local effects of changing patterns, the ruining, for example, of thousands of small businessmen with the rise of the mail-order house—such things will not be revealed by gross figures such as the GNP. When dealing with economic historians, we must be on guard against this possibility. But I do believe that their smooth curves, even with their limitations, give us a more realistic picture of the historical process than the historian of technology with his dramatic revolutions and his discontinuous leaps.

Economic historians have taken their penchant for continuity quite far—too far, at times, one might argue. Thus [Robert] Fogel used statistics, logic, and argument to demonstrate that the railroad innovation had no particular economic consequences. Others have shown that even great bursts of technological activity, such as those promoted by wartime conditions, can be contained in the smooth curves that existed independent of the bursts. The Civil War has recently suffered such a downgrading as a causal factor. Schmookler* came close to transposing the inventor—that eccentric, unpredictable, lonely individual of legend—into an economic man, rationally calculating relative advantage, assessing the market and inventing or not inventing, changing directions from railroads to electric shop motors in response to the same economic forces that the pawnbroker, the industrialist, or the merchant obeyed. Schmookler's key point was that invention was essentially an economic activity, and from this relatively simple concept a great deal followed.

Although I would not like to exchange one hackneyed stereotype for an-

---

*Editors' note: See below, p. 400.

other, there is a great deal to be said for Schmookler's picture of the inventor, and in studies carried out over the ten years before his recent death, he gathered a great deal of evidence to support it. He concluded that new goods and new techniques are unlikely to appear unless there is a preexisting demand in the society. In other words, more significant than a "social lag," there exists at any time a technological lag, a chronic tendency of technology to lag behind demand. This is the same point that Gilfillan made intuitively in 1935 and Friedrich Engels much earlier, but now Schmookler has brought what seems to me impressive evidence to support it. Studying a wide range of industries, he generalized that inventive effort varies directly with the output of the class of goods the inventive effort is intended to improve, with invention lagging slightly behind output. Increasing sales for one class of goods invariably produced an increase in inventions pertaining to that class; declining sales were followed by a decline in inventions.

The existence of the lag implies that causation could not have been the other way around. The recent business practice of setting research budgets at a fixed percentage of sales tends to assure this relation now, but it existed long before research became institutionalized and the practice became common.

Although the points are not nearly so well substantiated, Schmookler even believed that basic inventions establishing new industries are induced by these economic forces, and that the case holds even when a scientific discovery underlies an invention, for a discovery may contain the seeds of many potential applications, only some of which will be realized. Schmookler's view is in marked contrast with the earlier view of Ogburn, who insisted that an invention may answer no social need but may simply be a product of a scientific advance. The dispute between the two points of view is fundamental: in the one, technology is tied to society; in the other, it is tied to science. If we accept the contention of Price and others that science and technology develop independently of each other, we shall have to cast out Ogburn's view— although doing so will not yet establish the rival view.

If the pattern of invention thus depends in large measure upon socioeconomic change, we see once more why it is futile to attempt to trace social changes to technological innovations. Such broad social forces as urbanization, declining family size, changing status of women, increases in population and per capita income—the whole range of matters often attributed to technology—are themselves determinants of the direction of technological innovation. Technology, in a word, is used to help people do better what they were already doing for other reasons, and what they are doing for other reasons determines the nature of their future technology....

If there is anything that seems clear at this point in our knowledge, it is that the preferences of people do have a lot to do with the development of their technology. If this recognition forces historians of technology to consider intangible factors and assign a somewhat lesser role to technology than they customarily do, if it forces them, even, to learn a little history, it will at least have the virtue of making unnecessary the futile game of trying to find direct, mechanically operating connections between a technological innovation and social change. If we stop misdirecting our work, we may one day find out how to direct it.

JOHN A. HOSTETLER

Amish Society

The Amish are communities of people descended from German-Swiss religious groups formed during the sixteenth-century Protestant Reformation. Many Amish emigrated to the United States in the eighteenth century and settled in Lancaster County, Pennsylvania, from where some of their descendants migrated to farm areas across the United States. This selection by an anthropologist, John Hostetler, describes the traditions and values of the Amish people, who teach personal simplicity as a way of life. In some instances, Hostetler writes, the Amish reject change and tenaciously retain their customs. Nevertheless, they are very gradually accepting some modern technology, which requires a reorganization of their values.

Self-sufficiency in the economic life of the Amish people is associated with agrarianism and occupations associated with nature. Closeness to the soil, to animals, to plants and to weather are consistent with their outlook on life and with limited outside contact. Tilling the soil was not a tenet in Anabaptism but emerged as one of its major values when the movement was banished to survive in hinterlands. Hard work, thrift, mutual aid, and repulsion of city ways such as leisure and non-productive spending, find support in the Bible and are emphasized in day-to-day experience. With practical knowledge and hard work, a good living can be made from the soil; and here, the Amish contend, is the only place to have family life.

Woman's sphere and work is at home, not in the factory or in a paid profession. Cooking, sewing, gardening, cleaning, white-washing of fences, tending to chickens, and helping with the milking keeps her forever occupied.

Caring for the children is, of course, her principal work. She will never be a teacher outside the home, not even in her church or its formal activities. Her place in the religious life of the community is a subordinate one, though she has voting rights in congregational meetings or in nominating names for the ministry. An Amish woman's work, like the work of any American woman, is never done. But she is always with her children, and to break the monotony, there are weddings, quiltings, frolics, auction sales and Sunday services. For her satisfaction in life she turns to brightly colored flowers in the garden and in her house in the winter, to rug-making and embroidery work on quilts, pillowcases and towels, and to shelves full of colored dishes in her corner cupboard. These are her prized possessions, some the work of her hands, made not for commercial gain, but for the enjoyment of the household and her host of relatives. Within her role as homemaker she has greater possibility of achieving status recognition than the suburban housewife: her skill, or lack of it, has direct bearing on her family's standard of living. She sews all their clothes; plants, preserves, and prepares the food her family eats, and adds beauty to life with quilts, rugs, and flowers. Canning her own food, making her chowchow, and spreading the dinner table with home prepared food, are achievements that are recognized and rewarded by her society. . . .

Closeness to Nature

The little Amish community has a strong affinity for the soil and for nature. Unlike science, which is occupied with the theoretical reconstruction of the order of the world, the Amish view comes from direct contact with nature by the reality of work. The physical world is good, and in itself not corrupting or evil. The beautiful is apprehended in the universe, by the orderliness of the seasons, the heavens, the world of growing plants as well as the many species of animals, and by the forces of living and dying. While it is wrong to attend a show in a theater, it is not uncommon for an Amish family to visit the zoo or the circus to see the animals God has made.

The Amishman feels contact with the world through the working of his muscles and the aching of his limbs. In the little Amish community toil is proper and good, religion provides meaning, and the bonds of family and church provide human satisfaction and love.

The charter of Amish life requires members to limit their occupation to farming or closely associated activity such as operating a saw mill, carpentry, or mason work. In Europe the Amish lived in rural areas, always having a close association with the soil, so that the community was entirely agrarian in character. It is only in America that the Amish have found it necessary to make occupational regulations for protection from the influence of urbanism.

The preference for rural living is reflected in attitudes and in the informal relations of group life, rather than in an explicit dogma. For the Amish, God is manifest more in closeness to nature, in the soil and in the weather, and among plants and animals, than he is in the man-made city. Hard work, thrift, and mutual aid find sanction in the Bible. The city by contrast is held to be the center of leisure, of non-productive spending, and often of wickedness. The Christian life, they contend, is best maintained away from the cities. . . .

There are other moral directives in the little community but these form the essential core of what is viewed as right and wrong. The view of life and of man's place in the total scheme of things are determined by the sacred guides to life. These guides are: a biblical view of separation from the world, the vow of obedience, observance of the *Ordnung* [rules], upholding the true doctrine of shunning, and living close to the God-created environment. In all of these tradition plays an important part. The people of the little Amish community tend to regard the ways of their ancestors as sacred and to believe that these time-hallowed practices should be carefully guarded.

*   *   *

Land use is exceedingly intensive in the most densely settled communities. All available farm land is under cultivation, and in some regions where land is unsuited for cultivation, timber is kept for a supply of lumber and firewood. The typical farm includes a large garden and a larger tract or "truck-patch" for potatoes and other vegetables for home consumption. Orchards are common. In Pennsylvania, farmers follow a four or five year rotation that includes the production of wheat, oats, corn, and hay. In some areas this plan of rotation has been modified by cash crops such as tobacco, potatoes, tomatoes, or peas. The Amish people make much use of barnyard manure, as well as lime and commercial fertilizer. Feeding livestock and cattle are outstanding features of most Amish farms in the eastern United States. Cattle feeding is carried on as much for the sake of obtaining manure as for financial profit. Hog raising is a normal activity on midwestern farms but has diminished in the eastern part of the nation. For sanitary reasons, hog raising is kept to a minimum in areas where dairy farming has become a specialty. Farm income is supplemented by selling farm products to local and nearby city markets, especially in Pennsylvania.

Favorable markets for fluid milk in some areas have modified the Amish farm in recent years. Cow stables have been remodeled and equipped to meet the demands of milk inspectors. Milk houses and cooling systems, which are operated with gasoline engines, have been installed, and a considerable number of Amish people have purchased milking machines and bulk tanks. The size of the Amish dairy herd ranges from twelve to twenty or more.

A major difficulty encountered with the commercialization of the farm has

been the traditional observance of Sunday. Believing that Sunday is a holy day, a day on which there should be no business transactions, the Amish have refused to allow their milk to be picked up by trucks on Sunday. Some families solve this problem by using the weekend milkings for home consumption and by churning them into butter. One firm agreed to pick up the Sunday evening milk on Monday morning if proper cooling and storage were assured. But the large milk industries in most instances will not take on producers unless they agree to sell milk seven days per week. They will make no concession to Amish religion. This has resulted in a serious problem for the Amish. They have sought other ways to solve the problem, some by separating the milk and selling cream, and some by taking an inferior price. As a way out, the Amish have also established cheese factories in some areas. In this way they have an outlet for fluid milk, and the farmers who have not modernized their barns to meet the standards of state inspectors can sell their milk to the cheese house. . . .

Soil conservation practices such as contour farming and strip cropping are not generally practiced by the Amish. Such techniques were never a part of their Swiss and Palatinate agricultural background, and when contour farming was advocated by state agricultural colleges they never adopted it, but regarded it as "book farming," that is, theoretical rather than practical. The Amish are, however, equipped to restore fertility in depleted soils. . . .

Farming among the Amish requires no books. The intellectual know-how of farming is passed on from father to son orally and remains in the head, scarcely reaching the printed page. Perhaps the only exception to this is *Baer's Almanac* and Raber's *Calendar*. Both provide zodiac information and these signs are carefully studied by some, but not by all Amish people.

*   *   *

Changing a major rule requires time in Amish society. It may take decades, and a half century or more to observe even the slightest symbolic change. . . . Let us take the case of a Pennsylvania church which during the past decade, suddenly to the surprise of the wider community, allowed the automobile for its members. Automobile dealers in nearby towns experienced a sudden boom in sales. The news that automobiles "were allowed" soon spread through the entire region and the secular community became accustomed to seeing bearded fathers drive their automobiles on the highways. This Amish church had voted almost unanimously, according to one of its members, to allow automobiles.

Upon closer examination and upon reconstructing the sequence of events in this decision, it was not a decision as easy as taking a vote among the members. A number of events reaching back a half century and also contemporary conditions produced this extraordinary example of systemic linkage, or successful change, This church had already relaxed its rules by making changes from the strictest Amish in the community. This church itself split off around 1911 from the "strict" *Meidung* church, and although it retained the dress and in many respects retained the *Ordnung* generally, slight modifications were made over the years. Men began wearing buttons on work jackets, the hair was cut shorter, single women were permitted to work as cleaning maids for non-Amish people, and tractors were adopted for farming. With the availability of pneumatic tires and improved tractors with higher speeds, the Amish began using them on the road to pull wagons to town, to run errands to neighbors, and for the daily delivery of milk to market. Children and boys became completely familiar with the mechanics and skill of driving a tractor. . . .

Under the above circumstances the desire for the automobile became very apparent in the church among the young men and the married non-farming men who were employed in such occupations as milling, carpentry, masonry, and butchering. Farm hands ribbed their employers about the inconsistency of hitching up horses for road work when transportation was easier and more efficient with the tractor. Sons complained about the slowness of the horses, that they were too much trouble, and that it was dangerous to drive a carriage on the open highway. This informal conversation and "egging" by younger members to Amish landowners, some of them ministers, over a period of several years appears to have set the stage for a favorable nod in the church. That talk about a forbidden norm was permitted at all was important in creating an atmosphere of discussion. . . .

The only way the automobile could be discussed in church was if some member violated the rule of non-ownership. Early one spring a young single man, a baptized member of a respected family, purchased an automobile. The youngster had secured a learning driver's permit and drove his new possession to the farm of his parents. The whole family was shocked, especially the father and mother who were interested in maintaining "peace" in the church. Obedience to the church and godliness were more important to the family than any change of traditional norms. The father objected to having the automobile on his property. After much persuasion the son returned the automobile to the dealer with the hopes of regaining it later. The youth was not excommunicated since he acknowledged his transgression of the church rules.

A few days later a married man, employed at the town village, purchased an automobile. He did not make the mistake of driving it to his home on the farm, but kept it at his place of employment. He commuted to and from the village with his farm tractor. The church officials deliberated on a course of action. He was advised to "put it away," meaning to sell it, until the church could come to a unanimous decision. He refused the advice of the assembly

and was excommunicated. Meanwhile a third member, a young married farmer, purchased an automobile, but he too was excommunicated. Many members were in favor of automobiles but waited for a change in the *Ordnung*. With these offences committed, the officials had justification for bringing the question of ownership of automobiles before the church. Though the offenders had to be punished for their disobedience, there was still the question of whether or not to allow the automobile. It became obvious to the officials that they must arrive at some recommendation among themselves as to the appropriate action. . . .

Decision-making in the Amish church requires taking the *Rat* or counsel of the baptized members. This is done by each member voting on the recommendation that is placed before the assembly. As is customary, the two deacons in the said church polled the members' meeting, one taking the nod among the men and the other among the women. Members always remain seated and either affirm the decision of the ministers with a nod of the head, oppose it, or remain neutral on the question. The outcome of the *Rat* is usually expressed as: unanimous, practically unanimous, or not unanimous. The result of the automobile vote was practically unanimous with only four old persons not giving assent, and these soon joined the next most conservative Old Order Amish group in the community. Members were instructed to buy only black automobiles or to have them painted black. Within a few weeks most of the members came to church in automobiles and only a few of the older members came in carriages. Every Sunday from forty to fifty automobiles could be seen around the farm buildings where Sunday services were held. The acceptance of the automobile forced still other changes in this community of Amish. The young people who formerly courted with horse and carriage and in traditional ways now had many alternatives before them. The small community, which was bounded by the horse and carriage, was now expanded to include other Amish communities in and beyond the state. Families and young men may now travel a distance of one to five hundred miles on a weekend to see friends or relatives. One woman who opposed the automobile vote said, "Where will this lead to, if our young people are given the privilege of going wherever they want?" The forces of adjustment that were necessary in this small community within a few weeks, which in the secular society required a half century, gave rise to other intense pressure for accelerated change. . . .

Conditions Favorable to Change

. . . Change, as we have seen, is inevitable in Amish society. The general influences of American culture, both material and non-material, gradually find their way into segments of the Amish. The methods used to keep the community in bounds, described earlier, are not 100 per cent effective in keeping the outside out. The following changes have occurred in one or more communities: Ball bearings have been adopted on carriage wheels. Dairy barns have been remodeled to conform to standards required for selling fluid milk. The young men have changed from black to brown shoes. Hair is cut shorter than the previous generation. Mothers have changed from cotton to nylon material for some women's garments. Tractors for field work have been allowed. The trend from general to specialized farming is very apparent. Young men and women have become interested in education, in occupations other than farming, and in missionary work. Some Amish districts have gone so far as to allow electricity, ownership of automobiles, and telephones. Kitchens have been modernized with appliances. Bottled gas in lieu of electricity, milking machines run by small gasoline engines, and refrigerators operated with kerosene are still other changes. The adoption of such innovations requires reintegration of culture and reorganization of values.

KINGSLEY DAVIS

The Migrations of Human Populations

Since Neolithic times, according to demographer Kingsley Davis, technology has contributed to the migration of populations, making possible the expansion of the world's population. Even in ancient times, towns and cities were magnets, drawing in people from rural areas, and their prosperity attracted invaders. Emigration from Europe commenced on a large scale during the Industrial Revolution with people going to North and South America, South Africa, and Asia. Again, technological development in the newly settled areas stimulated population growth. Currently, Davis points out, there is a large migration from the underdeveloped countries to the developed countries spurred by technological inequality. The loss in numbers hardly makes a difference in the situation of the underdeveloped countries; what is serious, however, is that most of the emigrants are professionals and other highly trained people.

Human beings have always been migratory. Sometime between 10,000 and 400,000 years ago man's predecessor *Homo erectus* had spread from China and Java to Britain and southern Africa. Later, Neanderthal types spanned Europe, North Africa and the Near East; modern *Homo sapiens,* originating probably in Africa, reached Sarawak at least 40,000 years ago, Australia some 30,000 years ago and North and South America more than 20,000 years ago. Excluding Antarctica, Paleolithic man made his way to every major part of the globe. Except for species dependent on him, he achieved a wider distribution than any other terrestrial animal.

Since this propensity to migrate has persisted in every epoch, its explanation requires a theory independent of any particular epoch. My own view is that the abiding cause is the same trait that explains man's uniqueness in many other ways: his sociocultural mode of adaptation. As culture advanced and diversified, a profound and distinctly human stimulus to migration developed, namely technological inequality between one territorial group and another. At the same time the possibility of migration was increased by man's capacity to adjust culturally to new environments without the slow process of organic evolution.

Although the particular conditions of each epoch shaped migration, the underlying cause remained the same. Paleolithic man, for example, was a hunter and gatherer who naturally followed his prey and forage. Urging him on was the contrast between exploited territory and virgin territory. This tendency, inherent in any predatory animal, was augmented by the unique advantages his technology gave him in hunting itself and in adapting to environments into which his prey took him. With weapons and cooperation he could quickly skim the big game from an area and move on, and with fire, skins, shelters and tools he could adjust readily to the new climatic and dietary conditions he encountered. Soon, however, most areas (and eventually all of them) would be skimmed and occupied by humans. The thrill and above all the advantage of moving into an empty land would be gone; instead migration would involve confrontation between newcomers and earlier inhabitants. At that point the difference in technology between one group and another would replace the difference between exploited territory and virgin territory as the stimulus to migration. Men with superior techniques could invade and use more fully an area occupied by others.

Whatever the specific factors, the worldwide dispersion of Paleolithic man had significant consequences. By enlarging the resource base it enabled the human population to expand to a size otherwise impossible. Men remained sparse, to be sure, but they roamed everywhere. Migration also stimulated sociocultural evolution both by making environmental adjustments necessary and by diffusing innovations. Finally, since migration also involved interbreeding, it caused man, in spite of his worldwide dispersion and his adaptation to diverse environments, to remain a single species.

Offhand one might think that the coming of agriculture and animal husbandry some 10,000 to 12,000 years ago would have reduced migration by making people "sedentary." The evidence is to the contrary. Not only did some Neolithic practices, such as slash-and-burn agriculture and nomadic pastoralism, necessitate movement through a sizable territory but also the Neolithic transition as a whole created a gulf between peoples who had made the transition and those who had not. Furthermore, the Neolithic complex did not arise fully developed anywhere, nor did it ever cease developing; rather, technological improvements in production, weaponry and transport kept appearing, and that created inequality and hence migratory potential between one territory and another. Pastoralists or shifting cultivators could evict hunters and gatherers, because hunters and gatherers required more land per man and therefore could mobilize less manpower at any one spot. For the same reason permanent cultivators could evict migratory cultivators and herders, but they might be evicted in turn by pastoralists with superior weapons and greater mobility. . . .

With the rise of town-based and quasi-literate civilizations new kinds of inequality between one territory and another arose, generating migration. The civilized centers operated as magnets, drawing both peasants and artisans from the immediate hinterland and barbarians from beyond. The barbarians frequently came not as peaceful newcomers but as marauders or invaders. In eastern Europe and central Asia the vast steppes evidently allowed pastoralism and an increase in population but not much agriculture. From this region nomads (the word is Greek for pasturing) began their invasions; each tribe pushing the one before it. When the tribesmen learned to ride horses, by at least 1500 B.C., their rapid movement made possible the creation of empires stretching for thousands of miles. Each wave tended eventually to become sedentary itself, a target for a fresh wave of nomadic invaders.

The list of invaders from central Asia is bewildering. Among the best-remembered are the Hittites, who reached the Anatolian plateau by 2000 B.C., were masters of iron metallurgy by 1500 B.C. and succumbed to the Phrygians and others about 1200 B.C.; the Scythians, who drove and followed the Cimmerians into central Europe and raided Egypt in 611 B.C.; the Huns, who emerged in Mongolia and from the second century B.C. were the scourge of China and moved steadily westward, reaching the Volga around A.D. 250, Gaul and Italy the following century, and stopping in 453 with the death of Attila. The Roman Empire was finally subjugated by two sets of nomadic invaders; those from eastern Europe and central Asia (Goths, Vandals, Alani, Franks and Burgundians) and those from the Arabian peninsula. The latter expanded rapidly after A.D. 630, until by 750 the Islamic world extended from Spain to the Punjab. Much of the expansion was accomplished not by Arabs, however, but by nomads from central Asia. The Seljuk Turks, forced out by the resurgent Chinese of the Sung dynasty, overran Persia, Armenia,

Anatolia and Syria in the eleventh century. Two centuries later Mongol tribes under Genghis Khan conquered northern China, eastern Turkestan, Afghanistan, Persia, Russia, a large part of eastern Europe, Asia Minor, Mesopotamia, Syria and finally southern China. As a result the Ottoman Turks were pushed into Asia Minor in the fourteenth century and then to the Balkans, culminating with the conquest of Constantinople in 1453. The Turks ruled India from the eleventh to the sixteenth century, when the Moguls (offshoots of Genghis Khan's people) took over and ruled until the British arrived.

How much actual migration was involved in these conquests it is impossible to say, but it was clearly from sparsely settled territory to thickly settled and from less advanced societies to more advanced. It if had been the only form of movement into civilized centers, the centers could not have existed. A more normal type was the movement of peasants and artisans into the city to sell their wares or earn a wage. This, however, did not suffice. The rulers and entrepreneurs of the civilized world needed manpower under direct control, and they took it by force, mainly from the barbarian world. . . .

Since Europeans initiated the technological transformation, the key to modern migration is to be found in their relation to other peoples. In the sixteenth and seventeenth centuries, for the first time, the world as a whole began to be one migratory network dominated by a single group of technologically advanced and culturally similar states. Largely as a result of the European countries' use of this network, they eventually were able to start the Industrial Revolution and thus enormously enhance their world dominance. The subsequent spread of industrialism to other parts of the world made industrialism per se, not European culture, the main basis of technological inequality.

How did the Europeans, their armies, navies and economies honed by incessant warfare among themselves, deal with the world they had discovered? Their first impulse was to skim the cream, to obtain luxuries and precious metals by confiscation, all the while preventing their European rivals from doing the same, but this could not last. Soon they followed the ancient world's example by setting up trading posts and coastal fortifications, but they needed more control over indigenous production and therefore claimed entire territories. Their handling of each territory depended on its climate, accessibility and inhabitants. In these terms four types of territory can be distinguished.

The first type, inaccessible and sparsely inhabited (such as Tibet, central Africa and the eastern Andes), was left in abeyance and need not detain us. A second type, tropical or subtropical, sparsely inhabited and accessible by sea, was immediately exploited; a third type, also accessible and lightly populated but temperate, was eventually exploited; a fourth type, accessible but thickly populated, was handled more indirectly. . . .

Warm and accessible territories were of immense potential value, because their products complemented those of Europe. When sparsely peopled by aborigines, the land required only clearing. Hence in the region closest to Europe—the Caribbean and the Gulf of Mexico and the warm coasts of North and South America—the Europeans undertook the production of indigo, rice, cotton, spices, sugar, tobacco, coffee, tea and other tropical crops. For this they needed huge inputs of cheap labor, but Europeans themselves were too expensive and too ill-adapted to such work in a hot climate, and the original inhabitants were too few and too recalcitrant. To obtain the needed labor the European managers resorted to the same device the Greeks and Romans had used: slavery. According to estimates recently evaluated and summarized by the historian Philip D. Curtin, 9.6 million slaves were imported into slave-using areas between 1451 and 1870. Since mortality during the voyage was great—normally 10 to 25 percent for slaves—the total number enslaved probably exceeded 11 million, virtually all from Africa. In distance and number this movement transcended any other slave migration in history. . . .

Only with the introduction of a new and greater technological gap produced by the Industrial Revolution did European emigration take off. Although the continent was already crowded, the death rate began to drop and the population began to expand rapidly. Simultaneously urbanization, new occupations, financial panics and unrestrained competition gave rise to status instability on a scale never known before. Many a bruised or disappointed European was ready to seek his fortune abroad, particularly since the new lands, tamed by the pioneers, no longer seemed wild and remote but rather like paradises where one could own land and start a new life. The invention of the steamship (the first one crossed the Atlantic in 1827) made the decision less irrevocable.

Little wonder that the great period of voluntary overseas European migration was from 1840 to 1930, and that the mania moved across Europe along with industrialism. At least 52 million people emigrated during that period. This equaled a fifth of the population of Europe at the start and exceeded the number of Europeans already abroad after more than three centuries of settlement.

The prime destination was the nearest Temperate Zone land, North America, but the wave spilled over to Australia, southern South America, southern Africa and central Asia. The movement fed on itself, not only because the migrants wrote back to friends and relatives but also because the new lands underwent rapid development. They turned out crops and products that competed with those of Europe, worsening the plight of many Europeans and improving the prospects for migrants. By World War I, 65 years after the big wave had started, the New World countries already rivaled northwestern Europe economically.

The new lands were so vast that not all parts could be settled simultaneously. In Russia settlement began beyond the Urals, but elsewhere it hit the seacoasts first and worked its way inland. The moving frontier became a part of life and folklore.

What were the consequences of the migrations of slaves, indentured laborers and free migrants in the four centuries preceding the Great Depression? One was a steep rise in world population growth after 1750, because in the regions of origin (except in Ireland) the migrations did little to damp population increase, whereas in the regions of destination, after initial setbacks, they greatly stimulated it. The sending areas, by the standards of the time, were densely settled. Emigration therefore enabled them to postpone an inevitable change in birth or death rates. Comparative data show that in Europe the countries with the highest rates of emigration postponed longest the reduction in their birth rate. France, with little emigration, had the lowest birth rate; Ireland and Italy, with much emigration, had high birth rates. Ireland was the only country whose population declined; if it had had no migration but had exhibited the birth and death rates that actually existed, its population today would be nearly 12 million instead of about 3 million. In Europe as a whole emigration did little to hold down population growth; the population rose from 194 million in 1840 to 463 million in 1930—about double the rate for the world as a whole. Emigration had even less effect in Asia and Africa.

In contrast, in the areas of destination the effect was electric. Even the primitive peoples after initial decimation generally made a strong comeback, and the descendants of African, Asian and European immigrants multiplied so fast that they were widely cited as being an illustration of the biological maximum of human increase. The reason for the growth was that entire new continents were being transformed overnight from stone-age technology to modern technology. This was a much greater transition than what was happening in Europe itself; in fact, it was the most fantastic jump in cultural evolution ever known, and it took the lid off population growth. Between 1750 and 1930 the population of the main areas of destination increased 14 times, while the rest of the world increased only 2.5 times.

Another consequence of the migrations was a geographic redistribution of the world's population. In 1750 the new regions, which accounted for half of the world's land area, held fewer than 3 percent of its people; by 1930 they held 16 percent.

At the same time the world's racial balance was altered. Certain groups became extinct, others disappeared by hybridization and still others made great gains. Caucasians increased 5.4 times between 1750 and 1930, Asians 2.3 times and blacks less than two times.

Even more dramatic was the geographic displacement of races. By 1930 approximately a third of all Caucasians (and by 1970 more than half) did not live in Europe and more than a fifth of all blacks did not live in Africa. If all Europeans had stayed in Europe and had had the same natural increase that Europeans exhibited everywhere, there would have been 1.08 billion people in Europe in 1970 instead of 650 million. The earliest immigrants exercised a disproportionate influence on subsequent racial distribution because their natural increase lasted longer than that of later immigrants. Although the immigration of blacks into the U.S. was minuscule after 1850 compared with European immigration, they almost held their own by sheer excess of births over deaths. Blacks represented 15.7 percent of the American population in 1850 and 11.1 percent by 1970.

Although most of the migrations involved no drastic shift in climate, some of them did. In the U.S. there are now 11 million blacks outside the South and 50 million whites (mostly northwestern Europeans) in the South. In Queensland in Australia there are about 1.7 million whites and in sultry Panama about half a million.

As a result of the displacement and mixing of races there are more racial problems in the world today than at any time in the past. In nearly all immigrant countries, in the Americas, Southeast Asia and southern Africa, race is one of the most important bases of political division. In some countries particular hybrids have become separate groups, for example the "Coloureds" who comprise 9.4 percent of the population of South Africa and the "Creoles" who make up 35 percent of the population of Surinam. Among immigrant countries Australia has been most effective in excluding racial minorities. Australia's freedom from racial strife compares with that of Sweden or Denmark. . . .

The noted authority on European migration Eugene Kulischer compiled a table of population displacements in Europe from World War II to 1948. Omitting internal displacements, some of which were enormous, the total comes to 18.3 million. For the periods from 1913 to World War II and from 1948 to 1968 I have tallied exchanges and refugee movements totaling 28.7 million. If we add similar displacements in Asia, Africa and the Western Hemisphere, the grand total for the world during the period from 1913 to 1968 comes to 71.1 million. This number of migrants is considerably higher than the estimated 52 million who left Europe of their own free will in the heyday of the transatlantic movement from 1840 to 1930, in spite of the fact that the period is shorter (55 years compared with 90).

Clearly the amount of forced migration since 1913 belies predictions that world migration would diminish. But what about free migration? The answer is that it has not diminished either, but it has changed direction. Instead of flowing from the crowded industrial countries of Europe to open spaces in the New World, it has gradually shifted until it is now flowing toward developed countries everywhere. The nations of northwestern and central Europe, exporters of people for so long, are now net importers. The New World industrial countries, still relatively uncrowded, receive professional and highly qualified immigrants from industrial Europe, but increasingly their migration is from less developed countries in southern and eastern Europe, Asia, Latin America and Africa.

Evidence of the surge into developed nations is abundant. Four New World countries—Australia, Canada, New Zealand and the U.S.—received a net total of 13.9 million migrants between World War II and 1972. The U.S. alone, still admitting more foreigners than any other nation in the world, received 9.2 million during that period. More surprising is the tide of migrants into industrial Europe; for example, Sweden, for centuries a country of emigration, became a country of immigration after 1930. Other advanced countries in Europe have shown a similar reversal, some more sharply than Sweden. Data on seven such countries, including Sweden, reveal a net migration of 6.3 million between 1950 and 1972. Adding this figure to the one for the four New World countries gives a total of 20.2 million net migrants to 11 industrial countries during the period.

What explains the reversed migration into industrial Europe and the continued migration into New World industrial nations? In my view the driving force is the widening technological and demographic gap between the developed nations and the underdeveloped three-fourths of the world. The gap differs in several important respects from the former differences between Europe and the rest of the world. First, the developed nations are now scattered over the entire world instead of being concentrated in Europe. Second, the underdeveloped countries are no longer overwhelmingly colonies but rather are independent nations. Third, the technological gap has widened in absolute terms while commercial and intellectual communication has drawn the two classes of nations closer together. Fourth, the demographic contrast between the two groups has been reversed. Formerly the technologically advanced nations had the most rapid population growth; now it is the technologically backward nations, and their rates of growth are without precedent. The population of the 176 countries I classify as underdeveloped in 1950 increased by 1.04 billion from 1950 to 1972, while the population of the 47 developed countries increased by only 200 million. Originally more sparsely populated, the underdeveloped countries as a whole were already more densely settled than the developed ones in 1950. Their comparative density was still greater by 1970: 36.3 persons per square kilometer compared with 17.2 in the developed countries.

As a consequence of the gap as it is now constituted the advanced countries have on the average more resources per person, more workers in relation to dependents, more capital generated from savings and more investment and trade. They therefore have more jobs and offer higher wages. Their native populations have become so educated, comfortable and upwardly mobile that in times of labor shortage they refuse to fill low-paying, low-status or disagreeable jobs. Millions of workers in the bulging underdeveloped countries are eager to take those jobs, and employers are anxious to hire them. Hence legally or illegally the migrants come, their transit facilitated by modern means of travel and communication and even by government and international assistance.

The dichotomy between developed and underdeveloped is, of course, arbitrary. Special geographic and political circumstances aside, the general principle is that a nation tends to gain migrants from countries less developed than itself and to send migrants to countries more developed. When an underdeveloped nation is close to a developed one (as Mexico is to the U.S. or Greece is to Germany) or has special ties with a developed one (as Britain's ex-colonies have with her), the migratory pressure is very strong. In the U.S. the 1970 census counted 4.53 million people of Mexican origin, more than the number from the rest of Latin America combined but equivalent to only two years of current population increase in Mexico. In Germany the number of foreign workers, 167,000 in 1959, rose to 2,345,000 in 1973, 82 percent of whom came from six countries (Greece, Italy, Portugal, Spain, Turkey and Yugoslavia).

That the immigrant stream is being increasingly drawn from underdeveloped countries is easy to see. In the U.S. the current shifted from northwestern and central Europe to southeastern Europe, and then to Asia and Latin America. In European countries Africa and Asia are playing an increasing role in immigration. . . .

The most prosperous countries are the ones that have the largest net immigration. The influx not only gives them a large foreign population but also adds to their population growth. The growth comes in two ways: directly as a result of the net migration itself and indirectly as a result of the immigrants' natural increase after they arrive. The indirect effect is greater the longer the period of time under consideration is. (The entire population of the U.S. is the result of immigration at some time in the past.) For a period less than a generation the indirect effect is a function of the migrants' fertility and age-sex structure.

Normally, since young adults are more numerous among immigrants than among natives, their crude birth rate is higher. On the other hand, international migration ordinarily includes more males than females, thus depressing the crude birth rate. In recent years this rule has not held for the U.S., but in the receiving countries of Europe contemporary free migration, often called "labor migration," is largely composed of young males. In West Germany in 1970–1972, for example, foreign workers were 71 percent male. Finally, the indirect effect of immigration on population growth depends on the fertility of the immigrant women. Insofar as they come from underdeveloped countries, their fertility is high compared with that of native women. In the U.S. in 1970 the number of children ever born to women aged 40 to 44 was 4.4 per woman for those of Mexican origin and 2.9 for all women. In Sweden the fertility of foreign women in 1970 was 28.3 percent higher, age for age, than that of native women. An approximate calculation indicates that about 42 percent of Sweden's increase in population (1.04 million) between 1950 and 1970 was contributed by net immigration: 33 percent by the entry of the immigrants themselves and 9 percent by their natural increase during the period. Hence direct immigration accounts for a large share of the growth of the major industrial nations. . . .

Contemporary migration has drawbacks for the sending countries as well as the receiving ones. They derive chiefly from the fact that migration is inevitably selective. Although the quality of migrants may be lower on the average than that of natives in the developed countries, it is higher than that of natives in the underdeveloped nations. Since the developed countries cannot admit all who wish to come, they can pick and choose as the interests of their employers dictate. This means that the underdeveloped country does not simply lose untrained manpower but often loses trained manpower that is scarce and costly to produce. In the U.S. in 1972, of the immigrants admitted who had an occupation 31.1 percent were classified as "professional, technical and kindred workers," compared with only 14 percent in this category in the U.S. labor force. In 1971 there were 8,919 medical degrees conferred in the U.S.; in the same year 5,748 immigrant physicians were admitted and in the next year 7,143 were admitted. At the same time many thousands of American youths failed to gain admission to medical school. Even when, by the standards of the receiving country, the foreign workers are relatively unskilled, they are often on the average better trained than the ones who do not move. In any case emigration removes people of productive age and leaves children and old people, thereby raising the underdeveloped country's already high dependency ratio. The value of remittances sent back home may partly compensate for those losses, but the remittances are uncertain and subject to stoppage or control in times of crisis. Indeed, migration itself may be cut off and migrants may be returned precisely when the sending country is in its worst condition. How ominous that prospect can be is suggested by the fact that in Germany alone the number of Greek workers is equivalent to 8.4 percent of Greece's entire labor force and the number of Portuguese workers is equivalent to 4.7 percent of Portugal's labor force.

For the underdeveloped country emigration appears to be a stopgap allowing postponement of internal economic and demographic changes that would make emigration unnecessary. It is like borrowing money at high interest to pay off debts that one's income could not support in the first place. As the underdeveloped countries become still more crowded, there will be increasing pressure for greater admission to developed countries, on humanitarian grounds if for no other reason. By and large, however, the problems of underdeveloped countries are beyond solution by emigration. If developed nations tried to accept as migrants the excess population growth of the underdeveloped ones, they would currently have to receive about 53 million immigrants per year. This would give them a population growth of 5.2 percent per year, which added to their own natural increase of 1.1 percent per year would double their population every 11 years.

In the future the failure of international migration to solve problems will not necessarily prevent its happening. The present wave of voluntary movement from underdeveloped nations to developed ones may reach a maximum and be reduced, but if so, it will be replaced by other waves. Although particular migratory streams are temporary, migratory pressure is perpetual because it is inherent in technological inequality. In the past migration has helped to fill the world with people. That the world is now full is a new condition that complicates prediction. Another new condition is the degree to which nations can, if they wish, control their borders. Nations strong enough to prevent voluntary migration, however, are also strong enough to engender forced migration. Whether migration is controlled by those who send, by those who go or by those who receive, it mirrors the world as it is at the time.

NEAL F. JENSEN

The Food-People Problem

Following the agricultural revolution 10,000 years ago, the human population grew very slowly. An acceleration in the growth rate commenced about 1750 owing to changes in mortality and fertility, about which demographers are still puzzled. The increase was so marked that in 1798 the English economist Thomas Malthus was prompted to write his Essay on the Principle of Population, *expounding his theory that population grows faster than food supplies and that "inevitable famine stalks in the rear of misery and vice to limit the numbers of mankind." For many years Malthus' theory was discredited as the application of technology to agriculture increased the food supply and seemed to eliminate the threat of world-wide starvation. But technological improvements in sanitation, public health, and medicine contributed to population growth beginning in the nineteenth century by the reduction of mortality. In the twentieth century, developed societies are characterized by reduced fertility, whereas there has been a dramatic increase in the population of less developed countries, largely because of modern medical technology. As a result, "neo-Malthusians" have begun warning once again of the dangers of overpopulation.*

In this selection, Neal Jensen, an authority on plant breeding, asserts that the dramatic increases in wheat production that began in the 1930s in the United States will soon begin to level off. Scientific research in plant genetics and agricultural technology are approaching a ceiling, while environmental forces such as the weather could materially lower wheat yields. With absolute limits on food production, Jensen warns, world population must be stabilized.

Neal F. Jensen, "Limits to Growth in World Food Production." *Science*, Vol. 201, July 28, 1978, pp. 317–320. Copyright 1978 by the American Association for the Advancement of Science.

Could he return for a day, Robert Malthus would be amazed at how little things have changed in the last two centuries. His controversial discourse of 1798, arguing that the growth of populations occurs geometrically while the growth of food production occurs arithmetically, is still unproved and controversial. But some changes have occurred. For example, while the "passion between the sexes" continues unabated, there have been medical advances which now short-circuit the formerly close relation between the sex act and human births. Other scientific discoveries and technological applications have greatly increased the productivity of food crops and livestock.

The geometrical doubling growth of populations that Malthus postulated has, in fact, occurred, and it suggests that a critical test of his hypothesis is just a generation or so away. World population reached 4 billion people in 1976. Earlier predictions of 7 billion by 2000 are now being modified to reflect the recent slowing in population increase in some parts of the world. Uncertainty surrounds only the "when" of reaching 7 billion—there seems little "if" about it. A further uncertainty surrounds the eventual carrying capacity of the world, which someday might be called the normal population. It is generally conceded that world population cannot stabilize below 12 billion people and may go as high as 25 billion. Studies I made some years ago indicated that the range of experts' opinions of the eventual maximum world population included the figure of 50 billion.

Malthus has been somewhat discredited by technology. He did not foresee the Industrial Age and the geometric effect of technology upon economic growth. Technology, backed by cheap energy and vast resources, has enlarged food production capacity so much that today the supply often exceeds the world demand expressed in market terms, although for logistic and economic reasons the food may not get to the hungry people who need it. The result is that food production appears to have fared rather well in Malthusian terms. Indeed, my own data show that wheat yields in New York doubled between 1935 and 1975, just as world population doubled between 1930 and 1976. The current question is, what does technology in agriculture hold for the future of food production? I will attempt to assess the role of technology in the productivity of one crop, wheat, and draw some inferences for the future. . . .

The gain in wheat productivity for the United States as a whole has averaged 0.50 bushel per year over several decades. The New York gain of 20.1 bpa [bushels per acre] over the 40-year period, 1936–1975, amounts to 0.50 bpa per year, supporting the existence of a general phenomenon.

General Limits to Growth

The legacy of Malthus is that we cannot escape the question of limits to growth, whether it be growth of populations, economics, or food productivity. The question is particularly apt today as we see around us many constraints on economic growth—indeed, on living—related to our formerly "free" or inexpensive basic resources of land, water, air, minerals, and energy. Historically, the full costs of economic growth often have not been assessed or recognized within the same time frame. The result has been that we live in a world of continual corrective action for past transgressions against our life-support system. At the same time, our concern, insofar as we are wise enough, extends into the future. Today, solutions that may be technologically feasible may not be otherwise feasible. These are not solely economic matters—an environmental impact statement, for example, is basically a political instrument.

In agriculture, too, we have long operated under this generally accepted concept of unlimited growth. I think it is fair to assume that the technology of the future will not be as "free-wheeling" as the technology of the past. The constraints we now see on economic growth will also place limits on the way technology itself will evolve in the future. The farmer will not escape this. In fact, during the past few years the farmer has been caught in a whirlwind of bannings, withdrawals, and substitutes affecting fertilizers, herbicides, pesticides, livestock medicines and feeds, and many other supplies and operations.

Productivity versus Production: What Is Ahead?

Productivity and production, although related, are terms with vastly different meanings. Productivity is yield per area or the yield potential under a given set of environmental parameters. Production, however, is productivity realized, the sum of all harvested areas times their yields. Scientists work to increase wheat productivity. These increases make possible increased production, but the relation is not a direct one since many factors influence the actual production. For example . . . wheat productivity in New York [doubled] between 1935 and 1975 (from 19.2 to 39.3 bpa). The average yearly production of wheat for these two decades separated by 40 years, however, increased only 39 percent (from 4,868,300 to 6,778,200 bushels). The reason for this is that farmers reduced their wheat acreage from an average of 255,300 acres per year for the earlier decade to only 172,000 acres per year 40 years later.

The world food problem often translates into a scenario where we pose the question of maximum production to meet a crisis. How much wheat could New York (a state whose highest acreage was 790,000 in 1880 when 15,010,000 bushels were produced) produce today in all-out production? The answer would surely be disappointing. In my opinion it would be difficult today to find one half of the acreage available in 1880 for wheat—given the sizable irreversible losses to urbanization and the competing pressures for alternative uses even in a crisis. Even with a doubled productivity rate it would be difficult to attain the record of wheat production established a century ago.

What is the outlook for continued growth in productivity in wheat? The picture I have painted shows four decades of uninterrupted growth at a steady rate. The productivity can be attributed equally to genetic and other technological gain. The genetic gain potential still before us is undoubtedly large. The technological gain potential must also be considered sizable but is more speculative because technological directions will evolve from a mix of pressures and information yet to be discovered (how can one know the unknown?).

It has come as a shock to me to realize that my answer to the question posed in the previous paragraph is gloomier than the circumstances would suggest. It is the point of this article that, despite the favorable evidence presented, we are approaching the end of an epoch of research and of increases in wheat productivity. Our picture of the future is blurred because our viewpoint is within the time segment represented by the upward slope of the productivity line. It is self-evident that this slope cannot be sustained indefinitely. In fact, I believe this line will begin leveling off and this will be evident for the decade ending in 1985. I emphasize here that productivity will continue to grow, but at a slower rate. After a few decades of gradual slowing, the productivity line for New York wheat will assume a slope very much like the 1865 to 1935 segment, except, of course, at a higher yield level. It seems unlikely that any future combination of genetics, technology, or unknown factors will be able to generate a sustained rise in productivity from these high levels. Thus the entire past and future of wheat productivity in New York under rainfed conditions can be expressed as a zigzag line embodying the single dramatic upward slope that began in the mid-1930s.

The rate of productivity increase will become slower and will eventually become level for several reasons. First, improvements to the wheat plant through plant breeding become part of the productivity base. Continued incorporation of an improved character into new varieties results in a sideways movement in productivity rather than a new upward swing. Plant height is a good example. The heavy wheat head is positioned at the very tip of a slender, reedlike stem. It is like the handle of a lever whose fulcrum is the base of the wheat plant. The ability of the wheat head to remain standing until ripe for harvest through seasonal hazards such as wind and rain storms is dependent on three factors: (1) the firmness of the root attachment to the soil, (2) the length of the culm (leverage), and (3) the actual strength of the culm. Reduced height is a powerful way to increase resistance to lodging or falling over. With reduced height and increased lodging resistance, productivity can increase because the crop escapes the severe losses in quantity and quality which accompany downed grain. Also, the full effect of technological inputs, such as increments of fertilizer, can be realized.... Thus, height reduction represents a plant improvement whose impact already has been built into the productivity base of wheat.

Second, technology, other than genetic, already has identified and removed or modified as many bottlenecks to productivity as possible, given our present state of the art. For the future, technology will be under a tight rein of accountability to society for its impact on health, safety, environmental, and ecological concerns. These and energy considerations may sometimes dictate that possible solutions are not feasible. The predictable result is a slowing of productivity growth.

Third, as the facts which are bottlenecks to production of wheat are eliminated or modified through breeding and other technology the eventual residue will consist of forces over which man has less and less control. A good illustration is climate and weather. Although the slopes of the Kansas and North Dakota lines are similar to New York's, the level of production for both states is lower than that for New York. This is principally due to climate and weather, reflecting moisture available to the crops in the different areas. Climate, or weather, will play an ever important role in future crop production. While one may think of ways to increase production, particularly in a developing country, it is not so simple in a developed country where technology has long interacted within the competitive needs of society. In fact, with time it is easy to see factors conducive to shrinking agricultural production. For example, western U.S. irrigated agriculture faces gradual elimination as the pressures for higher priority water needs become evident.

Coming: Absolute Yield Ceilings

A productivity line with an undeviatingly upward slope ... must be a short-range phenomenon. To gain the perspective of the long run it is necessary to understand the nature of a yield ceiling. A ceiling has two faces, one relative, the other absolute. I think it fair to say that almost all productivity increases of the past have been made in the relative area; that is, the ceiling has been there but it has been possible to gradually raise it by bumping against it as we have discovered successive new inputs and favorable mixtures of technology. In technical terms the yield ceiling at a given time is the product of interaction between genotype (variety) and environment. Scientists have raised productivity by successfully modifying both the genotype and the environment. Nevertheless, somewhere down the road ahead of us lies the absolute ceiling, still somewhat flexible, with swings from year to year but basically buffered and resistant to upward change by the outside forces man has not yet mastered. What are these forces? I have mentioned weather. Others are the loss of land and irrigation to urban (people) demands. Another is economic— simply, it will cost too much to try and extract the last increment of possible yield. Agriculture is a high-risk enterprise where major economic costs are committed at the front end of the season and the whole enterprise is at hazard to many unknown forces of the season, including the market. In North Dakota during the drought years of the 1930s, decisions to harvest wheat fields in the fall often were made simply on the judgment as to whether the value of the wheat exceeded the remaining operations of binding (twine) and threshing— all previous costs of raising the crop already had been committed and were beyond recovery. As we approach the absolute yield ceiling for rain-fed wheat I believe we will see also the development of an altered strategy of production based on optimum rather than maximum returns. The input costs of the maximum strategy based on the hope of a bumper crop each year will have to be adjusted to a more moderate approach based on average pragmatic expectations and more in tune with the conservation of energy and resources of the future. . . .

Conclusions

In conclusion, I am aware that the favorable data on wheat productivity I have presented for New York through 1975 do not support my gloomy conclusions and prognosis for the future. Nevertheless, I strongly believe that my interpretation of an approaching yield ceiling is valid and that the Malthusian divergence of food production and people production rates will widen. I am not writing of the end of productivity gains—these will continue for an unknown time—but of a slowing in the rate. At the same time, agricultural production will inevitably decline so long as the urbanization and life-support pressure of people on the environment remains unchecked. We must remember, however, that a favorable or desired trend in population stabilization must be sustained for something like 70 years for the entire population to reach equilibrium throughout its age structure.

Foreign affairs of the future will be deeply affected by the outcome of the food-people problem. What I have presented here can be but a small input into the global mix and I hesitate to draw any conclusions because of the kaleidoscopic nature of the world food situation. For example, there are countries that have yet to reach the point at which agricultural yields "takeoff," and others that will never reach that point. Nevertheless, I suggest to those whose business it is to make projections on the world stage that absolute limitations to food production loom in the future. We have been surprised at the rapidity with which the energy crisis, the depletion of fossil fuel supplies, came upon us. It would be tragic indeed for this to be repeated with food. The bicentennial of Malthus's paper will be in 1998. Let us hope that by that date the problem, if not the solution, will be much clearer.

S. HUSAIN ZAHEER

India's Need for Advanced Science And Technology

Advances in medical technology that reduced infant mortality and permitted people to live longer and have more children survive into adulthood were responsible for the enormous population increase in the developing countries during the twentieth century. Now, it appears that scientifically-based technology is the only way that peoples in the developing countries can be saved from starvation. In this selection, one of India's top science administrators, Dr. S. Husain Zaheer, outlines the problems facing his country. The two major ones, he declares, are population control and an increase in agricultural productivity. To begin to solve these problems, he argues, India must develop advanced technology, and to accomplish this task, the industrialized nations must supply know-how and investment capital.

India had scientific traditions in the ancient and medieval periods. However, in spite of the existence of considerable scientific and technical potential, there was, due to historical reasons, a sharp break with earlier traditions in the growth of modern science. The limited growth of science in India before

S. Husain Zaheer, "Meeting National Needs through Science and Technology," *Government Science and International Policy*. A Compilation of Papers Prepared for the Eighth Meeting of the Panel on Science and Technology. (Committee on Science and Astronautics, U.S. House of Representatives, Washington, D.C., 1967.)

independence, therefore, has been without any deep roots in its history and in its society and its organization was patterned more or less on the "heroic concepts" of the nineteenth century based primarily on those of the United Kingdom and superimposed on the traditional and individualistic "guru and chela" (teacher and pupil) traditions of Indian society.

Since independence, there has been a major growth and, considering its limited resources, a comparatively substantial investment in science. A major problem in India, like in other developing countries, is the urgent need for a change in the outlook of scientists themselves toward society and of society toward science and scientists; i.e., the problem of integration of science and technology with society in order to remove its isolation as a foreign imposition. Science can no longer be looked upon merely as a bundle of various intellectual disciplines or isolated fields of specialization and a new awareness of the wider role of science and technology has got to be developed in order that science and technology may play their proper and due role in the economic, social, and political development of India in the same manner as science has done in other economically and socially more advanced societies. Consequently, there is an urgent demand for evolving a new era of science as part and parcel of the new society in India fully integrated with and not as an exotic imposition developing only under the impact of influences from other countries; i.e., the emergence of a new image of science freed from outdated and obscurantist traditions and recognizing its universal and international character. Progress toward this ideal is necessarily slow—being hampered by a large number of social and political factors. . . .

If it is accepted that it is necessary and possible for the developing countries to catch up with [technological] advancements, the first task would be to plan science and technology within the framework of clearly identified areas of development in the plans of industrial and economic growth. The conception of science and technology as something abstract must be discarded and planning of research and development made an integral part of economic growth. While in no way underestimating the non-economic returns by way of cultural advancement and satisfaction to creative minds, the sights must be kept clear in regard to the basic objective of science and technology in the developing economies. Science and technology must be in such circumstances far more closely and rigidly planned.

No developing country can have any hope of catching up, much less of becoming a leader in any single field, unless it starts from an advanced stage and develops specialist technology of its own thereafter. Transfer of technology from the advanced nations to the developing ones is one of the major planks for rapid and accelerated industrial growth and should figure prominently in the national economic and scientific policies of the developing countries. Advanced technology insures application of concentrated research, development, and experience at points of maximum return in the developing economy. Choice of technology from competing ones, economics of it, adaptation and orientation in keeping with the national resources of raw materials, and genius of the people should form the main task of science policy. It would hardly be wise or economic to make any effort to re-create technology or permit limited resources to be used to infructuous ends, while the transfer of advanced technology would itself require a sound technological base in the recipient countries to insure speedy adaptation and assimilation and further development. Planning of scientific and technical manpower and science education are a part and parcel of science policy and equally a part of planning for economic growth.

The planning of research and development must be such as to give greater emphasis on research sensitive areas. Use of latest techniques of social research such as operational research, management techniques, systems engineering, and programming through computers must be employed to forecast the requirements of scientific and technical manpower and work out systems

UPI PHOTO

Food for the hungry: at a resettlement camp in Bangladesh, women make jute mats in a United Nations "food for work" program in 1976.

in the developing economy where research can be employed to give maximum returns in terms of economic growth. . . .

Perhaps it would be inappropriate . . . to enumerate the growths which have taken place in the economic and the social service sector. . . . Suffice it, however, to say that there have been laudable increases but very much remains to be done. . . .

The people are much better educated now. . . . There has been in particular a remarkable growth in technical education of personnel. Expenditure on scholarships has increased . . . and new opportunities have been provided to students from poorer families. The people are also much more healthy. Malaria has been eradicated. The average expectation of life has increased from 32 years in the forties to 50 years today.

So far an attempt has been made to indicate (*a*) the realization by India that science and technology form a crucial element in its economic and social development and (*b*) the efforts and investment made so far for development and use science and technology. It would be noticed that even though in absolute terms the actual percentage of investment and the increase in the rate of investment are not satisfactory yet for a country like India taking into account its meager resources and that it started from almost zero, they can be considered as feasibly creditable. There are some obvious imbalances, as for example, in the life sciences like agriculture, investment in earth sciences and science education, and research in schools-universities, have been unsatisfactory. . . .

The economic indicators . . . indicate that steady improvements are taking place in India. It is, however, obvious that the rate of growth, especially when we take into consideration the rate of growth of population, is not satisfactory. Looking at the world picture we find that one of the most important international problems today is that the rich nations are getting richer and the poor nations are getting comparatively poorer—in fact the gap is growing wider. . . .

The problems that face India today are truly staggering—take, for example, the rate of populatiion growth. . . . India is waging an almost desperate struggle to control population growth. However, as is generally realized and accepted, success in these efforts is tied up with many complicated factors, and is closely related to rapid social betterment and economic development; i.e., with rapid increase in education and of economic growth.

Besides control of population, the second major problem facing India is to increase its agricultural productivity, specially of food grains. India again is making intensive efforts to increase . . . its agricultural production. . . . But here again, as is generally agreed, this is closely linked with (1) an enormous growth of industrial production in the fields of fertilizers and pesticides productions; (2) rapid development of engineering industries related to agriculture; (3) extensive programs for water conservation and distribution; and (4) scientific research and education in soil condition, production of better seeds, and improved methods of cultivation. Here again it is felt that success is within India's grasp. But both these problems, i.e., population control and increased agricultural production, need not only an enormous direct investment but also indirectly a very much increased rate of investment in science and technology, scientific research, and scientific education. . . .

India, therefore, is looking for international capital assistance and assistance by way of know-how and knowledge. . . .

Besides the active cooperation, dedication, and a crusading spirit of and by the scientists, engineers, and statesmen all over the world, in a report submitted by the author to the Prime Minister of India . . . it was estimated that such an international effort required a capital investment of about $20 to $30 billion for the next 20 years. This magnitude of capital could obviously be made available only if conditions were created to persuade the nations of the world to reduce their present total armaments bill. . . . It is for the statesmen of the nations of the world to answer the question if that is possible. . . . Of course, the major query still remains: will the nations of the world, even if they were persuaded and convinced to cut down their expenditure on armaments, have the will and foresight and the awareness to heed the warning sounded by the late President Kennedy: "The world cannot survive in peace—half rich, half poor."

To conclude it would be appropriate to quote what Roger Revelle said . . . :

> Science itself is essentially optimistic. That is why it is needed in all countries, and particularly in the developing countries—not because of the physical or mechanical things it can produce but because of its spirit of rationality and optimism; its faith that men can understand not only the world but themselves; its belief that changes in the human conditions can occur and that those changes can be guided by human beings; and its demonstration that there is a real unity of human thought, that men truly are a band of brothers.

WILSON CLARK

Intermediate Technology

The problems of the developing nations, most authorities agree, call for the application of industrial technologies as well as agricultural technologies. But there is widespread disagreement as to the kinds of technologies that are most suitable to conditions in these nations, which often lack capital, skilled manpower, and access to markets for their goods. Science writer Wilson Clark thinks that attempting to transfer modern industrial technologies to Third World nations is a mistake. He gives examples of "intermediate" technologies that have been successfully introduced into developing countries, and concludes that the use of small-scale technologies even in industrially developed nations would be beneficial in many instances.

O n the road to centralizing technology, many efficient, small-scale processes have given way to large-scale, inefficient systems. In agriculture, for example, the extensive use of machinery has resulted in a system which no longer produces a net output of energy, but instead consumes more British Thermal Units (BTUs) of oil than it yields in BTUs of food to the consumer. The magnitude of risk and waste has prompted an international debate over the scale of technology. It is one thing for productive nations like the United States to use energy-intensive agriculture to help fill total world needs, but is it appropriate for the less developed nations?

Assistance to developing nations by developed ones has concentrated on

industrialization and this has created a host of problems. The African writer
and technology specialist Jimoh Omo-Fadaka points out:

"Often aid creates a psychological dependence on getting still more aid. It
saps initiative and enterprise; again, it may foster—as it has in so many
nonindustrialized countries—a type of development wholly inappropriate to
circumstance, industrial plants are created, instead of basic water supplies
being improved. Aspirations are created that can never be fulfilled." On the
other hand, Omo-Fadaka indicates that the kind of development needed "is
one that will produce enough food to feed the whole of the population and
that will absorb the labor. In this type of development, agriculture should be
the key factor, not industry."

The long-term debt of the developing countries has increased from $40
billion to $130 billion since 1970. Fortunately, many of the technologies the
developing world is most in need of are not nearly as expensive as the technologies of the industrialized world.

A recent study of industries and employment in developing countries by the
International Labour Office (ILO) in Geneva concluded that new research
and development is needed to bring about a "shift to indigenous raw materials, small hand-operated machinery and the use of various waste materials
for the generation of energy in individual industrial plants." With funding
provided by Sweden, the ILO conducted research in developing countries.

They found, for example, that in comparison with modern, imported tin-
can-manufacturing processes, less expensive, labor-intensive processes in
Thailand employ more people for a smaller total investment. By investing in a
labor-intensive process, ten-year savings would amount to about $500,000.
Since local machine shops could manufacture some of the equipment, there
would be less revenue spent for foreign parts, and there would be more
growth in native industries.

In Kenya, the purchase of second-hand, rather than new, machines used in
the processing of jute might offer the advantage of increasing employment
and productivity. Machine shops of the East African Railway and small local
shops could be employed for "generating new types of technical knowledge in
the railway workshops, and... might contribute to making secondhand
equipment more viable in Kenya."

Although there has been too little effort by corporations and governments
to investigate the use of such appropriate techniques in the developing world,
a number of small, nonprofit organizations in Europe and America have
made great progress in this field for years. The English economist E. F. Schumacher coined the term "intermediate technology" to describe the kinds of
approaches and industries needed in the developing world. "I have named it
intermediate technology," he says, "to signify that it is vastly superior to the

primitive technology of bygone ages but at the same time much simpler, cheaper, and freer than the supertechnology of the rich." Schumacher, top economist and head of planning for the British Coal Board for 20 years, paid many visits to developing countries in the 1950s and '60s, observing—with despair—the limited usefulness of advanced technology in solving persistent problems of food and shelter. In 1966 he founded the Intermediate Technology Development Group (ITDG), with headquarters in London.

For example, at the request of the Zambian government, the ITDG developed an egg-tray production machine costing $19,500, which is scaled to the needs of the country. The cheapest machine available on the international market would have cost $390,000, and would produce a million trays a month, far more than Zambia needed. The small machine offers more employment opportunities and lower capital investment, and can be used to fabricate other items than egg trays.

In the United States, a number of nonprofit organizations provide similar services. VITA, or Volunteers in Technical Assistance, located near Washington, D.C., provides a consulting service which uses the expertise of more than 6,000 scientists and engineers who offer free advice on questions of technology and construction. VITA consultants have designed for use in developing countries machines and implements such as windmills, bicycle-powered pumps, spinning wheels and solar cookers. . . .

The developing countries need energy for small industries and home uses (such as lighting, cooking and pumping water). A new project of the United Nations' Environment Programme (UNEP) is testing the use of solar energy, wind power, and fuel production from manure and agricultural wastes for providing the energy needs of villages. . . .

A team from Oklahoma State, which has pioneered in wind-power research, will design a center for a village in southern Sri Lanka, and Brace Research Institute one for Senegal.

Each village center will use a combination of solar power, wind power and energy from agricultural wastes to provide lighting, pumping and cooking needs. The village energy centers will have the installed power equivalent of a 20- to 50-kilowatt electric generating plant.

In many developing countries these technologies will come none too soon: supplies of fuels—such as firewood and charcoal—are perilously low.

In Nepal, population pressures have caused farmers to seek more arable land, forcing a move to less fertile mountainous areas. Simultaneously, villagers seeking firewood have caused extensive deforestation, Erik Eckholm, of the Washington, D.C.–based Worldwatch Institute, describes the ecological havoc. "Topsoil washing down into India and Bangladesh is now Nepal's most precious export, but one for which it receives no compensation. As fertile soils slip away, the productive capacity of the hills declines, even while the demand for food grows inexorably."

To curb this growing problem, the Nepalese government is encouraging land and forest conservation, but the existing population is hard-pressed to find fuel for cooking and heating homes. As nearby sources of wood disappear, villagers have to travel hours to find firewood, or they burn cow dung. Since the dung is needed to restore fertility, burning it only exacerbates the dilemma of the people.

One answer to the fuel problem is methane digesters to convert cow dung into methane gas, used for cooking or heating, and produce fertilizer in the process. The basic technology required to treat manure for gas production has been used since 1900 in India. Methane gas is produced when rich animal manures (and other organic matter) are confined in an oxygen-free closed container (or "digester") at temperatures ranging between 59 and 122 degrees Fahrenheit (15–50 degrees Celsius); bacteria convert the wastes into methane gas at atmospheric pressure through a process similar to fermentation.

According to Brot Coburn, a young American Peace corps volunteer who has been bulding home-sized methane plants in Nepal since 1973, a digester designed to use the manure of five cattle will provide the cooking-fuel needs of a family of five. At the current cost of firewood in Pokhara, Nepal, the digester—costing about $400—would pay for itself in about five years. This assumes only the cost of scarce firewood, not the other by-product of the digester, a richer fertilizer than cow dung applied directly to the fields.

More sophisticated methane-digester technology is in practical, day-to-day use in the Peoples' Republic of China and in Taiwan. . . .

It is no coincidence that most of the projects described here were accomplished by individuals and small organizations, working independently and at low cost; this characterizes the goal of appropriate technologies. Although their application is often confined to developing countries, there is no compelling reason to limit the use of these technologies to countries where agrarian conditions and fuel scarcities are common. In fact, the increasing development of small-scale technologies in America, Japan and Europe indicates that this new industrial revolution may offset the historic tradition of industrial centralization. . . .

The idea that efficiency increases with size has rarely been challenged, and remains a key concept in economic theory, yet small businesses and organizations are often as efficient in their use of human and physical resources as large enterprises, if not more so. . . .

But the inertial mass of large capital investment works in favor of large organizations in our economy, and government subsidies and laws tend to encourage bigness to the detriment of small businesses and organizations. Even government funding for development of small-scale energy technologies, notably solar energy, has favored large corporations rather than small research groups and businesses which have already developed systems for heating homes and buildings. . . .

PART FOUR

Sources Of Technological Change

Introduction

There are numerous motivations stimulating technological innovations and change. For example, the desire to relieve pain during dental extractions—a humanitarian concern—led Horace Wells, a Connecticut dentist, to use nitrous oxide as an anesthetic in 1844. This trial was soon followed by the administration of ether and chloroform to patients during surgery to make them insensitive to pain. The spirit of adventure motivated Frenchmen in 1783 to make the first ascents in balloons filled with hot air. The desire to transform idleness into enjoyable leisure stimulated the invention of games the world over, in which balls are hit, thrown, or kicked in accordance with prescribed rules.

Among the more important sources creating technological innovation are the economic demands or needs of society, scientific knowledge that is applied to develop new technologies, engineering practices, military exigencies, and the stimulus of government. Critics have singled out one or more of these sources as responsible for the unchecked and reckless advance of technology, which they believe is now occurring. In this section, we shall explore each in a historical context and try to determine the specific nature of the critics' concerns.

In about the twelfth century, prosperous merchants, either by themselves or in partnership with others, began to finance distant trading ventures, which they hoped would return their money at a profit. Gradually, this system, which became known as capitalism, spread and was refined by the introduction of banking, credit, and insurance practices, and by the development of such institutions as joint stock companies, corporations, and stock exchanges. The capitalistic system has been a powerful stimulus to the development of technology. World-wide trade created a demand for raw materials and finished goods, and called for faster and more reliable methods of transportation and communication. As business and industrial activity increased, economists began to theorize that the accumulation of money or capital and its investment in completely new ventures—mines, factories, and buildings, which would return a profit—was the primary stimulus to economic growth. Growth, in turn, enhanced the standard of living of a society. Although some modern economists question whether continued growth is possible, most are convinced that growth is the keystone of an affluent society. Now, however, new technology has also become a major factor contributing to economic growth.

The selections by Meier, Wagar, Rosenberg, and Henderson focus on the question of the desirability of continued economic growth. Hugo Meier describes how nineteenth-century Americans welcomed the idea of growth and viewed technology as the principal element in their optimistic belief in progress. According to Alan Wagar, faith in continual growth was entirely suitable for last century's frontier mentality, but it is no longer tenable today. We must use our resources wisely, preserve the environment, stabilize the population, and reduce waste, he warns, in order to survive.

This idea of a static economy is rejected by Nathan Rosenberg. The disastrous consequences of growth, he writes, have been predicted since the time of Malthus. What the doomsayers fail to take into account, he asserts, is human ingenuity and ever-advancing technology, which make their prophecies irrelevant. Nonetheless, as Carter Henderson points out, a sizable number of Americans have opted for a simpler life style. Whether this phenomenon represents a basic shift in American attitudes cannot be determined right now, but it is surely a trend that bears watching.

The next group of selections focus on science, its application to develop new technologies, and criticisms of its present conduct. Derek Price documents the astonishing growth of science during the last three centuries and concludes that the current rate of growth cannot continue. Peter Drucker describes how the establishment of industrial research laboratories has accelerated the process of innovation, which in turn produces change in all areas of society. The process by which one major modern innovation, the transistor, emerged is described by Charles Weiner. The exchange of information and ideas between scientists working in government, university, and industrial laboratories, he shows, is a highly complex process, with the result that the "credit" for any innovation of any magnitude cannot be given to any single individual. Jerome Ravetz is highly critical of the present conduct of Big Science. It is no longer the idealistic pursuit of the truths of nature, he charges, but rather an enterprise designed to develop lethal weapons or profitable products, which may be socially harmful but for which the scientists disclaim responsibility. It is clear from these selections that the nature of the scientific enterprise has changed. Whether it is now a malevolent force, as Ravetz contends, is another matter, and the issue will be further discussed in Part Five.

Innovation may also come about through engineering; for example, by designing a product that can be mass produced efficiently and economically. Although, as Sir William Fairbairn points out, engineering has been practiced since ancient times, it only became a recognized profession in the nineteenth century. One of the most heralded engineering accomplishments of the early twentieth century was the mass production of labor-saving household appliances. According to Siegfried Giedion, the key development in this transformation of household work was the small electric motor, which powered washing machines, vacuum cleaners, and other appliances. But this revolution in household work, Ruth Cowan claims, did not just relieve drudgery. It completely changed the role of the housewife, and actually increased the number of hours that she worked.

By 1920, engineering had become a respected profession in the United States. Indeed, Thorstein Veblen wrote that engineers were much more qualified to control industrial production than were the capitalists. However, in the 1950s, when air and water pollution became common and poorly designed products threatened the lives and safety of people, the public attitude toward the engineering profession became critical. Ralph Nader describes the negligence with which General Motors' Corvair was designed, an experience that has been repeated frequently with other automobiles since he wrote.

Engineers, however, disclaim responsibility in such matters, declaring that their hands are tied by profit-minded managers. The best way in which the public interest can be served, Donald Marlowe asserts, is to give engineers more independence in decision-making and a greater voice in framing policies. Further, they should be permitted to apply their powerful methods in solving problems involving human values and aspirations. Marlowe's statement is essentially a modernized version of Veblen's position; that is, we should trust engineers with our future. But on this point many authorities would disagree, and the reader should refer once again to the selection by Roszak in Part Two.

The next group of selections addresses the subject of war, which has always stimulated technological innovation. As Quincy Wright shows, however, the effect of new weapons is not limited to questions of military strategy, but gradually involves increasingly large elements of society. Compulsory military service and conscription became necessary, and now the entire civilian population is affected. Strategic bombing resulting in civilian casualties and deaths was a feature of World War II, although this development was predicted two decades before the war by Giulio Douhet. Similarly, Einstein warned President Roosevelt, just before the outbreak of World War II, of the possibility of the development of the atom bomb. After its construction and use against Japan by the United States, the scientists and engineers who were involved in the project were reluctant to recommend the development of the "super" or hydrogen bomb. An exchange of missiles between Russia and the United States could produce such a level of radioactivity around the world that the entire human species would be endangered.

Their position, documented in the report of the Oppenheimer committee, was ignored, and both the United States and Russia began to make and test hydrogen bombs. As the Sakharov memoir testifies, Soviet scientists were also worried about the potential danger to human life of radioactive fallout from atmospheric nuclear tests, but Sakharov's protests were unavailing.

According to Herbert York and G. Allen Greb, the same complex process that led to the emergence of the transistor operates in the development of a new weapons system. There is no master plan; instead, ideas and innovations merge until the final product evolves. This kind of fortuitous process does result in the gaining of a temporary advantage over possible enemies. But it is a serious question whether this process of continually developing more dangerous and more costly weapons has become a way of life for the military planners and the scientists and engineers involved.

Throughout history, governments have encouraged technological innovations for peaceful as well as for military purposes. The final four selections in this section treat this subject. Reynold Wik describes the federal government's sponsorship of innovations in agricultural technology and its contributions in easing the burdens of farm life. Governments have also attempted to strengthen domestic industries by legislating protective tariffs and by establishing patent systems. Carroll Pursell gives an account of federal encouragement and promotion of scientific research and technological development during the twentieth century, and indicates how decisions relative to science and technology have now become major components of the political process.

Simultaneously, though, the federal government has had to exert its power to protect people from the unwanted by-products of technological change. John Burke's selection describes the establishment of the federal government's first regulatory agency. Newly created bureaus of this type are the Environmental Protection Agency, Consumer Product Safety Commission, and Occupational Health and Safety Administration. In our advanced society, Sanford Lakoff writes, government involvement with science and technology has become a necessity. Not only must the government promote innovation, he says, it must see to it that innovation serves social purposes.

In this broad historical survey of the five chief sources of technological innovation, it becomes clear that change is a complex process, and that the problems resulting from change cannot be attributed to any single factor. We cannot place blame only on the "growth ethic," or the "military-industrial complex," or the "scientific-technological elite," or the "system." Can we somehow make sense out of the complexity? Can we develop methods to channel technology in socially desirable directions? Can we agree on what is socially desirable? Can we, in other words, make sure that technology is used in the service of humanity?

HUGO A. MEIER

Technology And Democracy, 1800–1860

Americans in the early Republic, as historian Hugo Meier contends in this selection, had a steadfast faith in the power of technology to increase democratic economic and social opportunities. Progress through science and technology would yield abundance and permit all free citizens to have a generous share of the good things of life.

D emocratic economic opportunity . . . appears—like social opportunity— to have been advanced rather than retarded by the increasing influence of technology in the United States after 1800. The doubters appeared to be in a minority and remained so until mid-century.

Permeating the concept of democracy in the United States and closely linking its notions of social and economic opportunity with technology, was the current spirit of optimism which historians have labeled the "idea of progress." So closely were the technologist and his creations associated with progress in America during this period than believers in the relationship seemed often to share a faith more naïve than realistic. Not in terms of religion, nor of philosophy, nor in those of military glory or artistic achievement was the story of human progress to be told. Rather, progress must be identified with the

Hugo A. Meier, "Technology and Democracy, 1800–1860," *Mississippi Valley Historical Review*, XL (March 1957). Reprinted by permission of the *Journal of American History*.

sweeping advances which this age was making in applying science to the satisfaction of human wants. Enthusiastic support for this belief made technology a central element in the optimistic doctrine of progress in America, and by the time of Jackson was shaping for American engineers and inventors a solemn responsibility.

Indeed, by 1830 very many Americans had come to believe that their age was superior to any other, and the prospects of further progress greater, because of remarkable material advancements. Comfort, convenience, and good health were practical evidences of that progress, and constituted, too, a goal for future efforts. Machinery itself became a throbbing symbol of this notion of progress. . . . Edward Everett, then governor of Massachusetts and soon to become president of Harvard, declared in 1837 that "Mind, acting through the useful arts, is the vital principle of modern civilized society. The mechanician, not the magician, is now the master of life."

With so much confidence in the contemporary role of technology, it is not surprising that many Americans saw even greater future challenges for science and invention. Utopia, even, had a technological coloring. Not only did Owen's New Harmony *Gazette* lend ample space to descriptions of new improvements, but the Associationists invited to their ranks men with special scientific and mechanical experience, trusting in the principle of co-operation to control technology in the common interest. It was a modest but calculated surrender to the machine.

Nature's untamed might—steam, water power, electricity—fascinated these nineteenth-century adherents of social progress through technology. What wonders might man not perform by further mastery of wind and tide and the sun's own heat! But it was the glory of steam power, especially, which thrilled Americans, and ever greater seemed its promises as each puffing locomotive or steamboat passed by. Not even the compass or the printing press or gunpowder could equal in social significance the impact of steam power, insisted the Reverend James T. Austin in 1839. "It is to bring mankind into a common brotherhood; annihilate space and time in the intercourse of human life; increase the social relations; draw closer the ties of philanthropy and benevolence; multiply common benefits, and the reciprocal interchange of them, and by a power of yet unknown kindness, give to reason and religion an empire which they have but nominally possessed in the conduct of mankind.". . .

In the Crystal Palace exhibition in New York in 1853, Americans could witness personally the proud summing up of a half century of American technological and artistic achievement. Not alone the structure itself but the variety and excellence of the exhibits evoked enthusiasm—the fruits of American engineering and mechanical genius. Graceful form, beautiful outline, and poetic thought, remarked one observer, might now be traced not only in a Madonna or Venus but in "swan-like life-boats, those light and airy carriages, and highly finished engines—the epics of mechanics—and the host of tiny-handed operators for sewing, for card making, for spinning, for ornamental weaving, and a multitude of other works, including that wonderful distributor of thought, the printing press." To the Americans, therefore, the New York Crystal Palace exhibition, like the great London international exhibition two years earlier, offered an opportunity to display before the mid-nineteenth-century world a new conception of aesthetic expression. There remained much to be desired in the finish and polish of the samples of machinery and articles of manufacture produced by their own countrymen; but there was no question of the sincerity of purpose and the dedication of those products to the service of Everyman. Their very simplicity and humbleness suggested that.

In a much broader sense, the great exhibitions also summarized the general relationship of technology to the concept of democracy in America, and it seemed appropriate that Walt Whitman should have assisted at the dedication ceremonies in New York in 1853. There was no denying the tremendous technological progress that had been made since 1800—the constant improvement in engineering and the never ending multiplicity of new inventions. As it had begun, so the period ended, with American science and invention still in the service of the people. Popular needs still tended to determine the goals of engineers, inventors, mechanics, and their spokesmen. Perhaps inevitably, the materialistic approach of technology fitted admirably the tone of life of a bustling people who wanted to get things done quickly, who worshipped abundance, and who believed that every free citizen was entitled to a generous share of the things which brought physical comfort in their world. American technology in the years before the Civil War served those objectives well—a fact of which most Americans seemed fully aware.

J. ALAN WAGAR

Growth versus
The Quality of Life

J. Alan Wagar, a researcher with the U.S. Forest Service, argues that too often human progress has been equated with economic growth, change, and exploitation. Given the natural resources and the fertility of North America and the skills and political philosophy of the pioneers, the growth of the United States, Wagar feels, was inevitable. But he also thinks that we are going to have to adjust from a rapidly growing economy to one that is practically static. It is certainly not clear, he asserts, that people wish growth at the expense of the environment, and he warns that we should be prepared for the eventuality that we can no longer depend upon the resources of the underdeveloped nations. As Dubos noted in his article in the previous section, this concern for the consequences of unlimited economic growth is a phenomenon of the last two decades. Many of the arguments expressed by Wagar were given wide publicity here and abroad with the 1972 publication of The Limits to Growth *by Donella Meadows, Dennis Meadows, et al. Using a computer-simulation model, this controversial study predicted that civilization would soon outstrip the world's capacity to support it, and the book stimulated an international debate on the merits—and feasibility—of continued economic growth that has continued throughout the 1970s.*

I n economics, as in most other matters, past experience provides a major basis for current decisions, even though changing circumstances may have diminished the appropriateness of such experience. Such use of "conventional

J. A. Wagar, "Growth versus the Quality of Life," *Science*, Vol. 168, pp. 1179–1184, June 5, 1970. Reprinted by permission.

wisdom" may explain our continuing emphasis on economic and other types of growth despite the many problems created by such growth.

When the United States was sparsely populated, emphasis on growth made good sense. Growth of many kinds permitted exploitation of the rich environment at an accelerating rate and provided a phenomenal increase in wealth.

Growth still increases material wealth but has a growing number of unfortunate side effects, as each of us tries to increase his own benefits within an increasingly crowded environment. These spillover effects, which were of minor importance when settlement was sparse and neighbors farther apart, are now of major consequence. For example, a firm may make the most money from a downtown tract of land by erecting a tall office building there. Construction of the building will add to the gross national product, and the builders will be hailed for their contribution to "progress." However, the building will add to traffic congestion, exhaust fumes, competition for parking, the need for new freeways, and social disorder. These problems, which must be handled by someone else, become part of the "environmental mess" or "urban crisis.". . .

Growth is not an unmixed blessing, and the purpose of this article is to argue that growth is not longer the factor we should be trying to increase.

Unfortunately, growth is as deeply entrenched in our economic thinking as rain dancing has been for some other societies. In each case there is faith that results will come indirectly if a capricious and little-understood power is propitiated. Thus, instead of concentrating directly on the goods and values we want, we emphasize growth, exploit the environment faster, and assume that good things will follow by some indirect mechanism.

From time to time, the correlation between rainfall and rain dancing must have been good enough to perpetuate the tradition. Similarly, the correlations between exploitation of the environment, growth, and progress were usually excellent in our recent past. So great have been the successes of our economic habits that they have become almost sacrosanct and are not to be challenged.

However, here in the United States as in most of the world, the relationships between people and environment have changed drastically, and past experience is no longer a reliable guide. While we rush headlong through the present with frontier-day attitudes, our runaway growth generates noxious physical and sociological by-products that threaten the very quality of our lives. Although we still seem confident that technology will solve all problems as they arise, the problems are already far ahead of us, and many are growing faster than their solutions.

We cannot return to some golden and fictionally perfect era of the past, and we certainly should extend the knowledge on which not only our comfort but our very existence depends. However, to cope with the future, we may need a fundamental reanalysis of the economic strategy that directs our application of knowledge. Instead of producing more and more to be cast sooner and sooner on our growing piles of junk, we need to concentrate on improving our total quality of life.

If environmental resources were infinite, as our behavior seems to assume, then the rate at which we created wealth would depend mainly on our rate of exploitation, which is certainly accelerated by growth. However, the idea of an unlimited environment is increasingly untenable, in spite of our growing technological capacity to develop new resources.

[Economist Kenneth] Boulding has beautifully contrasted the open or "cowboy" economy, where resources are considered infinite, with the closed "spaceman" economy of the future. He has pointed out that, as the earth becomes recognized as a closed space capsule with finite quantities of resources, the problem becomes one of maintaining adequate capital stocks with the least possible production and consumption (or "throughput"). However, this idea of keeping the economic plumbing full, with the least possible pressure and flow, is still almost unthinkable. Experience to the contrary is still too fresh.

Cult of Growth

The economic boom of World War II, in contrast with the stagnation of the Great Depression, seemed to verify the Keynesian theory that abundance will follow if we keep the economy moving. As a result, continuing growth has been embraced as a cornerstone of our economy and the answer to many of our economic problems. At least for the short run, growth seems to be the answer to distribution of wealth, debt, the population explosion, unemployment, and international competition. Let us start with the distribution of wealth.

Probably no other factor has contributed as much to human strife as has discontentment or competition concerning wealth. Among individuals and nations, differences in wealth separate the "haves" from the "have-nots." The "have-nots" plot to redress the imbalance, and the "haves" fight to protect their interests and usually have the power to win. However, the precariousness of their position, if recognized, demands a more just balance. But, rather than decrease their own wealth, they find it much more comfortable to enrich the poor, both within a nation and among the nations. Only growth offers the possibility of bringing the poor up without bringing the rich down.

In our market society, the distribution of wealth has come to depend on jobholding, consumption, and, to an increasing extent, on creating dissatisfaction with last year's models. Unless this year's line of larger models can be sold, receipts will not be sufficient to pay the jobholders and assure further consumption. Inadequate demand would mean recession. We have therefore been urged: Throw something away. Stir up the economy. Buy now. And if there are two of us buying where there had been only one, wonderful! Rapid consumption and a growing economy help to distribute income and goods and have been accepted as part of "progress."

Problems of debt also seem to be answered by growth. To keep up with production, consumption may need to be on credit, or personal debt. But debt is uncomfortable. However, if we are assured that our income will grow, then we can pay off today's debt from tomorrow's expanded income. Growth (perhaps with just a little inflation) is accepted as an answer.

The same reasoning applies to corporate debt, the national debt, and the expansion of government services. As long as debt is not increasing in proportion to income, why worry? Debt is something we expect to outgrow, especially if we can keep the interest paid.

The population explosion is growth that is finally causing widespread concern. Yet many businessmen can think of nothing worse than the day our population stops growing. New citizens are the customers on which our economic growth depends. Conversely, economic growth can meet the needs of added people—if we are careful not to look beyond our borders.

Growth might also handle unemployment problems, and Myrdal has indicated that only an expanding economy and massive retraining can incorporate our increasingly structural "underclass" into the mainstream of American life.

Finally, there is the problem of international competition. In an era when our sphere of influence and overseas sources of economic health are threatened, strength is imperative. Yet our main adversary has grown from a backward nation to a substantial industrial and military power. To counter the threat, we expect to outgrow the competition.

The evidence suggests that growth is good and that we have always grown. Isn't it reasonable to believe that we always will? This question takes us from the short run to the middle and long run.

Dynamics of Growth

Viewed in the most general terms, growth will continue as long as there is something capable of growing and the conditions are suitable for its growth. The typical growth pattern starts slowly because growth cannot be rapid without an adequate base, be it capital, number of cells or organisms, or surfaces for crystallization. However, if other conditions are suitable, growth can proceed at a compound rate, accelerating as the base increases. But growth is eventually slowed or stopped by "limiting factors." These factors can include exhaustion of the materials needed for growth. They can also include lack of further space; the predation, disease, or parasitism encouraged by crowding; social or psychological disorganization; and concentrations of wastes or other products of growth. For example, the concentration of alcohol eventually limits the growth of yeast in wine.

Perhaps it is worth examining the U.S. economy within this frame of reference. Although its vigor has been attributed solely to free enterprise, or to democracy, or to divine grace, it fits the general growth model of a few well-adapted entities with growth potential (settlers) landing on an extremely rich and little exploited growth medium (North America).

Our settlers had, or soon acquired, the technological skills of Europe. They also had the good fortune to inherit and elaborate a political philosophy of equality, diffused power, and the right to benefit from one's own efforts. So armed, they faced a rich and nearly untouched continent. The growth we are still witnessing today is probably nothing more than the inevitable.

But the end of growth is also inevitable. In a finite environment no pattern of growth can continue forever. Sooner or later both our population growth and our economic growth must stop. The crucial questions are When? and How will it come about?

Malthus once saw food shortages as the factor that would limit population growth. At least half of the world lives with Malthusian realities, but the technological nations have so far escaped his predictions. To what extent can technology continue to remove the limiting factors? Will we use foresight and intelligence? Or will we wait until congestion, disease, social and psychological disorganization, and perhaps even hunger finally limit our growth?

Perhaps there is little time to spare. Many factors already in operation could stop or greatly curtail the economic growth of the United States within the next 10 to 30 years. Furthermore, the multiplier effect of many economic factors could transform an apparently low-risk decline into an accelerating downward spiral. If devastating results are to be avoided, the adjustment from a rapidly growing to a much slowed economy will take time, and we should examine the problems and possibilities far enough in advance to be prepared.

The Case for Pessimism

Some of the very problems we hope to outgrow result in part from growth. Certainly the rapid changes brought by a growing economy contribute strongly to unemployment, migration to the cities, and the uneven distribution of wealth. A great deal of our debt can also be attributed to growth, as people try to keep up with what is new. Even the population explosion may result in part from confidence that the future offers increasing abundance. By trying to inundate the problems with more growth, we may actually be intensifying the causes.

If there were no other powers in the world, technology might be sufficient to sustain our growth, replace our shortages, and keep us ahead of the problems. Boulding has suggested that we may have a chance, and probably only one, to convert our environmental capital into enough knowledge so that we can henceforth live without a rich natural environment.

But we are not alone.... We can expect competition in many places in a struggle for spheres of influence and the roots of power. The nation or bloc that can extend its influence can gain raw materials and markets and can deny them to its competitors.

It is doubtful that we can retain the hegemony enjoyed in the late 1940's, and technology cannot fully fill the breach. Our competitors have access to the same technology that we do, and, if they gain control of rich resources and markets while ours are declining, they can increase their power relative to ours.

Closely related to competition for spheres of influence are the rising nationalism and aspirations of the underdeveloped countries. Extractive economies have seldom made them wealthy, and they aspire increasingly toward industrialization. As elements in the global struggle for power, they can demand technological assistance by threatening to go elsewhere for it if refused. From their point of view, it would be rational to put their resources on the world market, to try to get enough for them to support aspirations toward technology, and to let us bid without privileged status.

The problem is compounded by rapid communication and increasing awareness by the aspiring nations that wealth and consumption are disproportionate. The United States, for example, has about 6 percent of the world's population and consumes about 40 percent of the world's annual production. Until such differences in wealth are substantially reduced, they will create constant tension and antagonism. While enduring the many frustrations and setbacks of incipient economic growth, the aspiring nations may be happy to do whatever they can to reduce our wealth. The possible effect is suggested by England's economic woes since she lost her empire and her control over vast resources and markets.

If the aspiring nations and the Communists are not enough to slow us down, perhaps our friends will add the finishing touch. Western Europe is becoming increasingly powerful as an economic bloc and will compete for many of the resources and markets we would like to have. From another quarter, we can expect increasing competition from the Japanese.

In addition to these external forces, there are processes within our own nation that could slow our rate of growth. One of them is the increasing recognition that the products of runaway growth can damage the quality of living, especially for adults who remember a different past. When our rivers are choked with sewage, our cities are choked with automobiles and smog, and our countryside is choked with suburbs, some people begin to wonder if "the good life" will be achieved through more growth and goods. When goods are so abundant and the environment so threatened, will people continue to want even more goods at the expense of environmental quality?

Even the growth promised by automation may be self-limiting. The machines used by "management" to replace "labor" are not going to engage in collective bargaining. However, labor outnumbers management at the ballot box and may well counter such threats by demanding government control of automation and the protection of jobs, even at the cost of slowing our economic growth.

We already have a rising number of permanently unemployed and unemployable people who probably threaten our domestic tranquillity far more than "have-not" nations threaten international stability. Our traditions of self-reliance seem increasingly inadequate now that jobholding depends largely on technological skills that are so much easier to acquire in some settings than in others.

In addition to such technological unemployment, Heilbroner has listed three other factors that may slow our growth. The first is the extent to which we now depend on defense expenditures to maintain growth and the likelihood that these outlays will eventually stabilize. His second point is that capitalism is inherently unstable, even though the factors that caused the Great Depression are now better understood and largely under control. His third point concerns the size of government expenditures that might be needed for anti-recession policy in the future. If investments in plant, equipment, and construction are all low in 1980, he has estimated that government expenditures of $50 to $75 billion per year may be required to maintain growth and that Congress may well balk at such appropriations.

Another factor that could slow growth was suggested by [Harrison] Brown. Growth can be slowed by the increasing amount of energy and organization required for subsequent units of output from resources of decreasing richness. So far, as we have used up the richest mineral resources, improved technology, imports, newly located deposits, and the redefinition of resources have kept us ahead of the problem. But, if the difficulty of extracting essential materials from the environment should ever happen to increase more rapidly than our technological efficiency, our economy could become static and then decline.

Perhaps of greater importance, Brown predicted that the level of organization needed for a very populous society would become so interdependent that failure at one point could trigger failures elsewhere until a chain reaction led to total collapse. In relation to his prediction, the chain reaction aspects of power failures in the Northeast, the Southwest, and elsewhere are sobering. Also sobering is the growing power of strikes to disrupt our economy.

As stated earlier, growth must inevitably stop, and the major uncertainties are When? and How? Despite these uncertainties, the factors examined above could limit our growth within the next few decades, and they merit careful thought. Because growth has become such an integral part of our economy, any sudden setback is greatly feared and could be disastrous. Nevertheless, transition from accelerating growth to some other economic pattern must eventually be made, and it is desirable that we make a smooth transition to something other than total collapse. . . .

Yet, true to the assumption that man is subservient to the economic system, we hear waste defended as necessary for our prosperity. Surely we can organize our economy efficiently enough to avoid having to throw things away to have more! Are we inescapably on such a treadmill? . . .

In its time the treadmill pattern of growth was progress enough and served us well. But as the relationships change between human numbers and the total environment, we must abandon unregulated growth before it strangles us.

The essential tasks ahead are to stabilize human population levels and to learn to recycle as much of our material abundance as possible. Ideally, the change to new ways would be by incremental, evolutionary, and perhaps experimental steps, although some writers believe an incremental approach may not work. But if steps of some kind are not started soon, they may well be outrun by the pace of events. Unless we can slow the treadmill on which we have been running faster and faster, we may stumble—and find ourselves flung irretrievably into disaster.

NATHAN ROSENBERG

Economic Growth, Technology, And Society

Nathan Rosenberg, an economic historian, largely discounts the prophets of gloom, who like Malthus predict disaster if growth continues. Natural resources, Rosenberg shows, are playing a role of declining importance in economic growth, because advancing technology has permitted substitutions. Further, Rosenberg argues, the models of doom fail to take into account the human capacity to adapt to novel situations and to modify technology accordingly.

Concern over the adequacy of natural resources has always been a central preoccupation of economists. . . .

Some of the basic insights of classical economics emerged out of this preoccupation with natural resource constraints. Malthus and Ricardo* were both pessimistic about the long-term prospects for economic growth in an economy experiencing substantial population growth and with only a fixed supply of land available for food production. . . .

The impact of the Malthusian conception is difficult to exaggerate. It not only forcefully focused upon the implications of limited resources for the

---

*Editors' note: David Ricardo (1772–1823), British economist.

Nathan Rosenberg, "Technological Innovation and Natural Resources: The Niggardliness of Nature Reconsidered," in *Perspectives on Technology,* Cambridge University Press, 1976. Reprinted by permission.

prospects for economic growth, but it specifically linked the problem, as it must be linked, to the rate of population growth. Malthus, it turned out, was not much of a demographer, but his formulation nevertheless has had an overwhelming influence upon the framework within which these matters have been discussed right up to the present day. . . .

Although Malthus and Ricardo had been primarily concerned with the adequacy of the supply of arable land to provide food for a growing population, the voracious appetite for natural resources of an industrializing economy shifted the locus of concern to other resources in the second half of the nineteenth century. As early as 1865 the distinguished English economist W. S. Jevons published his book, *The Coal Question,* in which he warned that the inevitably rising costs of coal extraction posed an ominous and urgent threat to Britain's industrial establishment. Having demonstrated the British economy's extreme dependence upon coal—especially in the form of its reliance upon steam power—Jevons went on to argue explicitly that there were no prospective reasonable substitutes for coal as a fuel. After examining recent trends in coal consumption he demonstrated the inevitability of rising coal costs associated with the necessity of conducting mining operations at progressively greater depths. As a result,

> I draw the conclusion that I think any one would draw, that *we cannot long maintain our present rate of increase of consumption; that we can never advance to the higher amounts of consumption supposed. But this only means that the check to our progress must become perceptible considerably within a century from the present time;* that the cost of fuel must rise, perhaps within a lfetime, to a rate threatening our commercial and manufacturing supremacy; and the conclusion is inevitable, that our present happy progressive condition is a thing of limited duration.

Even in the United States, a country of continental proportions that possesses a resources/man ratio far more favorable than that of Great Britain, expressions of concern over the adequacy of natural resources became widespread before the end of the nineteenth century. The Conservation Movement emerged into the consciousness of American political life at just about the time that Frederick Jackson Turner was announcing *The End of the Frontier* (1893). The movement was a significant force in American politial life from 1890 to 1920. Although the spokesmen of the Movement did not attack their concerns in an analytical or rigorous way, they did forcefully present a conceptualization of the problem which was to exercise a profound influence upon later thinking. In the words of [Gifford] Pinchot, Chief Forester of the United States, and an articulate leader of the Movement . . .

> The five indispensably essential materials in our civilization are wood, water, coal, iron, and agricultural products. . . . We have timber for less than thirty

years at the present rate of cutting. The figures indicate that our demands upon the forest have increased twice as fast as our population. We have anthracite coal for but fifty years, and bituminous coal for less than two hundred. Our supplies of iron ore, mineral oil, and natural gas are being rapidly depleted, and many of the great fields are already exhausted. Mineral resources such as these when once gone are gone forever.

The Conservationist view, therefore, was that nature contained fixed stocks of useful inputs for man's productive activities, which could be readily identified and measured in terms of physical units. These resources need to be husbanded carefully. Preference should be given to exploitation of renewable resources—agricultural crops, forests, fisheries—which should be operated on a sustained yield basis. Waste should be avoided in the exploitation of nonrenewable resources so as to pass on the largest possible inventory to future generations.

When, in the years immediately after the Second World War, economists turned their attention once again to the problems and prospects for long-term economic development, the Malthusian pressures once again figured prominently. Undeveloped countries all appeared to confront some serious demographic obstacle—either that of already high population densities, or rates of population growth which had sharply accelerated in the twentieth century, or both. Not surprisingly, Malthusian-type models made their appearance. In these models, that is to say, there existed a "quasi-equilibrium" such that temporary improvements which raised per capita income triggered off demographic changes which eventually restored the low level of per capita income at a higher absolute population size. Concern over the implications of population growth subsided in the United States and most high income countries by the late 1950's when the postwar "baby boom" appeared to have exhausted itself. However, it reappeared forcefully in the 1960's as a consequence of the growing concern over pollution and environmental deterioration; and its possible consequences have been more recently dramatized in the apocalyptic Meadows Report [*The Limits to Growth*], which attempts to focus attention on the long-term implications of continued growth of population and industrial production. In this model, exponential growth of population and industrial output confront a world of finite resources—mineral deposits, arable land, etc.—and limited capacity of the natural environment to absorb the growing pollution "fallout." The finite limits appear to guarantee, not merely a ceiling to future growth, but a precipitous and disastrous decline in approximately a century.

In assessing the Malthusian view and its implications over the past century or so, a convenient starting point is the recognition that Malthus and Malthus-like models have led to predictions which have been demonstrably and emphatically wrong. Indeed, it is difficult to understand the persistence of widespread attachment to a hypothesis which has been so decisively refuted by the facts of history. . . .

Classical models have failed to make relevant predictions primarily because they adopted an excessively static notion of the meaning of natural resources and because they drastically underestimated the extent to which technological change could offset, bypass or provide substitutes for increasingly scarce natural resources. A large part of what economists have had to say in recent years about the innovative process in relation to natural resources needs to be seen as a prolonged effort to break out of the restrictive conceptualizations inherited from Malthus and Ricardo.

The need to break out of this framework has been made increasingly apparent by numerous studies, each of which in its different way, has shown that natural resources have played a role of declining importance, at least within the favored circle of industrializing countries. Whereas classical economic models based upon fixed resources, population growth and diminishing returns, lead us to expect rising relative prices for extractive products and an increasing share of GNP* consisting of the output of resources, such features obviously have not dominated the growth experiences of industrial economies. The agricultural sectors, to begin with, have declined in their relative importance. A declining agricultural sector, as [Simon] Kuznets has shown, has characterized all economies which have experienced long-term economic growth. Indeed, perhaps the most distinctive characteristic of growing economies has been the complex of structural changes associated with the declining importance of the agricultural sector. Moreover, within agriculture itself, the implicit assumption that there were no good substitutes for land in food production has been belied by a broad range of innovations which have sharply raised the productivity of agricultural resources and have at the same time made possible the widespread substitution of industrial inputs for the more traditional agricultural labor and land—machinery, commercial fertilizer, insecticides, irrigation water, etc. . . .

Barnett and Morse (*Scarcity and Growth*) have attempted a quantitative test of the implications of the classical model that increasing natural resource scarcity should make itself apparent in the secularly rising unit cost of extractive products generally—agriculture, minerals, forestry and fishing. To do this they examined data for the American economy over the period from about 1870 to the late 1950's. Their findings are that unit costs of extractive

---

*Editors' note: Gross national product, or the total market value of all the goods and services produced by a nation.

products, as measured by labor plus capital inputs required to produce a unit of net extractive output, declined through the period 1870–1957. For agriculture, minerals and fishing the trends are persistently downward. Indeed, the time sequence of the trends is particularly damaging to the increasing scarcity hypothesis because in the cases of agriculture and minerals (which together account for about 90% of the value of extractive output) the rate of decline in unit costs was *greater* in the later portion of the period than the earlier. Forestry, in fact, is the only extractive industry in which the long-term trend in unit costs and relative prices has been upward since 1870. Since forestry has accounted for less than 10%, by value, of extractive output in the twentieth century, the influence of this sector has been swamped by the downward trends elsewhere. But, moreover, the rising cost of forest products seems to have induced a large-scale substitution of more abundant non-forest-based products for forest-based ones—for example, metals, masonry and plastics—which may account for the rough constancy since 1920 in the unit cost of forest products. . . .

In a suggestive article, "The Development of the Extractive Industries," Anthony Scott constructs a three-stage model for dealing with the sequence through which man has gone in the exploitation of his natural environment. In the third stage, where the application of science makes possible a sophisticated degree of control and manipulation over the environment, older scarcities are deliberately bypassed in the systematic search for substitutes from abundant and convenient sources—the classic early-twentieth-century example would be the Haber process of extracting nitrogen from the atmosphere. Scott makes the important point that there is no economic demand for specific minerals as such, but rather for certain *properties,* and that an advanced technology makes it possible to obtain these properties from materials available in great abundance.

> Demand for minerals is *derived* from demand for certain final goods and services. Therefore, certain properties must be obtainable from the raw materials from which such services and types of final goods are produced. Man's hunt for minerals must properly be viewed as a hunt for economical sources of these properties (strength, colour, porosity, conductivity, magnetism, texture, size, durability, elasticity, flavour, and so on). For example, there is no demand for "tin," but for something to make copper harder or iron corrosion-free. No one substitute for tin has been found, but each of the functions performed by tin can now be performed in other ways. Tin's hardening of copper (as in bronze) has been supplanted by the use of other metals. Food need no longer be packed in tin cans. Hence the immense capital investment that society might have been forced to undertake to satisfy its former needs for tin from the minute, low-grade quantities to be found in many parts of the world have been replaced by simpler investments in obtaining other materials. Chief of the replacements for tin is glass, made from apparently unlimited

quantities of sand and with little more energy than is needed to bring metallic tin to the user. Lead and mercury are being bypassed in similar fashion; zinc and copper may be next.

Thus, although there may be no close substitutes for tin in nature, there may be excellent substitutes for each of the properties for which tin is valued. The implications of this position are, of course, far-reaching. . . .

The current round of intense concern over natural resources, which began in the late 1960's, contains an important new element. Not only does it assert the inadequacy of the resource base to sustain continued growth of population (because of the limited supply of arable land) and industrial production (because of the limited supply of mineral resources). Insofar as models such as *The Limits to Growth* and its companion, Jay Forrester's *World Dynamics* (Cambridge, 1971) do this, they are essentially, for all their elaborate systems dynamics and computer methodology, Malthus in modern dress. That is, they deduce the conclusion, from certain restrictive behavioral assumption, that continued exponential growth is impossible in a world of finite resources. The new element in the present debate is the prominence, not to be found in Malthus, of concern over the pollution problem and the assertion of additional limits to growth imposed by the increasing incidence of environmental pollution.

Although an extended appraisal of the unique aspects of the new concerns is impossible here, it is important to stress that they are subject to the same criticisms which have been made of earlier static approaches to the resource problem. That is to say, all such extrapolations of recent trends which point to some collapse of the social system *t* years hence, fail fundamentally to take account of the human capacity to adapt and to modify technology in response to changing social and economic needs. Indeed, a central feature of such models (like the dog that did *not* bark in the night) is the total *absence* of any social mechanism for signalling shifting patterns of resource scarcity and for reallocating human skill and facilities in response to rising costs, either the rising direct costs of raw materials or the rising indirect costs associated with environmental pollution. One need not be a capitalist apologist to argue that the price mechanism does perform such a function, albeit imperfectly, or to believe that it might perform these functions even better under altered arrangements which no longer allowed private firms to treat our water courses and atmosphere as if they were free goods.

These models constitute a reversion to Malthusian Fundamentalism in another, closely related way. They allow no possibility that mankind's taste for offspring and therefore his reproductive practices will be altered *as a result of the growth experience itself.* In both the demographic and technological realms, therefore, such purely deductive models ignore the mass of dramatic evidence of the past 150 years or so that human behavior undergoes continuous modification and adjustment as a response to the changing patterns of opportunities and constraints thrown up by an industrializing society.

The thrust of these critical remarks is *not* that population growth, pollution, and increasing scarcity of key natural resources are unimportant problems, or that technology may be confidently relied upon to provide cheap and painless solutions whenever these key variables, and their interactions, begin to behave in problematic ways. Quite the contrary. The main complaints against such apocalyptic, neo-Malthusian models, built upon a global scale and incorporating naive behavioral assumptions, are first, that they define the problems incorrectly, and second, that they deflect attention from more "modest" but genuine questions which ought to be placed more conspicuously upon the agenda of social science research. Consider the following sample.

What are the determinants of human fertility? How is it likely to respond to future changes in mortality, income, urban densities, pollution, education levels, female employment opportunities, new goods (including new contraceptive techniques)? The very recent reductions in American fertility levels, which may indeed turn out to be a short-run phenomenon, indicate forcefully that demography is a subject about which our present state of ignorance is truly momentous. . . .

What sort of alterations in our system of property rights and our tax structure hold the greatest promise for the control of pollution? What are the technological possibilities that an altered incentive system might lead to the development of "cleaner" technologies, and at what cost?

What kinds of technological and social adaptations can we visualize if the cost of key resources, such as fuels, should increase substantially in the future?

What improvements can be made in our present mix of private and public institutions devoted to the "production" of useful knowledge?

Instead of the almost total preoccupation with supply considerations, is it possible to anticipate future shifts in income and social organization which may reduce the need for resources by shifting demand away from resource-intensive goods?

What can we find by opening up the "black box" of technological change? *How* responsive are human agents with the requisite talents, to the market forces which continually shift the prospective payoff to inventive and innovative activities? And—a very different question—what can be said in a systematic way about the responsiveness of nature? That is, given the incentives, what factors shape the prospects for success in overcoming different kinds of scarcities? Is there any meaningful sense in which it may be harder or easier to make resource-saving innovations than capital-saving innovations, or either of these innovations than labor-saving innovations? Is a dollar spent on agricultural invention likely to yield the same eventual social return as a dollar spent on manufacturing or on service? What of petrochemicals vs. textiles, or building materials vs. transportation, or pharmaceuticals vs. machine tools? Instead of computing the number of years before we "run out" of specific deposits, can we project reasonable estimates for the changing resource costs of alternative technologies?

Finally, of what relevance is the historical experience of the presently rich countries to the prospect for the presently poor countries? For one thing, the industrializing countries of the nineteenth and twentieth centuries were often able to overcome their own resource deficiencies by imports of primary products from the less developed world. When—or if—these regions are able to enter successfully into the stream of industrialization, what alternatives will they be confronted with in overcoming their own resource deficiencies? But, perhaps most fundamentally, what factors have determined the differential effectiveness with which societies have responded to the problems and constraints posed by their unique resource endowments? For, as Simon Kuznets has observed:

> It need not be denied that in the distribution of natural resources some small nations may be the lucky winners at a given time (and others at other times). But I would still argue that the capacity to take advantage of these hazards of fortune and to make them a basis for sustained economic development is not often given. In the 19th century, Brazil was commonly regarded as an Eldorado—and indeed enjoyed several times the position of a supplier of a natural resource in world-wide demand; yet the record of this country's economic growth has not been impressive, and it is not as yet among the economically advanced nations. The existence of a valuable natural resource represents a permissive condition, facilitates—if properly exploited—the transition from the pre-industrial to industrial phases of growth. But unless the nation shows a capacity for modifying its social institutions in time to take advantage of the opportunity, it will have only a transient effect. Advantages in natural resources never last for too long—given continuous changes in technology and its extension to other parts of the world.
>
> To put it differently: *every* small nation has some advantage in natural resources—whether it be location, coastline, minerals, forests, etc. But some show a capacity to build on it, if only as a starting point, toward a process of sustained growth and others do not. The crucial variables are elsewhere, and they must be sought in the nation's social and economic institutions.

CARTER HENDERSON

The Frugality Phenomenon

Skepticism about our nation's capacity for unlimited economic growth has now motivated five million Americans to choose to live a simpler life. While substantial numbers of Americans in the 1960s had begun to question the personal and societal rewards of a technologically advanced, materialistic society, the new advocates of a frugal life style seem motivated by their view of economic realism. Carter Henderson, co-director of the Princeton Center for Alternative Futures, Inc., describes this trend, which is growing at such a rate that estimates place the number who will have adopted lives of voluntary simplicity at about sixty million in the year 2000. Nor is the "frugality phenomenon" limited to the United States; similar movements are occurring in Canada, Great Britain, and Sweden. It is still too early, Henderson writes, to determine whether the frugality phenomenon represents a basic shift in Western economic attitudes. Nevertheless, the adoption by many Western Europeans and Americans of lower-consumption life styles must be taken into consideration by economists; and, should the movement continue to grow, it will, in the long run, undoubtedly affect the extent to which economic growth will continue to stimulate technological change.

What would happen if consumers decided to simplify their lives and spend less on material goods and services? This question is taking on a

Carter Henderson, "The Frugality Phenomenon." *Bulletin of the Atomic Scientists,* Vol. 34, No. 5 (May 1978). Reprinted by permission of the Bulletin of the Atomic Scientists, a magazine of science and public affairs. Copyright © by the Educational Foundation for Nuclear Science, Chicago, Illinois.

certain urgency as rates of economic growth continue to decelerate through-out the industrialized world, and as millions of consumers appear to be opting for more frugal lifestyles.

The Stanford Research Institute, which has done some of the most exten-sive work on the frugality phenomenon, estimates that nearly five million American adults are pursuing lives of "voluntary simplicity," and double that number "adhere to and act on some but not all" of its basic tenets.

The frugality phenomenon first achieved prominence as a middle-class re-jection of high-consumption life-styles in the industrialized world during the fifties and sixties. In *The Silent Revolution,* Ronald Inglehart of the University of Michigan's Institute of Social Research examined this experience in the United States and ten Western European nations. He concluded that a change has taken place "from an overwhelming emphasis on material well-being and physical security toward greater emphasis on the quality of life," that is, "a shift from materialism to post-materialism."

Inglehart calls the sixties the "fat years." Among their more visible trap-pings were the ragged and flamboyantly-patched blue jeans favored by the affluent young, reminding physicist-futurist Herman Kahn of the members of the Royal Court at Versailles who amused themselves by dressing up as milk-maids and woodsmen. Most of the retreat from materialism, however, was less visible. Comfortably-fixed Americans were going without, making things last longer, sharing things with others, learning to do things for themselves and so on. But while economically significant, it was hardly discernible in a U.S. gross national product climbing vigorously toward the $2 trillion mark.

Yet as the frugality phenomenon matured—growing out of the soaring six-ties and into the somber seventies—it seemed to undergo a fundamental transformation. American consumers continued to lose faith in materialism and were being joined by new converts who were embracing frugality because of the darkening economic skies they saw ahead. Resource scarcities, soaring energy prices, persistent inflation, high-level unemployment, alarming balance-of-trade deficits, the declining value of the U.S. dollar on foreign exchange markets forced consumers to look to their own resources. The one device which seemed most promising, the one over which they had the most control, was frugality—learning to live with less in a world where a penny saved was still a penny earned.

The Western democracies are now "in the midst of a revolution that we have only begun to perceive," former Secretary of State Henry Kissinger told a 1977 meeting of business, political and education leaders in Washington, D.C. The next decade, he added, will decide whether the industrial democ-racies will be able to manage their economic policies and maintain social peace "in the face of a probably lower long-term growth rate in the 1980s."

According to the Stanford Research Institute, the nearly five million Americans living lives of voluntary simplicity appear to be predominantly young (between the ages of 18 and 39), evenly divided between the sexes, almost exclusively white, from middle- or upper-class backgrounds, exceptionally well-educated, bimodal in income, politically independent, and largely urban. The reasons these men and women have chosen simple life-styles include the desire to live in a way that is "outwardly simple and inwardly rich," a "preference for smallness" as opposed to "complexity, anonymity, artificiality, dehumanization, manipulation, and wastefulness." Theirs is an "insistence upon living as naturally as possible," and a desire "to free the inner self for exploration" and to better cope with the "new scarcity."

The Stanford Institute authors projected that Americans pursuing lives of full voluntary simplicity would grow from 5 million to 25 million in 1987, and to 60 million in the year 2000; while those opting for partial voluntary simplicity life-styles would be from 10 million to 35 million in 1987, and 60 million by the turn of the century.

Paradoxically, the authors suggest that this growth in voluntary simplicity stands a good chance of being "perhaps the fastest growing consumer market of the coming decades: rising in value from about $35 billion today, to some $140 billion in 1987," and to "well over $300 billion in 2000 (all in 1975 dollars)." The reason for this paradox is that "the person living the simple life tends to prefer products that are functional, healthy, nonpolluting, durable, repairable, recyclable or made from renewable raw materials, energy-cheap, authentic, aesthetically pleasing, and made through simple technology."

The growth projected for both the number of Americans pursuing lives of voluntary simplicity, and the size of the new consumer markets this would generate, presupposes (1) a continuation of the pressures currently pushing people toward more frugal life-styles such as the prospect of chronic resource shortages, (2) that those choosing these life-styles will find them satisfying, and (3) that America's mass production/consumption economy will remain strong enough to avoid a severe depression and maintain decent living standards thereby avoiding widespread poverty leading to "social and economic revolution."

Many of the Stanford Research Institute's basic contentions about the trend toward more frugal life-styles are supported by a public opinion poll recently conducted by Louis Harris and Associates Inc. on the subject of America's unlimited economic growth.

"The American people," said Harris, "have begun to show a deep skepticism about the nation's capacity for unlimited economic growth, and they are wary of the benefits that growth is supposed to bring. Significant majorities place a higher priority on improving human and social relationships and the quality of American life than on simply raising the standard of living." Among the Harris Survey's more significant findings were that a large majority of the American public now prefers:

—"Teaching people how to live more with basic essentials" than on "reaching higher standards of living" (79 percent *vs* 17 percent);

—"Learning to get our pleasure out of non-material experiences," rather than on "satisfying our needs for more goods and services" (76 percent *vs* 17 percent);

—"Spending more time getting to know each other better as human beings on a person to person basis," instead of "improving and speeding up our ability to communicate with each other through better technology" (77 percent *vs* 15 percent); and

—"Improving those modes of travel we already have" rather than "developing ways to get more places faster" (82 percent *vs* 11 percent).

Harris sums up:

> Taken together, the majority views expressed ... suggest that a quiet revolution may be taking place in our national values and aspirations. Some of these attitudes reflect the energy crunch and the realization that the supply of raw materials is not boundless; others are a legacy of all those ideas young people pressed for in the 1960s that have now begun to take root in the 1970s.

Inevitable Limits

Many Americans got their first intimation of a less affluent future in 1972 when the Club of Rome published *The Limits of Growth,* a computer-based research study on "The Predicament of Mankind" done by a systems dynamics team at the Massachusetts Institute of Technology. The study concluded that the world economy is a single system doomed to collapse "If the present growth trends in world population, industrialism, pollution, food production, and resource depletion continue unchanged. ..."

At about the same time *A Blueprint for Survival* was being drawn up by the editors of the British magazine *The Ecologist.* "The principal defect of the industrial way of life with its ethos of expansion," said the *Blueprint,* "is that it is not sustainable. Its termination within the lifetime of someone born today is inevitable." The *Blueprint* report—endorsed by many of the United Kingdom's scientific/industrial elite—held that national survival depends on moving industrialized countries away from their mindless dedication to unending economic growth, and toward more realistic levels of production and consumption. "There is every reason to suppose," it concluded, that a "stable society would provide us with satisfactions that would more than compensate for those which, with the passing of the industrial state, it will become increasingly necessary to forgo."

Canada has released a study of its economic outlook which does not appear to share all the forebodings of *A Blueprint for Survival,* yet does foresee an urgent need for Canadians to change their high-energy, high consumption life-styles.

Canada's study, called the Conserver Society Report, is the end product of a project sponsored by 14 Canadian government departments and agencies which brought together an interdisciplinary team of professors from the universities of Montreal and McGill to investigate whether "uncontrolled, undirected industrial growth [is] a desirable option for Canada in the 1970s and 1980s" and whether "the Canadian ethic of perpetual acquisition and accumulation of things [is] conducive to a viable society with a high quality of life."

The university research group projected five scenarios of the future: *Status Quo* (doing more with more), *Growth with Conservation* (doing more with less), *High-Level Stable-State* (doing the same with less), *Buddhist* (doing less with less) and *Squander Society* (doing less with more).

The future scenario selected as the most rational and feasible was the Conserver Society whose "prime characteristic would be *a change of behavior without a radical change in the value system."* Its "goal would be *growth with conservation"* by striving to (1) "eliminate waste in the private and public sectors," (2) "direct our economic development, as far as possible, so that it harmonizes with nature," and (3) "to 'conserve' human and material resources by prolonging their useful life."

The Conserver Society report, whose theoretical framework was first published in July 1975, illustrates how Canada might be able to do more with less to "build a society where waste is reduced to a minimum" as seen in the following examples:

—Encourage the rental of infrequently-used goods such as vacation homes by several families as opposed to their outright ownership by just one;

—Use flexi-time scheduling so that vast sums of money and materials don't have to be wasted building peak-load highway systems and other facilities needed for only an hour or so a day;

—Stress conservation technology capable of creating recyclable products, substituting abundant and renewable energy sources such as solar power for disappearing reserves of fossil fuels, and cleaning up pollution;

—Reform Inefficient Consumption Habits (RICH) by turning off lights in empty rooms, driving less to save gas, insulating homes to save energy, discouraging overpackaging, etc.;

—Reduce the illusion of abundance through "full-cost pricing" which adds the cost of pollution control, depletion of non-renewable resources, etc., to the purchase price of goods;

—Coordinate the public and private sectors to they cooperate in the drive against waste rather than relying on market forces which are incapable of efficiently conserving the nation's resources.

While the Conserver Society Project has already generated ample controversy in Canada, it's been a veritable vicar's tea party compared with the furor touched off by a paper called "How Much is Enough?—Another Sweden" prepared by Goran Backstrand and Lars Ingelstam of the Swedish Government's Secretariat for Future Studies. This paper recommended that the Swedes, who are still among the world's wealthiest people with a physical quality of life second to none, put ceilings on their consumption of oil and meat, use buildings more economically, increase the durability of their consumer goods, and end the private ownership of automobiles. One Swedish men's magazine commented "it won't be any paradise, more like Hell. Is this really the way we want it to be? With thermostat spies and pork-chop detectives?"

In America frugality and hard work have been ingrained in our national character since the Pilgrims bequeathed us the Puritan ethic. Since then, we have been continually reminded of the joys of simple living and the discomforts of its antithesis by a succession of social commentators of whom Henry David Thoreau (*Walden*), Thorstein Veblen (*The Theory of the Leisure Class*), David M. Potter (*People of Plenty*) and Staffen Linder (*The Harried Leisure Class*) are among the more familiar. Only very recently, however, have we begun to believe that the transition from baroque to basic consumption patterns may be essential to our national economic survival.

One of the most fascinating examinations of the events which could encourage Americans to embrace voluntary frugality was a 1975 study done by the Stanford Research Institute for the U.S. Environmental Protection Agency. This study, entitled *Alternative Futures for Environmental Planning: 1975–2000* consists of ten scenarios describing what life may be like in the United States during the remainder of this century. Three of the scenarios depict "industrial success... within the limits of the industrial age paradigm," four scenarios focus on "industrial failure" in which "the industrial-age system functions but less efficiently and happily" than it does today, and the remaining four scenarios deal with "industrial transformation... in which the industrial-age paradigm... is drastically changed and, in effect, transcended."

Particularly interesting in the EPA study was its overall pessimism—seven out of the ten scenarios indicated a severe deterioration of the U.S. economy during the next quarter century, and it concluded that the more Americans living frugal life-styles, the better chance the U.S. economy will have to weather the economic storms ahead.

In examining frugality, and its potential as a meta-level resource for national economic survival, it is useful to look at the difference between those Americans currently turning to more frugal life-styles, and those who have been living in what is generally described as the "counter-culture."

While the two groups differ in many ways, the most significant difference would appear to be that frugal Americans largely accept the values of the industrial culture while the counter-culture Americans do not. The essence of the counter-culture life-style is its commitment to unhook from the consumption-driven mainstream economy.

The fact that more and more Americans are choosing to live lower-consumption life-styles is, of course, attracting attention throughout the country. Wall Street's Morgan Guaranty Trust Company was recently emboldened to recognize this "disconcerting" development in its monthly economic *Survey* (even though it did so in the non-alarming language of a fairy tale entitled "March of the Frugal Mice"). Outside the nation's financial capital, however, a lively family of publications such as *Mother Earth News, Whole Earth Catalog, Rain, Prevention,* and *Organic Gardening* are showing hundreds of thousands of readers how to actually create more self-reliant, wholesomely simple life-styles. Should these readers wish to pursue any particular aspect of frugal living they can do so by consulting hundreds of New Age books including *99 Ways to A Simple Life-stye, Handbook for Building Homes of Earth, Bicycle Transportation, How to Convert Your Auto to Propane/Methane, How to Grow More Vegetables, Sensible Sludge, Medical Self-Care* and *A Manual of Simple Burial.* . . .

Groups pursuing New Age life-styles are springing up all over the United States. A recent issue of the counter-culture magazine *Green Revolution,* for example, examined more than 200 communities, most of which got their start in the late 1960s or early 1970s. Begun in private homes, these communities are overwhelmingly populated by young people in their twenties and early thirties.

The emphasis on sharing and self-reliance in the growing New Age movement arises in part from its adherents' belief in the philosophy of "right livelihood" which might be broadly defined as engaging in work which does not threaten our human species, the planetary environment which sustains us, or future generations whose lives will be largely shaped by what we do today.

What seems to be happening in North America and the other market-oriented democracies is that a "counter-economy," more interested in psychic than material income, is taking firm root within our mainstream economy whose vitality is dependent on endlessly-growing production and consumption.

Those moving into the counter-economy are apparently convinced that the

Living with less: this spinner and weaver's home is built with second-hand materials, has no electricity, and is heated by wood only.

mainstream economy is now encountering limits to its previous exponential growth, and that those whose life-styles are attuned to *enough* rather than *more* will be far better equipped to get through the wrenching transition many see ahead.

Finally, and perhaps most important, is the suspicion now shared by so many people in the industrialized world that fulfillment on this planet cannot be found in the endlessly increasing consumption of material goods (even if this were possible in the future), but only through the life of the mind and spirit as taught by all the world's great religions.

Whether the frugality phenomenon represents a fundamental shift in Western economic attitudes, or something more transitory, remains to be seen. What does seem clear, however, is that the appearance of millions of Americans, Canadians, Britons, Swedes, and others willing to live more frugal lives could not have come at a more opportune time.

DEREK J. DE SOLLA PRICE

Little Science, Big Science

Through the analysis of a number of indices, Derek Price, a historian of science, describes in this selection how science has grown at an exponential rate since 1665. Because of the numbers of people who are now scientists, and because of the huge financial investment in the sophisticated instruments that modern science must employ, Price labels this enterprise "Big Science." He concludes that the growth of science cannot continue to accelerate, and that within the next century growth will cease. Since many experts believe that scientific knowledge is now the primary source of technological change, if scientific growth ceases, then technological innovation could be retarded.

Because the science we have now so vastly exceeds all that has gone before, we have obviously entered a new age that has been swept clear of all but the basic traditions of the old. Not only are the manifestations of modern scientific hardware so monumental that they have been usefully compared with the pyramids of Egypt and the great cathedrals of medieval Europe, but the national expenditures of manpower and money on it have suddenly made science a major segment of our national economy. The large-scale character of modern science, new and shining and all-powerful, is so apparent that the happy term "Big Science" has been coined to describe it. Big Science is so new that many of us can remember its beginnings. Big

Derek de Solla Price, ''Prologue to a Science of Science'' in *Little Science, Big Science*, Columbia University Press 1963, 1965. Courtesy of Professor Derek de S. Price and Columbia University Press.

Science is so large that many of us begin to worry about the sheer mass of the monster we have created. Big Science is so different from the former state of affairs that we can look back, perhaps nostalgically, at the Little Science that was once our way of life.

If we are to understand how to live and work in the age newly dawned, it is clearly necessary to appreciate the nature of the transition from Little Science to Big Science. It is only too easy to dramatize the change and see the differences with reckless naïveté. But how much truth is there in the picture of the Little Scientist as the lone, long-haired genius, moldering in an attic or basement workshop, despised by society as a nonconformist, existing in a state of near poverty, motivated by the flame burning within him? And what about the corresponding image of the Big Scientist? Is he honored in Washington, sought after by all the research corporations of the "Boston ring road," part of an elite intellectual brotherhood of co-workers, arbiters of political as well as technological destiny? And the basis of the change—was it an urgent public reaction to the first atomic explosion and the first national shocks of military missiles and satellites? Did it all happen very quickly, with historical roots no deeper in time than the Manhattan Project, Cape Canaveral rocketry, the discovery of penicillin, and the invention of radar and electronic computers?

I think one can give a flat "No" in answer to all these questions. The images are too naïvely conceived, and the transition from Little Science to Big Science was less dramatic and more gradual than appears at first. . . .

To get at this we must begin our analysis of science by taking measurements, and in this case it is even more difficult than usual to make such determinations and find out what they mean.

Our starting point will be the empirical statistical evidence drawn from many numerical indicators of the various fields and aspects of science. All of these show with impressive consistency and regularity that if any sufficiently large segment of science is measured in any reasonable way, the normal mode of growth is exponential. That is to say, science grows at compound interest, multiplying by some fixed amount in equal periods of time. Mathematically, the law of exponential growth follows from the simple condition that at any time the rate of growth is proportional to the size of the population or to the total magnitude already achieved—the bigger a thing is, the faster it grows. In this respect it agrees with the common natural law of growth governing the number of human beings in the population of the world or of a particular country, the number of fruit flies growing in a colony in a bottle, or the number of miles of railroad build in the early industrial revolution. . . .

Just after 1660, the first national scientific societies in the modern tradition were founded; they established the first scientific periodicals, and scientists

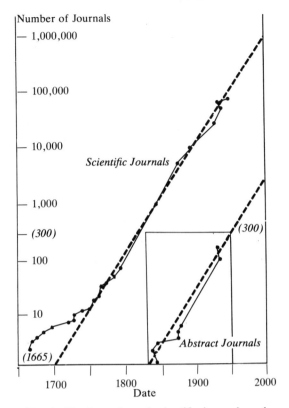

Fig. 1. Total number of scientific journals and
abstract journals founded, as a function of date.

Note that abstracts begin when the population of journals is approximately 300. Numbers
recorded here are for journals founded, rather than those surviving; for all periodicals containing any "science" rather than for "strictly scientific" journals. Tighter definitions might reduce
the absolute numbers by an order of magnitude, but the general trend remains constant for all
definitions. From Derek J. de Solla Price, *Science Since Babylon.* New Haven, Yale University
Press, 1961. Copyright © 1961 by Derek J. de Solla Price. By permission.

found themselves beginning to write scientific papers instead of the books that
hitherto had been their only outlets. We have now a world list of some 50,000
scientific periodicals (Fig. 1) that have been founded, of which about 30,000
are still being published; these have produced a world total of about six million scientific papers (Fig. 2) and an increase at the approximate rate of at
least half a million a year. In general, the same applies to scientific manpower.
Whereas in the mid-seventeenth century there were a few scientific men—a
denumerable few who were countable and namable—there is now in the
United States alone a population on the order of a million with scientific and
technical degrees (Fig. 3). What is more, the same exponential law accounts
quite well for all the time in between. The present million came through inter-

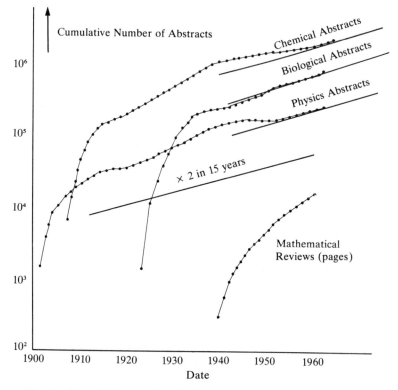

Fig. 2. Cumulative number of abstracts in various scientific fields, from the beginning of the abstract service to given date.

It will be noted that after an initial period of rapid expansion to a stable growth rate, the number of abstracts increases exponentially, doubling in approximately 15 years.

mediate stages of 100,000 in 1900, 10,000 in 1850, and 1000 in 1800. In terms of magnitude alone, the transition from Little Science to Big Science has been steady—or at least has had only minor periodic fluctuations similar to those of the stock market—and it has followed a law of exponential growth with the time rates previously stated.

Thus, the steady doubling every 15 years or so that has brought us into the present scientific age has produced the peculiar immediacy that enables us to say that so much of science is current and that so many of its practitioners are alive. . . .

A second clarification, one of crucial importance, must be made concerning the immediacy and growth of modern science. We have already shown that the 80- to 90-percent currency of modern science is a direct result of an exponential growth that has been steady and consistent for a long time. It follows that this result, true now, must also have been true *at all times in the past*,

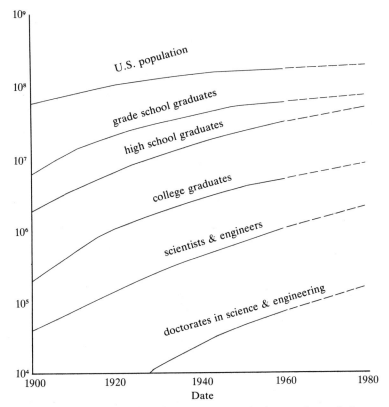

Fig. 3. Growth of scientific manpower and of general population
in the United States.

It may be seen that the more highly qualified the manpower, the greater has been its growth
rate. It will also be noted that there appears a distinct tendency for the curves to turn toward
a ceiling value running parallel with the population curve.

back to the eighteenth century and perhaps even as far back as the late seventeenth. In 1900, in 1800, and perhaps in 1700, one could look back and say
that most of the scientists that have ever been are alive now, and most of
what is known has been determined within living memory. In that respect,
surprised though we may be to find it so, the scientific world is no different
now from what it has always been since the seventeenth century. Science has
always been modern; it has always been exploding into the population, always
on the brink of its expansive revolution. Scientists have always felt themselves
to be awash in a sea of scientific literature that augments in each decade as
much as in all times before.

It is not difficult to find good historical authority for this feeling in all
epochs. In the nineteenth century we have Charles Babbage in England and
Nathaniel Bowditch in the United States bitterly deploring the lack of recog-

nition of the new scientific era that had just burst upon them. In the eighteenth century there were the first furtive moves toward special journals and abstracts in a vain attempt to halt or at least rationalize the rising tide of publications; there is Sir Humphrey Davy, whose habit it was to throw books away after reading on the principle that no man could ever have the time or occasion to read the same thing twice. Even in the seventeenth century, we must not forget that the motivating purpose of the *Philosophical Transactions of the Royal Society* and the *Journal des Sçavans* was *not* the publishing of new scientific papers so much as the monitoring and digesting of the learned publications and letters that now were too much for one man to cope with in his daily reading and correspondence. . . .

It is just possible that the tradition of more than 250 years represents a sort of adolescent stage during which every half-century science grew out of its order of magnitude, donned a new suit of clothes, and was ready to expand again. Perhaps now a post-adolescent quiescence has set in, and such exuberant growth has slowed down and is about to stop upon the attainment of adult stature. After all, five orders of magnitude is rather a lot. Scientists and engineers are now a couple of percent of the labor force of the United States, and the annual expenditure on research and development is about the same fraction of the Gross National Product. It is clear that we cannot go up another two orders of magnitude as we have climbed the last five. If we did, we should have two scientists for every man, woman, child, and dog in the population, and we should spend on them twice as much money as we had. Scientific doomsday is therefore less than a century distant.

PETER F. DRUCKER

Applied Science And Technology

For centuries, science and technology tended to be regarded as distinct fields, the former concerned with the pure pursuit of knowledge, the latter with making things for human purposes. With the exception of a few men like Sir Francis Bacon, the seventeenth-century British statesman and philosopher who predicted that scientific knowledge could lead to control over nature, most scientists did not consider their work to be of interest to technologists. In the nineteenth century, however, as Peter Drucker shows, science began to contribute to technology in a number of areas, and in the twentieth century this involvement has increased with the establishment of industrial research laboratories. Drucker makes a clear distinction between the technological work of the nineteenth century, which was "invention," and that of the present, which is "innovation." Current technological efforts involve deliberate attempts to bring about change in all areas of society, and therefore have enormously increased the impact of technology.

H and in hand with changes in the structure of technological work go changes in the basic approach to and methods of work. Technology has become science-based. Its method is now "systematic research." And what was formerly "invention" is "innovation" today.

Historically the relationship between science and technology has been a

---

From *Technology in Western Civilization*, Volume II: *Technology in the Twentieth Century*, edited by Melvin Kranzberg and Carroll W. Pursell, Jr. Copyright © 1967 by The Regents of the University of Wisconsin. Reprinted by permission of the Oxford University Press, Inc.

complex one, and it has by no means been thoroughly explored nor is it truly understood as yet. But it is certain that the scientist, until the end of the nineteenth century, with rare exceptions, concerned himself little with the application of his new scientific knowledge and even less with the technological work needed to make knowledge applicable. Similarly, the technologist, until recently, seldom had direct or frequent contact with the scientist and did not consider his findings of primary importance to technological work. Science required, of course, its own technology—a very advanced technology at that, since all along the progress of science has depended upon the development of scientific instruments. But the technological advances made by the scientific instrument maker were not, as a rule, extended to other areas and did not lead to new products for the consumer or to new processes for artisan and industry. The first instrument maker to become important outside of the scientific field was James Watt, the inventor of the steam engine.

Not until almost seventy-five years later, that is until 1850 or so, did scientists themselves become interested in the technological development and application of their discoveries. The first scientist to become a major figure in technology was Justus von Liebig, who in the mid-nineteenth century developed the first synthetic fertilizer and also a meat extract (still sold all over Europe under his name) which was, until the coming of refrigeration in the 1880's, the only way to store and transport animal proteins. In 1856 Sir William H. Perkin in England isolated, almost by accident, the first aniline dye and immediately built a chemical business on his discovery. Since then, technological work in the organic-chemicals industry has tended to be science-based.

About 1850 science began to affect another new technology—electrical engineering. The great physicists who contributed scientific knowledge of electricity during the century were not themselves engaged in applying this knowledge to products and processes; but the major nineteenth-century technologists of electricity closely followed the work of the scientists. Siemens and Edison were thoroughly familiar with the work of physicists such as Michael Faraday (1791–1867) and Joseph Henry (1791–1878). And Alexander Graham Bell (1847–1927) was led to his work on the telephone through the researches of Hermann von Helmholtz (1821–94) on the reproduction of sound. Guglielmo Marconi (1874–1910) developed radio on the foundation Heinrich Hertz (1857–94) had laid with his experimental confirmation of Maxwell's electromagnetic-wave propagation theory; and so on. From its beginnings, therefore, electrical technology has been closely related to the physical science of electricity.

Generally, however, the relationship between scientific work and its technological application, which we today take for granted, did not begin until after the turn of the twentieth century. As previously mentioned, such typically modern devices as the automobile and the airplane benefitted little from purely theoretical scientific work in their formative years. It was World War I that brought about the change: in all belligerent countries scientists were mobilized for the war effort, and it was then that industry discovered the tremendous power of science to spark technological ideas and to indicate technological solutions. It was at that time also that scientists discovered the challenge of technological problems.

Today technological work is, for the most part, consciously based on scientific effort. Indeed, a great many industrial research laboratories do work in "pure" research, that is, work concerned exclusively with new theoretical knowledge rather than with the application of knowledge. And it is a rare laboratory that starts a new technological project without a study of scientific knowledge, even where it does not seek new knowledge for its own sake. At the same time, the results of scientific inquiry into the properties of nature—whether in physics, chemistry, biology, geology, or another science— are immediately analyzed by thousands of "applied scientists" and technologists for their possible application to technology.

Technology is not, then, "the application of science to products and processes," as is often asserted. At best this is a gross oversimplification. In some areas—for example, polymer chemistry, pharmaceuticals, atomic energy, space exploration, and computers—the line between "scientific inquiry" and "technology" is a blurred one; the scientist who finds new basic knowledge and the technologist who develops specific processes and products are one and the same man. In other areas, however, highly productive efforts are still primarily concerned with purely technological problems, and have little connection to science as such. In the design of mechanical equipment—machine tools, textile machinery, printing presses, and so forth—scientific discoveries as a rule play a very small part, and scientists are not commonly found in the research laboratory. More important is the fact that science, even where most relevant, provides only the starting point for technological efforts. The greatest amount of work on new products and processes comes well *after* the scientific contribution has been made. "Know-how," the technologist's contribution, takes a good deal more time and effort in most cases than the scientist's "know-what"; but though science is no substitute for today's technology, it is the basis and starting point.

While we know today that our technology is based on science, few people (other than the technologists themselves) realize that technology has become in this century somewhat of a science in its own right. It has become research—a separate discipline having its own specific methods.

Nineteenth-century technology was "invention"—not managed or organized or systematic. It was, as our patent laws, now two hundred years old, still define it, "flash of insight." Of course hard work, sometimes for decades, was usually required to convert this "flash" into something that worked and could be used. But nobody knew how this work should be done, how it might be organized, or what one could expect from it. The turning point was probably Edison's work on the electric light bulb in 1879. As his biographer Matthew Josephson points out, Edison did not intend to do organized research. He was led to it by his failure to develop through "flash of genius" a workable electric light. This forced him, very much against his will, to work through the specifications of the solution needed, to spell out in considerable detail the major steps that had to be taken, and then to test systematically one thousand six hundred different materials to find one that could be used as the incandescent element for the light bulb he sought to develop. Indeed, Edison found that he had to break through on three major technological fronts at once in order to have domestic electric lighting. He needed an electrical energy source producing a well-regulated voltage of essentially constant magnitude; a high vacuum in a small glass container; and a filament that would glow without immediately burning up. And the job that Edison expected to finish by himself in a few weeks required a full year and the work of a large number of highly trained assistants, that is, a research team.

There have been many refinements in the research method since Edison's experiments. Instead of testing one thousand six hundred different materials, we would today, in all probability, use conceptual and mathematical analysis to narrow the choices considerably (this does not always work, however; current cancer research, for instance, is testing more than sixty thousand chemical substances for possible therapeutic action). Perhaps the greatest improvements have been in the management of the research team. There was, in 1879, no precedent for such a team effort, and Edison had to improvise research management as he went along. Nevertheless, he clearly saw the basic elements of research discipline: (1) a definition of the need— for Edison, a reliable and economical system of converting electricity into light; and (2) a clear goal—a transparent container in which resistance to a current would heat up a substance to white heat; (3) identification of the major steps to be taken and the major pieces of work that had to be done—in his case, the power source, the container, and the filament; (4) constant feedback from the results of the work on the plan; for example, Edison's finding that he needed a high vacuum rather than an inert gas as the environment for his filament made him at once change the direction of research on the container: and finally (5) organization of the work so that each major segment is assigned to a specific work team.

These steps together constitute to this day the basic method and the system of technological work. October 21, 1879, the day on which Edison first had a light bulb that would burn for more than a very short time, therefore, is not only the birthday of electric light; it marks the birth of modern technological research as well. Yet whether Edison himself fully understood what he had accomplished is not clear, and certainly few people at the time recognized that he had found a generally applicable method of technological and scientific inquiry. It took twenty years before Edison was widely imitated, by German chemists and bacteriologists in their laboratories and in the General Electric laboratory in the United States. Since then, however, technological work has progressively developed as a discipline of methodical inquiry everywhere in the Western world.

Technological research has not only a different methodology from invention; it leads to a different approach, known as innovation, or the purposeful and deliberate attempt to bring about, through technological means, a distinct change in the way man lives and in his environment—the economy, the society, the community, and so on. Innovation may begin by defining a need or an opportunity, which then leads to organizing technological efforts to find a way to meet the need or exploit the opportunity. To reach the moon, for instance, requires a great deal of new technology; once the need has been defined, technological work can be organized systematically to produce this new technology. Or innovation can proceed from new scientific knowledge and an analysis of the opportunities it might be capable of creating. Plastic fibers, such as nylon, came into being in the 1930's as a result of systematic study of the opportunities offered by the new understanding of polymers (that is, long chains of organic molecules), which chemical scientists (mostly in Germany) had gained during World War I.

Innovation is not a product of the twentieth century; both Siemens and Edison were innovators as much as inventors. Both started out with the opportunity of creating big new industries—the electric railway (Siemens), and the electric lighting industry (Edison). Both men analyzed what new technology was needed and went to work creating it. Yet only in this century—and largely through the research laboratory and its approach to research—has innovation become central to technological effort.

In innovation, technology is used as a means to bring about change in the economy, in society, in education, in warfare, and so on. This has tremendously increased the impact of technology. It has become the battering ram which breaks through even the stoutest ramparts of tradition and habit. Thus modern technology influences traditional society and culture in underdeveloped countries. But innovation means also that technological work is not done only for technological reasons but for the sake of a nontechnological economic, social, or military end.

Scientific discovery has always been measured by what it adds to our understanding of natural phenomena. The test of invention is, however, technical—what new capacity it gives us to do a specific task. But the test of innovation is its impact on the way people live. Very powerful innovations may, therefore, be brought about with relatively little in the way of new technological invention.

A very good example is the first major innovation of the twentieth century, mass production, initiated by Henry Ford between 1905 and 1910 to produce the Model T automobile. It is correct, as has often been pointed out, that Ford contributed no important technological invention. The mass-production plant, as he designed and built it between 1905 and 1910, contained not a single new element: interchangeable parts had been known since before Eli Whitney, a century earlier; the conveyor belt and other means of moving materials had been in use for thirty years or more, especially in the meat-packing plants of Chicago. Only a few years before Ford, Otto Doering, in building the first large mail-order plant in Chicago for Sears, Roebuck, used practically every one of the technical devices Ford was to use at Highland Park, Detroit, to turn out the Model T. Henry Ford was himself a highly gifted inventor who found simple and elegant solutions to a host of technical problems—from developing new alloy steels to improving almost every machine tool used in the plant. But his contribution was an innovation: a technical solution to the economic problem of producing the largest number of finished products with the greatest reliability of quality at the lowest possible cost. And this innovation has had greater impact on the way men live than many of the great technical inventions of the past.

CHARLES WEINER

How the Transistor Emerged

The transistor is the heart of all modern solid-state devices such as the small but powerful computers that guide space vehicles to the moon and other planets. According to historian Charles Weiner, the transistor was not "invented," but instead "emerged" as a result of group research at Bell Telephone Laboratories. Weiner describes the environment of modern industrial research laboratories in which scientific knowledge is applied to develop new and important technologies. It is interesting to note that the Nobel Prize–winning research group did not participate in the commercial development of the transistor. This task was assigned to another team—a development group—at Bell.

The transistor's genesis, its immediate impact, and its subsequent significance will be celebrated through a stream of words and pictures. And no wonder! When only a few years old, the device already was regarded by many as a major invention of the century. Indeed, its emergence is a contender as the most-studied happening in the history of twentieth-century science and technology.

Why has the transistor story achieved almost mythic stature in so short a time? A major reason is that it symbolizes the goals and organization of post–World War II science and technology. . . . The emphasis in this article will be on the social communication processes that linked what might be called the microenvironment at the Laboratories to the macroenvironment in which it thrived—the broad academic and scientific community outside. Glimpses of these links are provided by the personal letters, notebooks, and other original source materials that have been made accessible for historical study in recent years. This documentation includes interviews that have recorded the recollections of some of the individuals involved in these events.

These sources show that a group of social inventions in the two decades before 1945 played a vital role in the emergence of the transistor. The social inventions, which include international fellowships, specialized review publications, and even modes of organizing research efforts, facilitated both formal and informal communication between the Bell industrial laboratory, the universities, and the international scientific community. They contributed in a major way to the Laboratories' awareness of developments in basic physics and to its ability not only to utilize the results of the academic research and education system, but to contribute to them as well.

The birth date of any event is often considered to be the date of its announcement to the public, rather than the date of its "private" birth. Thus, the public birth date of the point-contact transistor was determined by the press announcement of July 1, 1948. The last of ten short blurbs in *The New York Times* under the heading, "The News of Radio,". . . consisted of eight sentences beginning: "A device called a transistor, which has several applications in radio where a vacuum tube ordinarily is employed, was demonstrated for the first time yesterday at Bell Telephone Laboratories. . . where it was invented.". . .

The first transistor had been made some six months before—December 23, 1947—and it was recorded in this entry in the laboratory notebook of Walter Brattain: "This circuit was actually spoken over and by switching the device in and out, a distinct gain in speech level could be heard and seen on the scope presentation with no noticeable change in quality."

An Authorization for Work

If the public announcement was in July 1948 and the private birth was in December 1947, than conception took place almost 2½ years earlier in the summer of 1945 with a remarkable internal Bell Laboratories document, which began as follows:

> **Authorization for work**
>
> *Subject:* Solid State Physics—the fundamental investigation of conductors, semiconductors, dielectrics, insulators, piezoelectric and magnetic materials.
> *Statement:* Communication apparatus is dependent upon these materials for most of its functional properties. The research carried out under this case has as its purpose the obtaining of new knowledge that can be used in the development of completely new and improved components and apparatus elements of communication systems.

> We have carried on research in all of these areas in the past. Large improvements in existing types of apparatus and completely new types have resulted. Thermistors, varistors, and piezoelectric network elements are typical examples of new types. The quantum physics approach to structure of matter has brought about greatly increased understanding of solid-state phenomena. The modern conception of the constitution of solids that has resulted indicates that there are great possibilities of producing new and useful properties by finding physical and chemical methods of controlling the arrangement and behavior of the atoms and electrons which compose solids.

The document, which goes on for several pages, was approved by Harvey Fletcher and James Fisk, and signed by Mervin J. Kelly, Executive Vice President.

This document formally set in motion the Laboratories solid-state research program and a few months later led to the establishment of the now-famous semiconductor group from which the transistor emerged. An analysis of its first paragraphs provides an interesting point of entry into the transistor story because they reveal something of the earlier processes of growth and interaction involved in the invention's history. The last two sentences, in particular, are significant; they refer to the increased understanding of solids made possible through quantum physics and the consequent possibilities of producing new and useful properties in materials of interest for communications technology. These developments had been set in motion two decades earlier in Europe when the new concepts inspired by the quantum theory were proposed, debated, and tested by an international group of physicists who traveled from one university to another to exchange ideas and communicate the results of recent research. Their mobility was aided in a major way by the first of the social inventions important to the transistor story—international fellowships from the Rockefeller Foundation. These fellowships provided modest travel funds and stipends, thus enabling the bright young physicists of the period to come together with colleagues for the face-to-face discussions that are so crucial at a time when new ideas are brewing. In such encounters the basic concepts of the new quantum theory of matter were hammered out between 1925 and 1927, providing physicists with powerful tools to describe electronic and atomic systems and to explain the cohesion of atoms in solid crystals.

In 1928 Arnold Sommerfeld, Felix Bloch, and others applied these concepts to the theory of metals, building on the work of several major contributors to the new quantum theory—including Erwin Schrödinger of Austria, Wolfgang Pauli of Switzerland, P. A. M. Dirac of England, and Enrico Fermi of Italy. By 1931 A. H. Wilson of England had developed a quantum mechanical theory that related what had been learned about the motion of electrons in metals to a comprehensive theoretical explanation of insulators and semiconductors.

The 1931 publication of Wilson's theoretical model of a solid semiconductor coincided with an increasing interest in semiconductors for use in electrical communications. . . .

One of the physicists concerned with semiconductors during this period was Walter H. Brattain, who had joined the Bell Telephone Laboratories in 1929. In 1931, he and J. A. Becker turned their attention to copper oxide, a well-known semiconductor of interest to Bell because of its possible use as a modulator. Let us take a look at Brattain's background, since he was later to become a key member of the semiconductor research group set up in 1945 and a coinventor of the transistor. . . .

Brattain [had done his] graduate work at Minnesota, where his thesis advisor was John Tate, who had taken his degree in Berlin under the renowned physicists Franck and Hertz. While at Minnesota, Brattain heard lectures by visiting European physicists, including Franck, Sommerfeld, and Schrödinger. He also learned quantum mechanics from John H. Van Vleck, one of the leading U.S. theorists, with special interest in the application of quantum mechanics to the solid state. . . .

In April [1929] Brattain attended a meeting of The American Physical Society in Washington, where Tate introduced him to Becker, who was "looking for a man," and he got his job as a physicist at Bell. Brattain's original interest in the Laboratories had been sparked during his undergraduate days [at Whitman College] when, as he recalls, "I had become aware of the Bell Telephone Laboratories because Professor Brown had received the *Bell System Technical Journal,* and among other things, there appeared from time to time articles by Darrow, trying to explain the newer things in physics in his gorgeous language."

A Communications Link

The *Bell System Technical Journal* proved to be an important communications link between the staff of the Laboratories and the outside physics community. The *Technical Journal* was started in 1922, and a year later, at the suggestion of the editor, Karl K. Darrow began contributing a series of articles on contemporary advances in physics. . . .

Darrow's contribution was an important one, not only as mediator between the immediate advancing part of basic physics research and the industrial laboratory, but in synthesizing and abstracting for the academic world as well.

The *Bell System Technical Journal* served a badly needed function and also helped to increase the mobility of talented physicists within the overlapping domains of academic and industrial research. Darrow synthesized the results of current basic research published in the physics journals and reported at the physicists' meetings, and made them accessible to a wider audience; at the same time, the basic physics literature was being enriched through contributions resulting from research at Bell Labs.

One such contribution was the 1927 paper by Bell physicists C. J. Davisson and L. H. Germer on electron diffraction, a discovery that resulted in a Nobel prize for Davisson in 1937. It was Davisson's presence, in part, that attracted William Shockley to the Laboratories in 1936, immediately after M.I.T. had awarded him the Ph.D. in physics. Shockley had studied under John Slater, who had made major contributions to the development of quantum mechanics and its applications to the solid state. One of Shockley's fellow physics graduate students at M.I.T. was James Fisk, who joined Bell Labs in 1939, headed physics research there in the postwar period, and ultimately became president. At Cambridge in the mid-1930s Shockley and Fisk met another young physicist, John Bardeen, who studied at Harvard with Van Vleck after completing his doctoral work under Eugene Wigner at Princeton. Wigner, like Van Vleck and Slater, played a key role in the development of solid-state physics in the U.S. in the 1930s. . . .

Bardeen had . . . pursued graduate work at Wisconsin and gained further industrial experience with the Gulf Research and Development Corp. as a geophysicist. Van Vleck was at Wisconsin at the time, and Bardeen learned quantum mechanics from him, just as Brattain had done earlier at Minnesota. Bardeen eventually returned to an industrial research environment when he joined the new semiconductor research group at Bell Labs after the war. Thus, by the mid-1930s the career patterns of many of those who were later to play key roles in the development of the transistor at Bell Labs—Brattain, Shockley, Fisk, and Bardeen—were set, influenced significantly by the communication channels of the academic and industrial research communities.

Scientists within the Laboratories, like those in the universities, also developed their own informal approach to keeping abreast of current developments in basic science. The "journal clubs" and colloquia characteristic of U.S. academic physics departments in the 1930s had their counterparts in an informal discussion group of some eight Bell Labs scientists, including Brattain and Shockley, in the late 1930s. One night a week they read and discussed basic work in the quantum mechanics of the solid state. . . .

The Materials Research Effort

Against this background of the diffusion of knowledge about solid-state physics, there was renewed interest in crystal detectors in the 1930s, which led to developments in materials research that also proved essential to the transistor development. For example, George Southworth at Bell wanted a crystal detector for microwaves because vacuum tubes would not work at those frequencies. He was an old radio man familiar with the "cat's-whisker" rectifiers and asked the chemists and metallurgists at Bell Labs if they could produce silicon pure enough that its properties could be more predictable and controllable when the material was used as a crystal detector. In 1930, Cornell University chemists had launched efforts that resulted in their development of methods for preparing large quantities of purified germanium, and by the late 1930s research at industrial and academic laboratories in the U.S. and Europe produced results that meant vast improvement in the purity and uniformity of these semiconductor materials.

The 1945 "Authorization for Work" noted that "there are great possibilities of producing new and useful properties, by finding physical and chemical methods of controlling the arrangement and behavior of the atoms and electrons which compose solids." These possibilities had greatly increased since the late 1930s through joint efforts of a number of industrial and academic laboratories during World War II. The development of radar in the late 1930s provided a focus for efforts to produce crystal detectors that were predictable in their electrical properties, and were stable and sensitive. This work was able to proceed within a theoretical framework that had been established in the 1930s by A. H. Wilson, Schottky, and others.

For example, at the University of Pennsylvania Frederick Seitz launched an effort in collaboration with the Du Pont laboratories to produce high-purity silicon, achieving spectroscopic purity above 99.9 percent. Further work by the Pennsylvania group added significantly to the knowledge of silicon as a semiconductor.

Another center for materials research was in the Physics Department at Purdue University, where a research group under the leadership of Karl Lark-Horovitz launched a systematic study of germanium in 1942. The group's investigations showed how to produce germanium in a relatively pure form, and demonstrated that its electrical properties were predictable from its impurity content. . . .

These were some of the results obtained through a wartime semiconductor crystal study project that involved Purdue, the University of Pennsylvania, M.I.T., the General Electric Company, Bell Laboratories, and other institutions in a joint effort for the M.I.T. Radiation Laboratory, which had been established in 1940 for radar research and development. Thus both academic and industrial laboratories were joined in the same research and development field. The communication among these institutions helped create the mutual awareness of the potential of semiconductor materials that contributed to the ultimate success of the Bell effort. . . .

The significance of the wartime semiconductor developments in setting the stage for the invention of the transistor cannot be overemphasized....

The wartime effort not only helped to bring institutions together on common problems but it also had an unsettling effect on individual scientists who suddenly found themselves working in new research environments, with new colleagues, and, in many cases, in new fields. This mixing of people, ideas, institutions, and missions gave scientists an increased awareness of different research styles and traditions, and provided them with an opportunity to supplement their knowledge of other branches of physics and of new experimental techniques.

Another important feature of wartime research was that the work was oriented toward the creation of practical devices such as amplifiers and detectors. This experience certainly remained in the consciousness of physicists, even though their postwar work may not have been directly aimed at producing devices.

Within the Physics Community

It is also important to recall that the few years immediately preceding the Bell Labs decision to launch a concentrated solid-state research effort were ones in which the physics discipline was taking cognizance of new needs and opportunities. In 1943 the first steps were taken within The American Physical Society to organize more effectively in order to facilitate communication among those physicists concerned with metals and other solids....

By 1947 a separate division of the society devoted to solid-state physics had been formed, thus providing formal recognition of the increasing importance of solid-state studies as a central field within the discipline.

One of the signers of the 1943 statement urging efforts to aid communication among solid-state physicists was William Shockley, who also played a role in the Bell Labs discussions leading to the launching of their solid-state program in July 1945. In 1945 Shockley talked with Mervin J. Kelly, who directed research at Bell Labs, about the desirability and opportunity for the Laboratories to mount an intensive solid-state physics research program in the postwar period. The increased theoretical understanding of the physics of solids, the availability of new methods of producing pure semiconductor materials with specified properties, and the wartime applications of crystal detectors all pointed to the need for further efforts to understand the basic physics of semiconductors.

All of these developments were reflected in the "Authorization for Work," which carried the names of Fletcher, Fisk, and Kelly. These men were all physicists, and because of the effective communication channels established during the preceding decade they were well aware that solid-state physics had advanced to a point where a program of fundamental research would have a

Microminiature computer circuit (enlargement of a transparency).

reasonably good chance of paying off for the Bell Telephone System. . . .

There is yet another kind of awareness reflected in the memo, however, and that is the awareness shown in the very carefully directed research policy of Bell Laboratories. The first paragraph of the Authorization justifies the solid-state research program as follows: "The research carried out under this case has its purpose the obtaining of new knowledge that can be used in the development of completely new and improved components and apparatus elements of communication systems."

Thus, although some have looked to Bell Laboratories as a paradigm for nondirective basic industrial research, we can see clearly from this that their philosophy was to support research in those fields of current basic science that seemed to have the greatest relevance to the mission of the Bell Telephone System. . . .

In mid-1945, as the war was drawing to an end, Kelly recognized Bell's long-term communications needs and the relevant developments in the macro-environment within which the Laboratories functioned. Thus he and others at Bell could clearly identify solid-state physics as a potentially important pay-off area, one in which to authorize a concentrated research effort.

The policy of the Laboratories was further reinforced by not expecting groups concerned with fundamental research to pursue the development of information they might turn up. In other words, their work did not spill over into the development stage. This was true of the transistor, for once it was discovered, a separate development group was formed, and it points up the awareness by the Laboratories' management of the scientist's need to maintain his orientation and his relation to his primary research mission.

Thus, if Bell Laboratories in its research policy was to show itself more structured in managing research than some may have imagined, it was also to show itself far freer than practically any other industrial research organization at that time.

The issuance of the authorization memo set in motion the process by which a solid-state physics group was set up under the leadership of Shockley and S. O. Morgan. Among the subgroups that were established was one on semiconductors. The aim of this subgroup was to explore the essential physics of semiconductors in order to determine how and why they worked. Management's hope was, of course, that this knowledge could be useful in developing semiconductor devices of interest to communications technology. The solid-state amplifier was one such device. Shockley recalls how his prewar conversation with Kelly regarding the possibilities of using electronics means to solve problems in relays and telephone exchanges "led me to see if one could not find other ways of accomplishing the same sort of things that vacuum tubes accomplished. I went through a series of trials of this sort prior to World

War II and, in fact, Walter Brattain and I worked together trying to make a version of a transistor in late 1939 or early 1940."...

The semiconductor group was truly interdisciplinary, including experimental and theoretical physicists, a physical chemist, and a circuit expert. There was also a close collaboration with the metallurgical groups. This kind of interdisciplinary focus on a single basic research project would have been difficult to achieve in an academic environment at that time because of the traditional departmental structures of most universities.

An important aspect of the group's work was the ability to communicate across the disciplinary boundaries and, according to the participants, this was achieved to a remarkable degree....

The take-off point for the actual experimental work that led to the point-contact transistor in December 1947 came in January 1946, when the group made the first of two major decisions: to focus on the simplest semiconductor materials—silicon and germanium. As Bardeen pointed out in his Nobel Lecture:

"Of great importance for our research program was the development during and since the war of methods of purification and control of the electrical properties of germanium and silicon. These materials were chosen for most of our work because they are well suited to fundamental investigations with the desired close coordination of theory and experiment."

The second major step was the decision to focus on the surface properties of the semiconductor as well as on its bulk properties. Briefly, Shockley—based on his knowledge of quantum mechanics and his application of it to semiconductors—had performed calculations that indicated it should be possible to modulate the resistance of a thin layer of semiconductor by applying strong electric fields. When resulting experiments failed to find the predicted field effect, Bardeen came into the situation to see what was wrong. The Schottky theory of the formation of a space-charge layer was a major influence. Bardeen suggested that electrons trapped at the surface of the semiconductor might be acting to cancel out the electric fields needed to produce the field effect. His "theory of surface states" provided the basis for deciding to proceed with research on surface phenomena.

During the remainder of 1946 and 1947, many experiments were carried out by Bardeen, Shockley, Brattain, and others at the Laboratories to test the surface-state theory and, eventually, to study the now observable field effect. The experimentation grew increasingly sophisticated until, finally, in December 1947, the transistor effect was observed. Brattain's description of the final stages of the experimental work was intriguing, both for what it reveals about the style of the work in general as well as for its echoes of Brattain's youthful background in dismantling and reassembling engines.

"After discussions with John Bardeen we decided that the thing to do was to get two point contacts on the surface sufficiently close together, and after some little calculation on his part, this had to be closer than 2 mils. The smallest wires that we were using for point contacts were 5 mils in diameter. How you get two points 5 mils in diameter sharpened symmetrically closer together than 2 mils without touching the points, was a mental block.

"I accomplished it by getting my technical aide to cut me a polystyrene triangle which had a small narrow, flat edge and I cemented a piece of gold foil on it. After I got the gold on the triangle, very firmly, and dried, and we made contact to both ends of the gold, I took a razor and very carefully cut the gold in two at the apex of the triangle. I could tell when I had separated the gold. That's all I did. I cut carefully with the razor until the circuit opened and put it on a spring and put it down on the same piece of germanium that had been anodized but standing around the room now for pretty near a week probably. I found that if I wiggled it just right so that I had contact with both ends of the gold that I could make one contact an emitter and the other a collector, and that I had an amplifier with the order of magnitude of 100 amplification, clear up to the audio range."

Brattain's actual account of the first successful demonstration on December 23, 1947, of the amplifier circuit and, on the next day, of an oscillator, can be followed in his laboratory notebook entries. These notes mark the private birth of the point-contact transistor, and although the event triggered an explosive increase in related research at the Laboratories that led to the junction transistor in 1951 and, eventually, to a whole new industry, it also marks the end of the story recounted here.

It is a story in which, when viewed in terms of the relationship between the microenvironment of Bell Laboratories and the outside macroenvironment, the transistor clearly emerges from a complex interaction of individuals, ideas, and institutions. It seems reasonable to assume that this interaction would not have taken place without a number of social inventions: fellowships that enabled physicists to come together at a crucial time in the development of powerful new concepts; publications that provided a needed synthesis of new developments in physics for scientists within industrial laboratories as well as for the entire physics discipline, and that provided links between the two communities; the organization of more effective patterns of communication for scientists specializing in new fields of work within their professional societies; a research philosophy that enabled Bell Laboratories to attract talented people and use them effectively, and so on. These institutional inventions were essential in stimulating fundamental research in physics and eventually translating the results into a practical device of immense social consequence.

JEROME R. RAVETZ

Social Problems of Industrialized Science

Jerome Ravetz, a mathematician and philosopher of science, asserts that science, at the present time, is qualitatively different from what it was in the past. This change has come about, he argues, because the focus has shifted from the pursuit of truth to a concentration on the application of scientific knowledge. Big Science, as Price terms it, requires enormous resources, so that much of the research and development costs are supported by the state in one form or another. There are four visible abuses in the conduct of science at the present time, Ravetz claims. It is shoddy, entrepreneurial, reckless, and dirty. To pursue a career in science, one has to make accommodation to these conditions.

The obsolescence of the conception of science as the pursuit of truth results from several changes in the social activity of science. First, the heavy warfare with "theology and metaphysics" is over. Although a few sharp skirmishes still occur, the attacks on the freedom of science from this quarter are no longer significant. This is not so much because of the undoubted victory of science over its early contenders as for the deeper reason that the conclusions of natural science are no longer ideologically sensitive. What people, either the masses or the educated, believe about the inanimate universe or the biological aspects of humanity is not relevant to the stability of society, as it was once thought to be. The focus of sensitivity is now in the social sciences; and the techniques of control by those in authority vary in subtlety in accordance with local requirements and traditions. Hence the leaders of the community of natural science no longer need to hold and proclaim this sort of ideological commitment as a rallying slogan for their followers and potential recruits. Also, the experience of modern scientists in their work, seeing the rapid rate of obsolescence of scientific results, makes the vision of the pursuit of truth not so much wrong as irrelevant. But, more important, the attention of the general educated public has shifted away from the problem of the nature of pure science and its relation to philosophy and religion. It is now concentrated on the visible triumph of technology based on applied science. Applied science has now become the basic means of production in a modern economy. The prosperity, and economic independence, of a firm or of a nation does not rest so much in its existing factories as in the "research and development" laboratories, where the industry of the future is being created and the competition of the future is being met. Thus, industry has been penetrated by science. . . .

Influences from Runaway Technology

The relations of science with industry will not be uniformly close in all fields; the connections will be strongest with the most modern, rapidly developing technologies, where innovation depends entirely on large-scale, sophisticated "research and development." It is these areas, such as aerospace, electronics, and parts of biological engineering, where the pace of development is so rapid, and the ecological and social effects so unpredictable and dangerous, that have been the focus of public conern in the menace of "science" to humanity. Those who take the decisions to plunge into ever greater "progress" in this work are not afflicted by any special wickedness or even irresponsibility. They are merely continuing the attitudes and practices inherited from the industry of the past, which might be called "myopic engineering." Provided that a particular development was technically viable and not at risk of penalties under the law, then so long as it seemed likely to make a profit, it would be adopted with no further thought of its consequences. Hitherto, the effects of such a policy, however disastrous, were localized to the region where they were put into practice; thus the rural South of England knew little, and generally cared less, of what was being done to the Midlands and North by the Industrial Revolution. And it is undeniable that the generally short-sighted and ruthless men who created the industry of the nineteenth century laid the material foundations for the prosperity of the present.

But the engine of innovation and production which we now possess is qualitatively different in many respects. Its effects are so pervasive that there is no place to hide from them. Also, the work of innovation in the advanced technologies is now a large industry in its own right. Its projects have some special features, which make them very different from the work of inventors and scientific consultants in the past. First, the investigation of any technical problem of development requires the prior commitment of enormous resources, both in funds and skilled manpower. Also, such problems are necessarily speculative, in several ways. It cannot be guaranteed in advance that the research will produce a device which works at all. Moreover, even if it works, there is the risk that during its years of development, either a change in the technical or commercial context, or the appearance of a better suited product from a competitor, will deprive it of its market. Perhaps most significant of all, the concept of "profit" has been transformed. For much of the sophisticated technological work is done for the State, for use outside the market sector, as for war. Although the prospect of foreign sales of a device is part of the calculation of "benefit," the basic component of benefit is assessed through a scientific study of its potential uses. Even where a key industry is nominally in the private sector, the State will take responsibility for its continued prosperity, through research contracts, guaranteed purchases, and other techniques. Thus a particular innovation may be recognized as risky from the technical point of view, dubious from the commercial point of view, of very slight use to anyone at all, even the State, and a potentially serious nuisance to the public and source of legal and political difficulties, and yet still receive enormous sums from the State because of its contribution to national prestige and its importance for maintaining employment and morale in a key industry. The Anglo-French supersonic transport is a perfect case in point of this phenomenon. At the extreme, national prestige may become so involved in a project that all considerations of cost, benefit, and profit (except, of course, to the private firms doing research and manufacture) are cast to the winds, and a glamorous technical project, such as the moon race, absorbs resources on a gargantuan scale, all in the name of "science."

Although this new industry of "R. and D." employs many scientists (indeed, the bulk of graduates in science and technology go there rather than into teaching or university research), its working ethics are descended from industry, private and state-supported, rather than from academic science. In America, the enormous defence and aerospace industries carry on in the time-honoured American tradition of "boondoggling" on Government funds; the most effective path to super-profits being to keep the relevant Government agencies for cost-accounting and quality-control either remote, or weak, or complaisant.

Thus we can speak of this new technology as "runaway" in several respects. In calculating cost and benefit, it ignores all those costs of a project for which it cannot legally be called to account: in particular, the degradation of the natural and human environment. Since the combined effect of the present and future technological developments is likely to be catastrophic, this rush onwards can truly be considered as out of control. And in its internal workings, the absence of that traditional discipline, crude and frequently distorted but in the last resort effective, of the test of a commercial market, makes the category of "profit" an artificial one, to be determined by the judgement of men in State agencies, in co-operation with the promoters themselves.

It is in the borderland between science and this sort of technology that we find some significant pathological phenomena. The first occurs when a contractor (individual or institutional) develops a really big enterprise, which is most likely to be on some mission-oriented research in a field where money is plentiful and not too many questions are asked. There then develops a research business, making its profit by the production of results in the fulfilment of contracts. The director of such an establishment is then truly an entrepreneur, who juggles with a portfolio of contracts, prospective, existing, extendable, renewable or convertible, from various offices in one or several agencies. The business is precarious, of course, for his only capital is in his friendly contacts with those who decide on the allocation of funds. In such a research factory, conditions are not usually conducive to the slow, painstaking, and self-critical work which is necessary for the production of really good scientific results. Hence much, most, or even all the work can be shoddy; but the entrepreneur does not operate in the traditional market of independent artisan producers who evaluate work by consensus. So long as he can keep his contacts happy, or at least believing that they personally have more to lose by exposing themselves through the cancellation or non-renewal of contracts than by allowing them to continue, his business will flourish.

It is in such circumstances that a man of high prestige and real talent will produce a stream of shoddy work. Too busy to do any thinking himself, and yet requiring a steady stream of publications as a proof of his continued competence, he will toss off pieces, either alone or with his associates, which will produce a list of titles of the necessary length. Although large-scale science is more exposed to the risk of invasion by entrepreneurs, size is not the determining factor. Whenever a research contractor, however modest his plant, sets the goals of his establishment to be the renewal and extension of contracts rather than the achievement of worthwhile results, he is an entrepreneur.

Even when scientific work of good quality is being done, the style of runaway technology can infect a field; the old, diffuse ideal of material benefit gives way to something more sharply defined and intoxicating: the possibility of the creation of new technical powers. The patent dangers of some of these powers, in the present state of civilization, have been brushed aside as of little consequence, or as the responsibility of someone else. It is so many generations since people in our civilization believed that there are "secrets too powerful to be revealed," that a scientist of our age cannot conceive himself as being in the position of the sorcerer; and yet he is. Thus "reckless science," as a special product of the technical and social conditions of scientific inquiry in our time, must be identified and controlled, for the safety of humanity in the long run and for the preservation of science in the short run.

Finally, the demands of military technology in particular provide opportunities for employment of scientists on research projects whose intended application lies beyond the pale of civilized practice and morality. The weapons called "ABC"—atomic, biological, and chemical—are each, in their own ways, morally tainted. Research and development of such weapons can be plausibly justified in terms of defence and deterrence; but the experience of the scientists on the original atomic bomb project shows that once the weapon is available, the tender consciences of the scientists who created it will not have much influence on the decisions on its use.

These four abuses, shoddy science, entrepreneurial science, reckless science, and dirty science, are distinct in their natures, but there will be tendencies for them to overlap in practice. Also, each of them arises from conditions inherent to the situation of contemporary science; and there is no clear line of demarcation, in the results of the work or in the attitudes of the scientist, whereby one can condemn one man and exonerate the next. But because they are more closely related to the demands of modern technology, and sometimes more easily popularized as exciting than traditional research, they will tend to attract a lion's share of the available funds, thereby providing the most attractive career prospects for recruits, and drawing into their ambit those who could not otherwise carry on their research. This effect can be seen in the United States, where the total budget for "science" is enormous, but where all save a small fraction is allocated to military R. and D. and the space-race. Even there, it can be argued that Congress would never allocate more than it does for the direct support of the esoteric and peculiar activity of pure research; but the result is that the "scientist" is seen as costing a lot of money to the taxpayers, and if he wants to use some of it, he is under pressure to make his accommodation with those who control these branches of runaway technology.

SIR WILLIAM FAIRBAIRN

The Engineering Profession

Fairbairn was one of England's most noted nineteenth-century engineers. Here, he describes the great engineering works of antiquity and the Middle Ages and sketches the beginnings of modern engineering practice. When Fairbairn wrote this account in 1877, engineering had become a recognized and respected profession.

The progressive improvements of the last hundred years, whether in our means of communication, in the spread of our knowledge, in the position of our science, in our arts and manufactures, in our provisions of war, or in our personal and domestic comforts and enjoyments, have been largely dependent on the work of the engineer. In some cases he has been almost the sole author of the progress made; in scarcely any would such progress have been possible, unaided by the mechanical design and constructive art which it is his province to supply.

The profession of engineering has, indeed, now taken such a high position in the economy of modern life, and its members are called on to exercise such important functions in the community, that the nature of their occupation cannot but be a matter of general interest....

The term *engineer*, as defining an occupation, is an old one; but it was originally applied only to persons in the military profession, and does not appear to have been used by civilians until the middle of the last century....

William Pole, editor, *The Life of Sir William Fairbairn, Bart.* London, Longman's Green and Co., 1877.

The term *engineer* comes more directly from an old French word in the form of a verb—*s'ingénier* ... and thus we arrive at the interesting and certainly little known fact, that an engineer is, according to the strict derivation of the term, not necessarily a person who has to do with engines, but anyone who seeks in his mind, who sets his mental powers in action, in order to discover or devise some means of succeeding in a difficult task he may have to perform.

It would be impossible to give a nobler or more appropriate description than this, of the manner in which our greatest engineering works have been produced, or the nature of the qualifications by which the greatest men in the profession have acquired their renown. . . .

In 1588 a curious work was published, in Italian and French, by a Capitano Agostino Ramelli, who styled himself "Ingegniero del Christianissimo Rè di Francia e di Pollonia." It is a description of various ingenious constructive devices for both military and civil use; and here, therefore, we have an early identification of the term engineer with precisely the kind of work that modern engineers are engaged in. . . .

The use of the term in England can be traced back to the thirteenth century. In the wardrobe account of King Edward I, A.D. 1300, occur the following passages:—

> To Master Reginald, engineer, for going by the King's order from Berwick-upon-Tweed to Newcastle-upon-Tyne, to the Sheriff of Northumberland, to procure and chuse timber for the making of machines for the castle of Berwick, for seventy-eight days' expenses in going and returning, and for hackneys for riding, &c., 2*l.* 8*s.* 0*d.* (Another article charges his pay at 6*d.* per diem.)
>
> Brother Thomas, of Bamburgh, and brother Robert de Ulm, master engineers, retained in the King's service for the Scottish wars, with Alan Bright, carpenter; Robert at 9*d.* per diem, Alan at 4*d.*
>
> Gerard de Mayek, engineer, and Gaillard Abot, carpenter, employed by the King to make the pele of Dumfries at 6*d.* per diem each. . . .

... Down to a recent period the title engineer was unknown in any application except its military one. It was not applied to the constructors of similar works in civil life. And yet the construction of such works generally has existed from time immemorial. One of the earliest fables of antiquity—the destruction of the Hydra by Hercules—is supposed to have referred to what we should now call the engineering work of draining the low lands of Argos, and damming up the sources of the inundations. And when we come to the more trustworthy records of history, we find that the most ancient civilised nations occupied themselves practically with works of an engineering character, and on a very large scale.

In Mesopotamia there must have been, thousands of years ago, men who possessed considerable mechanical knowledge and much constructive skill, and traces of their occupations still remain. The Phœnicians, too, constructed harbour and other engineering works with great ability.

The ancient works of Egypt are celebrated, not only for the colossal magnitude of the buildings, but also for the ingenious and useful character of the hydraulic arrangements. It is in Egypt that we find the invention of the arch, the first rudiments of which may be traced back, it is said, to the time of Amunoph the First, 1540 B.C. The original canal across the Isthmus of Suez was made under an Egyptian dynasty.

The Greeks, independently of their skill in building generally, must, from the extent of their coasts, have been well occupied in hydraulic constructions; but it is to their successors, the Romans, we may turn for the most remarkable examples of ancient engineering.

The immense extent of roads constructed by this nation, their durability, and the skill shown in surmounting the obstacles of marshes, lakes, and mountains, have excited astonishment and admiration. Twenty-nine great military roads centred in Rome, some of which were carried to the extreme points of the vast empire; and the whole of the roads were estimated as measuring 52,964 Roman miles. Many of the more important of these were admirable specimens of construction, abounding in excellent detail. The bridges, built in great numbers, and many of great size, were remarkable for their solidity. Trajan's Bridge over the Danube, built about A.D. 120, was the most magnificent in Europe; it consisted of twenty arches, each 180 feet span.

In hydraulic constructions the Romans also excelled. The works for supplying water to cities were often of great magnitude, and laid out with much skill. For Rome alone many conduits were used, one of which, the Aqua Claudia, was nearly fifty miles long. The quantity of water brought into the city was very large, and in addition to the great and numerous public fountains, the houses had water laid on.

Aqueducts of Roman construction still exist in many parts of Europe; among these, the Pont du Gard, near Nismes, is one of the most celebrated. It is 560 feet long and 160 feet high, and is supposed to have been executed by Agrippa, who was governor of Nismes in the time of Augustus, and was declared *curator perpetuus aquarum*. The aqueduct of Segovia, in Spain, is 2,220 feet long, and was built by Trajan. That of Lisbon has thirty-five arches, and is 263 feet high.

Canal works were common in ancient Italy. The Etruscans had cut many for drainage purposes, and one in the Pontine Marshes was executed by the Romans 162 B.C. Pliny mentions several "useful and magnificent" works of the kind constructed by Trajan.

The drainage of Lake Albano, 400 B.C., and that of Lake Fucino, A.D. 52, were great works, showing high skill and enterprise.

The town drainage of Rome, by the Cloaca Maxima, was also an engineering achievement that deserves mention.

From the nature of these works we may be fully convinced that they were designed by men well acquainted theoretically with the principles of natural philosophy current in their era; and, as a matter of practice, how excellently they were done is testified by the manner in which they have stood the ravages of time. We may indeed predict that there are few engineering works of our day which will, at the end of thousands of years, make as favourable an appearance as those of the ancients do now.

We have not much information as to who were the actual designers of such works; probably, however, the architect, who has in all ages been a well-defined practitioner, took on himself the responsibility of building constructions generally. Brunelleschi, who built the great dome of Florence, and Michael Angelo, who designed St. Peter's at Rome, acted as architects, but really did also the work of engineers. . . .

After the fall of the Roman Empire, we still find occasional examples of fine constructive works, as, for example, the great aqueduct of Spoleto, which was built by Theodoric King of the Goths A.D. 741. It has ten large Gothic arches, each seventy feet span, and is 328 feet high above the valley it crosses. It remains to this day in good condition, and still supplies water to the town.

About the twelfth century attention became strongly directed, in France, to the improvement of the internal communications of the country, and an association was formed under the name of the "Frères Pontiers" (Brethren of the Bridge), with the object of building bridges wherever rivers were dangerous or difficult to ford. This society, really a society of civil engineers, extended its branches over all parts of Northern Europe, and executed great numbers of important works, some of which are still in existence, as, for example, the Old Bridge at Lyons, and the celebrated one over the Rhone lower down, at St. Esprit, which was nearly half a mile long. The first stone London Bridge was also erected about the same date by the same body. The Ponte Vecchio at Florence, having three segmental arches of ninety-five feet span, was built in 1345, and the first stone bridge in Paris dates 1412. The Rialto at Venice, with a single arch ninety-seven feet span, was built by Michael Angelo in 1578. . . .

It was . . . about the middle of the eighteenth century before engineering in England may have been said to begin in earnest, by the employment of James Brindley to construct a large system of canals in Lancashire. Brindley was by trade a wheelwright and millwright, and, having naturally a mechanical turn of mind, he had acquired great skill in millwork and mechanical construction generally. The Duke of Bridgewater, having obtained in 1759 an Act empowering him to make a canal to convey his coals from Worsley to Manchester, about ten miles, and having heard of Brindley's ingenuity, resolved to employ him. In a few years the canal was completed, and Brindley afterwards executed many more, in that district as well as in other parts of the country, altogether about 360 miles in length, and involving engineering works of considerable magnitude and variety. He died in 1772.

Brindley has been usually held in great honour as an engineer; but it must be recollected that his works appear of higher merit because of the extremely backward state of engineering knowledge and practice in this country at that time. Hydraulic constructions, including the formation of canals and all appertaining works, were really in an advanced state on the continent before Brindley's day, and there was probably little done by him that had not been anticipated there. But he was an uneducated man, and even if the hydraulic information published by the Italians and French had penetrated to this country (which is very doubtful), it could hardly have been intelligible to him. Hence, he deserves credit for having, by his own unaided and unlettered practical intelligence and skill, accomplished so much in the face of what were no doubt great difficulties.

Another eminent man, who lived about the same time, John Smeaton, was in a very different position; as, to practical talents not inferior to Brindley's, he added the advantages of a good education and considerable scientific knowledge. He was, like Brindley, occupied at first with mechanical pursuits. He was apprenticed to a mathematical instrument maker, and afterwards went into that business on his own account. But he was fond of science, and he made several communications to the Royal Society, who, in 1753, elected him a Fellow, and in 1759 awarded him their gold medal. In 1756, the Eddystone lighthouse having been destroyed by fire, Smeaton was applied to, on the recommendation of the President of the Royal Society, to rebuild it. He had just before made a careful study of the great engineering works of Holland and Belgium, during a tour in those countries, and he felt confidence in undertaking the task. The new lighthouse was completed in 1759, and its construction, ably described by himself, has commanded universal admiration.

Smeaton was afterwards engaged, down to his death in 1792, in many other engineering works—river and canal navigations, drainage and reclamation of lands, harbours, roads, bridges, water supplies, pumping-engines, and machinery. His reports, which have been collected and published, are admirable models of what such documents should be. He did not execute any works of the gigantic character which has more lately so impressed the popular mind;

but considering his accurate and extensive scientific knowledge, his good education and position in society, his great practical skill and experience, his literary ability, his logical and sound judgment, and the zealous and conscientious care and attention he bestowed on whatever he undertook, he is admitted by all competent judges to hold the very highest rank as an engineer. . . .

The profession being thus fairly launched and named, and an impulse given to the demands upon it by the improvements in the communications and trade of the country, many practitioners followed. . . .

If we consider . . . the effect that this work has had on trade, on commerce, on finance, on government, on every branch of industry, and indeed on every possible aspect of human interests, we cannot hesitate to admit that the profession of engineering has become truly a great power.

SIEGFRIED GIEDION

Engineering
The Household

One of the most revolutionary changes resulting from engineering practice after the turn of the twentieth century was the mechanization of the household, which is described by Siegfried Giedion, a scholar of architecture and mechanization, in this selection. Although inventors worked on household appliances beginning in the 1850s and 1860s, it was not until Nikola Tesla developed the small electric motor and electric power networks came into existence that vacuum cleaners, dishwashers, washing machines, and other household tools became common.

The small electric motor, ranging from the size of a billiard ball to that of a football, can be inconspicuously built in an moved wherever needed. While demanding very little care and upkeep, it is the most adaptable of prime movers. It meant to the mechanization of the household what the invention of the wheel meant to moving loads. It set everything rolling. Without it, mechanical comfort in the house could have advanced little beyond its condition in the 'sixties.

From mid-century on, labor-saving tools were designed with a surprisingly sure hand. . . . The principles of the vacuum cleaner, of the dishwasher, of the washing machine, were discovered almost instantaneously. But their success and assimilation had to await the mechanical mover.

The electric motor likewise went through a long incubation period. To trace its hopeful beginnings, its fumblings and false starts, more numerous than for any other mover, would lead us away from our topic. We shall merely give a few co-ordinates by which to locate it in time. The first electric motor was built by Michael Faraday after his discovery of inducted currents (1831). It consisted of a copper disk rotating between the poles of a powerful magnet. The galvanic current arising in the disk could easily be canalized. But Faraday was not interested in the problem of its practical application. His was the attitude of the eighteenth-century scientist, interested only in discovery. Regarding himself as a natural philosopher, he left the industrial exploitation to others.

Many obstacles stood in the way of immediate solution. Over half a century had to pass before the electric motor, from the small dimensions in which Faraday had conceived it, after passing through the gigantic, was again condensed into the small and reliable instrument; and almost a century before its ubiquitous use was taken for granted.

The coming onto the market of the small electric power unit is closely linked with the name of Nicola Tesla, although it was far from being the main achievement of this master of high-frequency currents and of the multiple-phase motor that first made possible the economical transmission of energy. In the spring of 1889, almost immediately after his pioneer patent for the multiple-phase motor, Nicola Tesla, together with the Westinghouse Company, put on the market a 1/6 horsepower alternating current motor directly driving a three-blade fan. This motor could not be regulated in speed or direction. But the simple appliance, easily portable from one room to another, marks the starting point of innumerable moves to punctuate the house with local power units. The year 1889 saw many other patents drawn up for fans driven by electric motors. But Tesla's simple apparatus owes its significance to the fact that it did not remain a mere idea, but was commercially manufactured and put on the market.

The hot and damp American summer inevitably brought attempts to substitute mechanical devices for the hand fan—something, at any rate, that should free the hand of its waving to-and-fro motion. Such mechanisms "operated by means of a single lever and cord" might be pedal-driven or harnessed to a rocking chair. "When a person is sitting in the chair and rocks it ..." the fan placed overhead "is vibrated by the slightest motion of the chair." If in the 'sixties one wanted a truly automatic fan, it would have to be a clockwork one. "The case containing the works is attached to the ceiling and provided with revolving wings." There were also at this time clockwork fans for table-top use, adjustable in speed.

Nicola Tesla's electric fan of 1889 led the development by about a quarter century. Electric current in the 'nineties in Europe as in America, was a luxury. There were no electrical networks. The first large-scale generating plant was projected in 1891, but did not begin operation until 1896. Erected by the Westinghouse Company near Niagara Falls to supply the near-by city of Buffalo with current, it consisted of three Tesla A.C. dynamos, each of 5000 H.P. Theaters such as the Paris Opera, department stores, factories, generated their own current. Most of the power then consumed was for propelling electrical streetcars. There were devices by which current for dental offices might be tapped directly from the 500-volt lines "without the least danger for the operator and the patient."

The question whether electricity should be made available to the masses was discussed everywhere in the 'nineties. At London's exclusive Society of Arts, ... a lecturer, Crompton, concluded that electric current was "too expensive to become general." Wherever the question was raised, in Philadelphia or in London, the experts were dubious. Only the great inventors saw its possibilities. Nicola Tesla, at the start of the 'nineties, foretold that electricity would soon be used as casually as water.

RUTH SCHWARTZ COWAN

The "Industrial Revolution" In the Home

Historian Ruth Schwartz Cowan describes a very significant and unanticipated change resulting from the introduction of labor-saving household appliances in the early decades of the twentieth century. Although women were relieved from drudgery, the role of the middle-class housewife actually expanded. She worked longer hours and was expected to do jobs that she had not performed previously. Cowan suspects that this role change may explain certain facets of the Women's Liberation Movement of the 1960s and 1970s.

When we think about the interaction between technology and society, we tend to think in fairly grandiose terms: massive computers invading the workplace, railroad tracks cutting through vast wildernesses, armies of women and children toiling in the mills. These grand visions have blinded us to an important and rather peculiar technological revolution which has been going on right under our noses: the technological revolution in the home. This revolution has transformed the conduct of our daily lives, but in somewhat unexpected ways. The industrialization of the home was a process very

Ruth Schwartz Cowan, "The Industrial Revolution in the Home: Household Technology and Social Change in the Twentieth Century," from *Technology and Culture* 17 (1976). Published by the University of Chicago Press, Chicago 60637. © 1976 by the Society for the History of Technology. All rights reserved. Reprinted by permission.

different from the industrialization of other means of production, and the impact of that process was neither what we have been led to believe it was nor what students of the other industrial revolutions would have been led to predict. . . .

*   *   *

The *Ladies' Home Journal* has been in continuous publication since 1886. A casual survey of the nonfiction in the *Journal* yields the immediate impression that the decade between the end of World War I and the beginning of the depression witnessed the most drastic changes in patterns of household work. Statistical data bear out this impression. Before 1918, for example, illustrations of homes lit by gaslight could still be found in the *Journal;* by 1928 gaslight had disappeared. In 1917 only one-quarter (24.3 percent) of the dwellings in the United States had been electrified, but by 1920 this figure had doubled (47.4 percent—for rural nonfarm and urban dwellings), and by 1930 it had risen to four-fifths. If electrification had meant simply the change from gas or oil lamps to electric lights, the changes in the housewife's routines might not have been very great (except for eliminating the chore of cleaning and filling oil lamps); but changes in lighting were the least of the changes that electrification implied. Small electric appliances followed quickly on the heels of the electric light, and some of those augured much more profound changes in the housewife's routine.

Ironing, for example, had traditionally been one of the most dreadful household chores, especially in warm weather when the kitchen stove had to be kept hot for the better part of the day; irons were heavy and they had to be returned to the stove frequently to be reheated. Electric irons eased a good part of this burden. They were relatively inexpensive and very quickly replaced their predecessors; advertisements for electric irons first began to appear in the ladies' magazines after the war, and by the end of the decade the old flatiron had disappeared; by 1929 a survey of 100 Ford employees revealed that ninety-eight of them had the new electric irons in their homes.

Data on the diffusion of electric washing machines are somewhat harder to come by; but it is clear from the advertisements in the magazines, particularly advertisements for laundry soap, that by the middle of the 1920s those machines could be found in a significant number of homes. The washing machine is depicted just about as frequently as the laundry tub by the middle of the 1920s; in 1929, forty-nine out of those 100 Ford workers had the machines in their homes. The washing machines did not drastically reduce the time that had to be spent on household laundry, as they did not go through their cycles automatically and did not spin dry; the housewife had to stand guard, stopping and starting the machine at appropriate times, adding soap, sometimes attaching the drain pipes, and putting the clothes through the wringer manually. The machines did, however, reduce a good part of the drudgery that once had been associated with washday, and this was a matter of no small consequence. Soap powders appeared on the market in the early 1920s, thus eliminating the need to scrape and boil bars of laundry soap. By the end of the 1920s Blue Monday must have been considerably less blue for some house-wives—and probably considerably less "Monday," for with an electric iron, a washing machine, and a hot water heater, there was no reason to limit the washing to just one day of the week....

* * *

... The change from the laundry tub to the washing machine is no less pro-found than the change from the hand loom to the power loom; the change from pumping water to turning on a water faucet is no less destructive of traditional habits than the change from manual to electric calculating. It seems odd to speak of an "industrial revolution" connected with housework, odd because we are talking about the technology of such homely things, and odd because we are not accustomed to thinking of housewives as a labor force or of housework as an economic commodity—but despite this oddity, I think the term is altogether appropriate.

In this case other questions come immediately to mind, questions that we do not hesitate to ask, say, about textile workers in Britain in the early nineteenth century, but we have never thought to ask about housewives in America in the twentieth century. What happened to this particular work force when the technology of its work was revolutionized? Did structural changes occur? Were new jobs created for which new skills were required? Can we discern new ideologies that influenced the behavior of the workers?

The answer to all of these questions, surprisingly enough, seems to be yes. There were marked structural changes in the work force, changes that in-creased the work load and the job description of the workers that remained. New jobs were created for which new skills were required; these jobs were not physically burdensome, but they may have taken up as much time as the jobs they had replaced. New ideologies were also created, ideologies which reinforced new behavioral patterns.... Middle-class housewives, the women who must have first felt the impact of the new household technology, were not flocking into the divorce courts or the labor market or the forums of political protest in the years immediately after the revolution in their work. What they were doing was sterilizing baby bottles, shepherding their children to dancing classes and music lessons, planning nutritious meals, shopping for new clothes, studying child psychology, and hand stitching color-coordinated curtains....

The significant change in the structure of the household labor force was the disappearance of paid and unpaid servants (unmarried daughters, maiden aunts, and grandparents fall in the latter category) as household workers— and the imposition of the entire job on the housewife herself. Leaving aside for a moment the question of which was cause and which effect (did the disappearance of the servant create a demand for the new technology, or did the new technology make the servant obsolete?), the phenomenon itself is relatively easy to document. Before World War I, when illustrators in the women's magazines depicted women doing housework, the women were very often servants. When the lady of the house was drawn, she was often the person being served, or she was supervising the serving, or she was adding an elegant finishing touch to the work. Nursemaids diapered babies, seamstresses pinned up hems, waitresses served meals, laundresses did the wash, and cooks did the cooking. By the end of the 1920s the servants had disappeared from those illustrations; all those jobs were being done by housewives— elegantly manicured and coiffed, to be sure, but housewives nonetheless. . . .

As the number of household assistants declined, the number of household tasks increased. The middle-class housewife was expected to demonstrate competence at several tasks that previously had not been in her purview or had not existed at all. Child care is the most obvious example. The average housewife had fewer children than her mother had had, but she was expected to do things for her children that her mother would never have dreamed of doing: to prepare their special infant formulas, sterilize their bottles, weigh them every day, see to it that they ate nutritionally balanced meals, keep them isolated and confined when they had even the slightest illness, consult with their teachers frequently, and chauffeur them to dancing lessons, music lessons, and evening parties. There was very little Freudianism in this new attitude toward child care: mothers were not spending more time and effort on their children because they feared the psychological trauma of separation, but because competent nursemaids could not be found, and the new theories of child care required constant attention from well-informed persons—persons who were willing and able to read about the latest discoveries in nutrition, in the control of contagious diseases, or in the techniques of behavioral psychology. These persons simply had to be their mothers.

Consumption of economic goods provides another example of the housewife's expanded job description; like child care, the new tasks associated with consumption were not necessarily physically burdensome, but they were time consuming, and they required the acquisition of new skills. Home economists and the editors of women's magazines tried to teach housewives to spend their money wisely. The present generation of housewives, it was argued, had been reared by mothers who did not ordinarily shop for things like clothing,

*Housewife's workload, 1947: a suburban housewife stands amid a typical week's load
of housework and supplies.*

bed linens, or towels; consequently modern housewives did not know how to shop and would have to be taught. Furthermore, their mothers had not been accustomed to the wide variety of goods that were now available in the modern marketplace; the new housewives had to be taught not just to be consumers, but to be informed consumers. Several contemporary observers believed that shopping and shopping wisely were occupying increasing amounts of housewives' time.

Several of these contemporary observers also believed that standards of household care changed during the decade of the 1920s. The discovery of the "household germ" led to almost fetishistic concern about the cleanliness of the home. The amount and frequency of laundering probably increased, as bed linen and underwear were changed more often, children's clothes were made increasingly out of washable fabrics, and men's shirts no longer had replaceable collars and cuffs. Unfortunately all these changes in standards are difficult to document, being changes in the things that people regard as so insignificant as to be unworthy of comment; the improvement in standards seems a likely possibility, but not something that can be proved.

In any event we do have various time studies which demonstrate somewhat surprisingly that housewives with conveniences were spending just as much time on household duties as were housewives without them—or, to put it another way, housework, like so many other types of work, expands to fill the time available. A study comparing the time spent per week in housework by 288 farm families and 154 town families in Oregon in 1928 revealed 61 hours spent by farm wives and 63.4 hours by town wives; in 1929 a U.S. Department of Agriculture study of families in various states produced almost identical results.... Just after World War II economists at Bryn Mawr College reported... 60.55 hours spent by farm housewives, 78.35 hours by women in small cities, 80.57 hours by women in large ones—precisely the reverse of the results that were expected. A recent survey of time studies conducted between 1920 and 1970 concludes that the time spent on housework by non-employed housewives had remained remarkably constant throughout the period. All these results point in the same direction: mechanization of the household meant that time expended on some jobs decreased, but also that new jobs were substituted, and in some cases—notably laundering—time expenditures for old jobs increased because of higher standards. The advantages of mechanization may be somewhat more dubious than they seem at first glance.

*   *   *

The housewife is just about the only unspecialized worker left in America— a veritable jane-of-all-trades at a time when the jacks-of-all-trades have disappeared. As her work became generalized the housewife was also proletarianized: formerly she was ideally the manager of several other subordinate workers; now she was idealized as the manager and the worker combined. Her managerial functions have not entirely disappeared, but they have certainly diminished and have been replaced by simple manual labor; the middle-class, fairly well educated housewife ceased to be a personnel manager and became, instead, a chauffeur, charwoman, and short-order cook. The implications of this phenomenon, the proletarianization of a work force that had previously seen itself as predominantly managerial, deserve to be explored at greater length than is possible here, because I suspect that they will explain certain aspects of the women's liberation movement of the 1960s and 1970s which have previously eluded explanation: why, for example, the movement's greatest strength lies in social and economic groups who seem, on the surface at least, to need it least—women who are white, well-educated, and middle-class.

Finally, instead of desensitizing the emotions that were connected with household work, the industrial revolution in the home seems to have heightened the emotional context of the work, until a woman's sense of self-worth became a function of her success at arranging bits of fruit to form a clown's face in a gelatin salad. That pervasive social illness, which Betty Friedan characterized as "the problem that has no name," arose not among workers who found that their labor brought no emotional satisfaction, but among workers who found that their work was invested with emotional weight far out of proportion to its own inherent value.

THORSTEIN VEBLEN

The Role of
The Engineers

By 1920, mass-produced automobiles were pouring off the assembly line, and manufacturers were beginning to apply mass production methods to other consumer durable goods—stoves, washing machines, dishwashers, and refrigerators. At the time, the sociologist Thorstein Veblen, whose scathing attacks on mercenary big businessmen had caused outrage in many quarters, believed that the major flaw in the economic system of the United States was to permit capitalists to be in control of production. Instead, Veblen argued, the engineer's logical and objective approach to problems uniquely fitted him to take command of industrial production. In this task, engineers should be aided by economists. In many ways, his position presaged the frequently voiced modern attitude of "leave it to the experts."

In more than one respect the industrial system of today is notably different from anything that has gone before. It is eminently a system, self-balanced and comprehensive; and it is a system of interlocking mechanical processes, rather than of skilful manipulation. It is mechanical rather than manual. It is an organization of mechanical powers and material resources, rather than of skilled craftsmen and tools; although the skilled workmen and tools are also an indispensable part of its comprehensive mechanism. It is of an impersonal nature, after the fashion of the material sciences, on which it constantly draws. It runs to "quantity production" of specialized and standardized goods and services. For all these reasons it lends itself to systematic control under the direction of industrial experts, skilled technologists, who may be called "production engineers," for want of a better term.

This industrial system runs on as an inclusive organization of many and diverse interlocking mechanical processes, interdependent and balanced among themselves in such a way that the due working of any part of it is conditioned on the due working of all the rest. Therefore it will work at its best only on condition that these industrial experts, production engineers, will work together on a common understanding; and more particularly on condition that they must not work at cross purposes. These technological specialists whose constant supervision is indispensable to the due working of the industrial system constitute the general staff of industry, whose work it is to control the strategy of production at large and to keep an oversight of the tactics of production in detail.

Such is the nature of this industrial system on whose due working depends the material welfare of all the civilized peoples. It is an inclusive system drawn on a plan of strict and comprehensive interdependence, such that, in point of material welfare, no nation and no community has anything to gain at the cost of any other nation or community. In point of material welfare, all the civilized peoples have been drawn together by the state of the industrial arts into a single going concern. And for the due working of this inclusive going concern it is essential that that corps of technological specialists who by training, insight, and interest make up the general staff of industry must have a free hand in the disposal of its available resources, in materials, equipment, and man power, regardless of any national pretensions or any vested interests. Any degree of obstruction, diversion, or withholding of any of the available industrial forces, with a view to the special gain of any nation or any investor, unavoidably brings on a dislocation of the system; which involves a disproportionate lowering of its working efficiency and therefore a disproportionate loss to the whole, and therefore a net loss to all its parts.

And all the while the statesmen are at work to divert and obstruct the working forces of this industrial system, here and there, for the special advantage of one nation and another at the cost of the rest; and the captains of finance are working, at cross purposes and in collusion, to divert whatever they can to the special gain of one vested interest and another, at any cost to the rest. So it happens that the industrial system is deliberately handicapped with dissension, misdirection, and unemployment of material resources, equipment, and man power, at every turn where the statesmen or the captains of finance can touch its mechanism; and all the civilized peoples are suffering privation together because their general staff of industrial experts are in this way required to take orders and submit to sabotage at the hands of the statesmen and the vested interests. Politics and investment are still allowed to decide matters of industrial policy which should plainly be left to the discretion of the general staff of production engineers driven by no commercial bias. . . .

In effect, the progressive advance of this industrial system towards an all-inclusive mechanical balance of interlocking processes appears to be approaching a critical pass, beyond which it will no longer be practicable to leave its control in the hands of business men working at cross purposes for private gain, or to entrust its continued administration to others than suitably trained technological experts, production engineers without a commercial interest. What these men may then do with it all is not so plain; the best they can do may not be good enough; but the negative proposition is becoming sufficiently plain, that this mechanical state of the industrial arts will not long tolerate the continued control of production by the vested interests under the current businesslike rule of incapacity by advisement.

In the beginning, that is to say during the early growth of the machine industry, and particularly in that new growth of mechanical industries which arose directly out of the Industrial Revolution, there was no marked division between the industrial experts and the business managers. That was before the new industrial system had gone far on the road of progressive specialization and complexity, and before business had reached an exactingly large scale; so that even the business men of that time, who were without special training in technological matters, would still be able to exercise something of an intelligent oversight of the whole, and to understand something of what was required in the mechanical conduct of the work which they financed and from which they drew their income. Not unusually the designers of industrial processes and equipment would then still take care of the financial end, at the same time that they managed the shop. But from an early point in the development there set in a progressive differentiation, such as to divide those who designed and administered the industrial processes from those others who designed and managed the commercial transactions and took care of the financial end. So there also set in a corresponding division of powers between the business management and the technological experts. It became the work of the technologist to determine, on technological grounds, what could be done in the way of productive industry, and to contrive ways and means of doing it; but the business management always continued to decide, on commercial grounds, how much work should be done and what kind and quality of goods and services should be produced; and the decision of the business management has always continued to be final, and has always set the limit beyond which production must not go.

With the continued growth of specialization the experts have necessarily had more and more to say in the affairs of industry; but always their findings as to what work is to be done and what ways and means are to be employed in production have had to wait on the findings of the business managers as to what will be expedient for the purpose of commercial gain. This division between business management and industrial management has continued to go forward, at a continually accelerated rate, because the special training and experience required for any passably efficient organization and direction of these industrial processes has continually grown more exacting, calling for special knowledge and abilities on the part of those who have this work to do and requiring their undivided interest and their undivided attention to the work in hand. But these specialists in technological knowledge, abilities, interest, and experience, who have increasingly come into the case in this way— inventors, designers, chemists, mineralogists, soil experts, crop specialists, production managers and engineers of many kinds and denominations—have continued to be employees of the captains of industry, that is to say, of the captains of finance, whose work it has been to commercialize the knowledge and abilities of the industrial experts and turn them to account for their own gain. . . .

Hitherto, then, the growth and conduct of this industrial system presents this singular outcome. The technology—the state of the industrial arts—which takes effect in this mechanical industry is in an eminent sense a joint stock of knowledge and experience held in common by the civilized peoples. It requires the use of trained and instructed workmen—born, bred, trained, and instructed at the cost of the people at large. So also it requires, with a continually more exacting insistence, a corps of highly trained and specially gifted experts, of divers and various kinds. These, too, are born, bred, and trained at the cost of the community at large, and they draw their requisite special knowledge from the community's joint stock of accumulated experience. These expert men, technologists, engineers, or whatever name may best suit them, make up the indispensable General Staff of the industrial system; and without their immediate and unremitting guidance and correction the industrial system will not work. It is a mechanically organized structure of technical processes designed, installed, and conducted by these production engineers. Without them and their constant attention the industrial equipment, the mechanical appliances of industry, will foot up to just so much junk. The material welfare of the community is unreservedly bound up with the due working of this industrial system, and therefore with its unreserved control by the engineers, who alone are competent to manage it. To do their work as it should be done these men of the industrial general staff must have a free hand, unhampered by commercial considerations and reservations; for the production of the goods and services needed by the community they neither need nor are they in any degree benefited by any supervision or interference from the side of the owners. Yet the absentee owners, now represented, in effect, by the syndicated investment bankers, continue to control the industrial experts and limit their discretion, arbitrarily, for their own commercial gain, regardless of the needs of the community.

Hitherto these men who so make up the general staff of the industrial system have not drawn together into anything like a self-directing working force; nor have they been vested with anything more than an occasional, haphazard, and tentative control of some disjointed sector of the industrial equipment, with no direct or decisive relation to that personnel of productive industry that may be called the officers of the line and the rank and file. It is still the unbroken privilege of the financial management and its financial agents to "hire and fire." The final disposition of all the industrial forces still remains in the hands of the business men, who still continue to dispose of these forces for other than industrial ends. And all the while it is an open secret that with a reasonably free hand the production experts would today readily increase the ordinary output of industry by several fold, — variously estimated at some 300 per cent to 1200 per cent of the current output. And what stands in the way of so increasing the ordinary output of goods and services is business as usual. . . .

As a matter of course, the powers and duties of the incoming directorate will be of a technological nature, in the main if not altogether; inasmuch as the purpose of its coming into control is the care of the community's material welfare by a more competent management of the country's industrial system. It may be added that even in the unexpected event that the contemplated overturn should, in the beginning, meet with armed opposition from the partisans of the old order, it will still be true that the duties of the incoming directorate will be of a technological character, in the main; inasmuch as warlike operations are also now substantially a matter of technology, both in the immediate conduct of hostilities and in the still more urgent work of material support and supply.

The incoming industrial order is designed to correct the shortcomings of the old. The duties and powers of the incoming directorate will accordingly converge on those points in the administration of industry where the old order has most signally fallen short; that is to say, on the due allocation of resources and a consequent full and reasonably proportioned employment of the available equipment and man power; on the avoidance of waste and duplication of work; and on an equitable and sufficient supply of goods and services to consumers. Evidently the most immediate and most urgent work to be taken over by the incoming directorate is that for want of which under the old order the industrial system has been working slack and at cross purposes; that is to say the due allocation of available resources, in power, equipment, and materials, among the greater primary industries. For this necessary work of allocation there has been substantially no provision under the old order.

To carry on this allocation, the country's transportation system must be placed at the disposal of the same staff that has the work of allocation to do; since, under modern conditions, any such allocation will take effect only by use of the transportation system. But, by the same token, the effectual control of the distribution of goods to consumers will also necessarily fall into the same hands; since the traffic in consumable goods is also a matter of transportation, in the main.

On these considerations, which would only be reinforced by a more detailed inquiry into the work to be done, the central directorate will apparently take the shape of a loosely tripartite executive council, with power to act in matters of industrial administration; the council to include technicians whose qualifications enable them to be called Resource Engineers, together with similarly competent spokesmen of the transportation system and of the distributive traffic in finished products and services. With a view to efficiency and expedition, the executive council will presumably not be a numerous body; although its staff of intelligence and advice may be expected to be fairly large, and it will be guided by current consultation with the accredited spokesmen (deputies, commissioners, executives, or whatever they may be called) of the several main subdivisions of productive industry, transportation, and distributive traffic.

Armed with these powers and working in due consultation with a sufficient ramification of subcenters and local councils, this industrial directorate should be in a position to avoid virtually all unemployment of serviceable equipment and man power on the one hand, and all local or seasonal scarcity on the other hand. The main line of duties indicated by the character of the work incumbent on the directorate, as well as the main line of qualifications in its personnel, both executive and advisory, is such as will call for the services of Production Engineers, to use a term which is coming into use. But it is also evident that in its continued work of planning and advisement the directorate will require the services of an appreciable number of consulting economists; men who are qualified to be called Production Economists.

The profession now includes men with the requisite qualifications, although it cannot be said that the gild of economists is made up of such men in the main. Quite blamelessly, the economists have, by tradition and by force of commercial pressure, habitually gone in for a theoretical inquiry into the ways and means of salesmanship, financial traffic, and the distribution of income and property, rather than a study of the industrial system considered as a ways and means of producing goods and services. Yet there now are, after all, especially among the younger generation, an appreciable number, perhaps an adequate number, of economists who have learned that "business" is not "industry" and that investment is not production. And, here as always, the best is good enough, perforce.

"Consulting economists" of this order are a necessary adjunct to the personnel of the central directorate, because the technical training that goes to make a resource engineer, or a production engineer, or indeed a competent industrial expert in any line of specialization, is not of a kind to give him the requisite sure and facile insight into the play of economic forces at large; and as a matter of notorious fact, very few of the technicians have gone at all far afield to acquaint themselves with anything more to the point in this connection than the half-forgotten commonplaces of the old order. The "consulting economist" is accordingly necessary to cover an otherwise uncovered joint in the new articulation of things. His place in the scheme is analogous to the part which legal counsel now plays in the manœuvres of diplomatists and statesmen; and the discretionary personnel of the incoming directorate are to be, in effect, something in the way of industrial statesmen under the new order.

RALPH NADER

Unsafe
At Any Speed

About the time Veblen praised engineers, they began to come under suspicion. In the 1920s waste from industrial plants produced the first noticeable water pollution, and planned obsolescence came under attack. In the 1950s several cities started to have serious smog problems. Instead of being praised for their contribution to technology, engineers were blamed for the damaging and un-wanted side effects of innovation. Perhaps more than anyone else, Ralph Nader focused on the shortcomings of engineers. In 1965 he published Unsafe at Any Speed, *which accused automotive engineers of disregarding ethical principles and of ignoring public safety. The publicity given to his critical analysis and Nader's own crusade spurred the consumer movement, which has become a powerful social and political force today.*

Ever since the Corvair was introduced, General Motors' official reaction to criticisms has been silence. The handling hazards of Corvairs did not proceed from engineering mysteries or the prevalence of one technical "school of thought" over another. The Corvair was a tragedy, not a blunder. The tragedy was overwhelmingly the fault of cutting corners to shave costs. This happens all the time in the automobile industry, but with the Corvair it happened in a big way. What was there for General Motors to say?

The tragedy of the Corvair did not begin that thirtieth day of September in 1959 when it went on display in dealer showrooms. Nor did it begin when

Ford test drivers got hold of two Corvairs somewhat prematurely from a dealer in early September and lost control of them at the company's test track. It began with the conception and development of the Corvair by leading GM engineers—Edward Cole, Harry Barr, Robert Schilling, Kai Hansen and Frank Winchell.

Cole, now a General Motors executive vice-president, provided the managerial ignition. He was an old devotee of rear-engined cars and right after World War II became involved with a short-lived experimental Cadillac having a rear engine. A prototype, ponderously bedecked with dual tires at the rear for stability, was soon shelved. To Cole, however, the idea of a rear-engined car remained attractive and he carried it over with him to Chevrolet and developed a project proposal as he rose in that division's hierarchy. In 1955, as chief engineer of Chevrolet, Cole saw a market for a small, "compact" car. Already an unpretentious import with a rear, air-cooled engine and independent suspension was "pre-testing" the American market with rising commercial success. But Cole and his associates were not in any mind merely to produce an American stereotype of the Volkswagen. This was to be a brand new kind of car utilizing the lessons of past models and the advances of the latest automotive technoloy. When he rose to head Chevrolet division in the summer of 1956, Cole put some of his finest engineering talent to work on preliminary design work. In the spring of 1957, Barr, Schilling, and Hansen made formal presentations before the top-level GM engineering policy committee and the executive committee. It was then that the official go-ahead to build the Corvair was given to Chevrolet. Kai Hansen was made head of the project.

A small, light car project naturally would look to the European experience. This is what Hansen and his associates did before coming up with the Corvair design. To aid in such an evaluation, they had the benefit of one of GM's most creative engineers, Maurice Olley. Originally hailing from Rolls Royce, Olley was a prolific inventor with over twenty-five U.S. patents issued in his name and assigned to General Motors. His field of specialization was automobile handling behavior. In 1953 Olley delivered a technical paper, "European Postwar Cars," containing a sharp critique of rear-engined automobiles with swing-axle suspension systems. He called such vehicles "a poor bargain, at least in the form in which they are at present built," adding that they could not handle safely in a wind even at moderate speeds, despite tire pressure differential between front and rear. Olley went further, depicting the forward fuel tank as "a collision risk, as is the mass of the engine in the rear." Unmistakably, he had notified colleagues of the hurdles which had to be overcome.

Hansen's group was familiar with the risks of its appointed task. Its members knew well the kinds of priorities which would force them to dilute their engineering standards. First, the new automobile had to sell well and make a "target rate of return" on investment, according to GM's unique and well-established policy of guaranteed profits. The way to do this, General Motors' management decided, was to make a small, lighter car, with fuel economy, which would seat six passengers comfortably and give a ride comparable to a standard Chevrolet passenger sedan. Given the goal of designing a much lighter vehicle, this was no routine task. If these objectives could be achieved, the quest for profit maximization would have reached new frontiers. An automobile representing a reduction of 1,332 pounds of material, or more than one-third the weight of a standard 1960 Chevrolet, that could sell for only about $200 less than standard models would constitute a marvel of production cost efficiency and sales ingenuity.

In January 1960, Hansen told a meeting of the Society of Automotive Engineers: "Our first objective, once the decision was made to design a smaller, lighter car, was to attain good styling proportions. Merely shortening the wheel base and front and rear overhang was not acceptable. To permit lower overall height and to accommodate six adult passengers, the floor hump for the drive shaft had to go. Eliminating the conventional drive shaft made it essential then that the car have either rear-engine, rear-drive or front-engine, front-drive. Before making a decision, all types of European cars were studied including front-engine, front-drive designs. None measured up to our standards of road performance."

Chevrolet engineers decided that the best and most "esthetically pleasant" utilization of passenger space dictated the use of a rear-engine, rear-drive design. This decision presented the problem, according to Hansen, of successfully applying the arrangement to a chassis that combines stability with a good ride and easy handling qualities. Hansen's job was to get the various factors working for safer handling—principally, front and rear weight-distribution, tire-pressure differentials and tire design, suspension geometry, and relative dynamic behavior in the front and rear—and still keep a soft ride and maximum cost reduction possible.

Hansen and his fellow engineers could not have been under any misapprehension as to the magnitude of the handling challenge before them. They had to deal with by far the heaviest rear-engined automobile in the western world, having between sixty and sixty-three per cent of its weight on the rear wheels. This fact alone posed handling problems considerably in excess of those afflicting the smaller and lighter rear-engined European cars. Ocee Ritch describes the consequences of this weight and size difference between rear-engined cars by way of simple analogy: "If you swing a bucket at the end of a short rope and accidentally hit your brother in the head, is he more apt to suffer a concussion if the bucket is empty or full? Similarly, if you increase the length of the rope and swing it at the same speed, will it cause more damage? Right on both counts. The more weight or the longer the arm, the more force is generated. In the case of the automobile, deviating from a straight line is the equivalent of swinging the bucket."

Automotive engineers will say, in defending their performance, that every car is a compromise with economic and stylistic factors. This statement, if true, is also meaningless. For the significant question is, who authorizes what compromises of engineering safety? Hansen has never publicly revealed what choices he would have preferred to take had he been given more authority against the erosive demands of the professional stylists and the cost department. The secret world of the automobile industry does not encourage free and open engineering discussion of alternative courses of action. But on occasion there is an exposition of what was actually done. Before a meeting of the Society of Automotive Engineers on April 1, 1960, in Detroit, Charles Rubly, a Chevrolet engineer who worked on the Corvair, gave his colleagues the practical considerations: "One of the obvious questions is: 'If you wish more of the roll couple to be taken on the front wheels, why did you leave the stabilizer off?' First, we felt the slight amount of gain realized did not warrant the cost; secondly, we did not wish to pay the penalty of increased road noise and harshness that results from use of a stabilizer. Another question that no doubt can be asked is why did we choose an independent rear suspension of this particular type? There are other swing-axle rear suspensions, of course, that permit transferring more of the roll couple to the front end. Our selection of this particular type of a swing-axle rear suspension is based on: (1) lower cost, (2) ease of assembly, (3) ease of service, and (4) simplicity of design. We also wished to take advantage of coil springs . . . in order to obtain a more pleasing ride . . ."

Mr. Rubly's four reasons could be reduced to one: lower cost. Having made such concessions, the Corvair engineers had to compensate for the strong oversteering tendency of the design. This was done by recommending to the Corvair owner certain critical tire pressure differentials which he should maintain between front and rear wheels. Corvair buyers received this advisory near the end of the owner's manual: "Over-steer problems may also be encountered with incorrect pressures. Maintain the recommended inflation at pressures at all times."

No definition of "over-steer" is given in the manual. The recommended pressures are fifteen psi (pounds per square inch) on the front wheels and twenty-six psi on the rear wheels when cold (defined as "after car has been parked for three hours or more or driven less than one mile") and eighteen psi front and thirty psi rear when "hot." According to the Chevrolet division, such pressure differences promote vehicle stability by introducing proper steer characteristics.

It is well established that cornering stability can be improved with any weight distribution, front or rear, by manipulating tire inflation pressures. (Equally inflated tire pressures, front and rear, says Professor Eugene Larrabee of the Massachusetts Institute of Technology, makes the Corvair dangerous to drive.) But any policy which throws the burden of such stability on the driver by requiring him to monitor closely and persistently tire pressure differentials, cannot be described as sound or sane engineering practice. The prominent automotive engineer Robert Janeway expressed a deeply rooted technical opinion in engineering circles when he evaluated the use of this human expedient: "Instead of stability being inherent in the vehicle design, the operator is relied upon to maintain a required pressure differential in front and rear tires. This responsibility, in turn, is passed along to service station attendants, who are notoriously unreliable in abiding by requested tire pressures. There is also serious doubt whether the owner or service man is fully aware of the importance of maintaining the recommended pressures.". . .

Ocee Ritch states flatly that the "suggested fifteen psi front, twenty-six psi rear (cold) or eighteen psi and thirty psi (hot) is far too low for high-speed driving or cornering." His recommended course of action requires a constant attention by the operator which proper vehicle design should have rendered wholly unnecessary. "Our prolonged experiments indicate that pressures should be increased until the tires begin to lose adhesion, then reduced slightly . . . a trial-and-error process, since production units are fitted with at least three brands of tires, each differing slightly from the other, plus the fact that loading and any suspension changes make significant differences."

The Corvair driver becomes puzzled on confronting such a range of advice. If he writes to the Chevrolet division for clarification, he receives a reply assuring him that the manual's recommendations are the optimum tire pressures and were derived after exhausting research and testing. But clearly a more heavily loaded Corvair, such as one with five passengers, requires different tire pressures to minimize differences in tire deflections front and rear. Corvair engineers knew about this problem and considered raising the recommended rear tire pressures. Once again, however, they succumbed to the great imperative—a soft ride. Rubly recounts it plainly enough: "The twenty-eight psi would reduce the rear-tire deflection enough but we did not feel that we should compromise ride and add harshness because under hot conditions tire pressures will increase three to four psi." Remarks such as these make it difficult to give full credence to company claims and advisories dealing with automotive safety. For behind the façade of engineering authority is the reality of the "trade-off"—auto industry cant for the bare-bones concessions to the cost and style men. The engineering assurances can not be taken at face value in such a context of undisclosed adulteration.

Another area intimately related to Corvair stability is the load-carrying capacity of the tires. According to the Tire and Rim Association, a tire industry standards group, the maximum permissible load capacity of the size tire used on the Corvair is 835 pounds per tire at twenty-four psi. These maximum permissible loads are derived after compromise between the tire manufacturers and the automobile companies. Yet even under these less than stringent standards, the rear tires of the Corvair are ordinarily overloaded with two or more passengers. . . .

It would not be fair to say that the Corvair engineers designed a vehicle but forgot about the driver. They knew the risks in a design where the car usurps the driving task under certain expected stresses of highway travel. These stresses occur not just in high-speed emergency conditions but in ordinary driving situations within legal speed limits. The combination of factors which leads to the critical point of control loss may occur with a statistical infrequency, but the traditional integrity of automotive design has been to embrace just such situations. Stability limitations, for example, must be evaluated under the most unfavorable loading conditions—six passengers with luggage in the case of the Corvair even though most mileage registered by Corvairs will be with fewer passengers.

When it serves their promotional interests, the automobile manufacturers show great concern about the most infrequently occurring situations. A continuing illustration is the elaborate defenses which they make for producing vehicles with up to four hundred horsepower and a speed capability reaching 150 miles per hour. Is such power and speed hazardous? Not at all, claim the companies, for they provide an important margin of safety in emergency conditions. Apparently emergency conditions include speeds up to and over 100 miles an hour.

The men who headed the Corvair project knew that the driver should be given a vehicle whose handling is both controllable and predictable. They knew that impossible demands could be placed upon the driver by an inherently oversteering vehicle. For the past thirty-five years, American cars have been designed so as to be basically understeering. The Corvair was the first mass-produced exception. Dr. Thomas Manos, the highly respected automotive engineering professor of the University of Detroit, is not teaching Hansen's group anything they do not know when he states his judgment about oversteering automobiles: "The driver must become aware that he has to continually fight the wheel or continuously correct because he is the factor which makes the vehicle a stable piece of equipment." Hansen's group didn't forget about the driver. They did not have the professional stamina to defend their engineering principles from the predatory clutches of the cost-cutters and stylists. They did not forget about the driver; they ignored him.

DONALD E. MARLOWE

Public Interest— First Priority in Engineering Design?

Donald Marlowe, a former president of the American Society of Mechanical Engineers, fights back at critics of the engineering profession. Engineering, he says, is the primary source of technological innovation and economic growth. The best way to ensure that the public interest will be served in engineering design, Marlowe argues, is to quantify—i.e., to assign numerical values to— all of the non-economic values inherent in engineering problems. Only engineers and economists, Marlowe asserts, are capable of conceptualizing such large-scale problems. Engineers should oppose the criticisms of those who act on intuition, he argues, and re-emphasize the power of the engineering method.

The engineering profession inherits Shakespeare's warning—"the evil lives after, the good is buried with the bones." Robert Hutchins urges "Stamp Out Engineering Schools." Kenneth Boulding has written, "The domination of almost all of our resources policy by engineers is utterly disastrous." Yet we define historical ages in terms of their technology (Mesolithic, Bronze Age, Iron Age, and so on), and any attempt to study culture divorced from its technology is destined to be fruitless. Darwin, Marx, and Bessemer were contemporaries—which one of them had the greatest influence on today's civilization?

Donald Marlowe, "Public Interest—First Priority in Engineering Design?" ASME Paper 68 WA/AV-3. By permission of the American Society of Mechanical Engineers.

As far as we can see at the present time, it appears that no nation will succeed in meeting the aspirations of its people if it does not adopt a modern, western, technology, unless, of course, such a country isolates itself like the monks of Shangri-la.

The absence of engineering students from the social turmoil represented by student lock-ins or sit-ins is not to be attributed to their lack of social awareness, but is rather that they know that their engineering education is relevant to the current situation and is perhaps the best preparation for the real problems of the future. As Kranzberg has said, "the engineering schools have found no ways of predicting the future, but they are associated with its command." They embody, as Gabor has said, "that magnificent spirit by which a poor animal almost toothless and clawless, has raised itself gradually to the status of modern man."

Engineers' Choice

We might well ask why Engineering wears such different masks to its different viewers—from Boulding's "utter disaster" to Gabor's "magnificent spirit." One might suspect that Boulding and Gabor and all of the followers of each would agree that the choice of masks is too important to be ignored—that technology has the potential of either saving or destroying the world. If it can do either, why the difficulty of choice? Why not simply choose to save?

Probably few engineers ever deliberately chose the philosophy that they have no concern for the uses to which their developments are put. They are not anarchists. It is true, however, that engineering has prospered in an atmosphere of entrepreneurship, and that engineers are more often political reactionaries than they are liberals.

In another setting I have stated that the last surviving Jeffersonian democrat will be an engineer. The basic criterion for choice among several possible solutions to a problem has always been the economic choice, as far as most engineers are concerned. And, indeed, in a developing non-affluent society, it may well be that the economic choice provides the most efficient distribution of resources. In the painful process of capital accumulation, there may well have been times when slavery, sweatshops, company unions and political repression had their economic place. The engineer in an entrepreneurial economy could then have the satisfaction of serving his technology and serving his society by the same acts.

Source of Economy

Those who bemoan the narrowness of the engineer's vision would do well to recognize that this very concentration of attention has been the first source of our current productive economy. Even with such a constrained view of engineering—that is, the rigorous solution of technological problems to the criterion of economic choice—we must recognize that, at the stage of economic development attained in the United States at the turn of the century, even this limited concept of engineering was probably in the Public Interest *at that time*. Indeed, until the United States had its "take off" assured, and the prospect of an exponentially growing economy before it, it might well have been frivolous of the engineering profession to have taken any other point of view.

However, even if we dispose of Hutchins as the last representative of a tradition wherein "The crisis in the humanities is that they have incorrectly defined the soul of our culture so that they now have relatively less to tell us about the total cultural setting in which man finds himself" (W. D. Maxwell), we cannot so lightly dispose of Boulding, for after all, the methods of economists more closely approach the methods of engineers with every passing year. Perhaps one good definition of an affluent society would be a society *which can afford to* consider values other than economics, after it has passed the stage wherein it has the potential to supply all of the necessities of all its people. This situation has not been a common one, and the world has had little experience with handling it.

For the engineer, bereft of economic choice, the first question to raise its head is one of "values." That is, if economics alone is not to rule, how are other values introduced into the analysis? If beauty, or comfort, or esthetics, or race, or spirit, are to be thrown into the scales, how are these values to be quantified?

Quantification of Values

These are not entirely new questions. Certainly, since the early 1930s men engaged in multiple-use hydroelectric power development have struggled with such questions. How does one quantify recreational values, wild-river preservation, the Colorado River Gorge?

While some steps have been taken, many questions remain. Particularly, as one moves into urban engineering, the choice of values becomes most obscure. How does one value the coherence of an ethnic neighborhood, the division of a school district, the effect of a zoning change or the impact of a public park? Yet these are far from inconsequential matters. In fact, the location of a modern urban highway is far more likely to be determined today by the number of poor families displaced than by the traditional criterion of minimizing earthmoving.

One might argue that this problem of quantification of values is not the responsibility of the engineering profession; that sociology, or philosophy, or psychology should bear this load.

Right here, however, one comes upon one of the hard realities of the world. Vision is sharply divided between those who are familiar with mathematical analysis and those who are not. And the former group is rather small, consisting largely of the mathematicians, scientists and engineers, and in the last few years, the economists. Of all of these, only the engineers are very numerous—about 500,000 in number.

Importance of Analysis

Why is this familiarity with analysis important? The world (or its component parts) is difficult enough to study when the conceptual model which represents it is simplified, linearized, and reduced to simple economic choice.

The complaints of men like Hutchins and Boulding have been that the model has been too simple; that analysis of this simplified model produces results which ignore the human dimension. But if the human dimension is to be included, the model becomes infinitely more complex, and its analysis will require mathematical abilities of an even higher level. One could, of course, discard modeling as a technique, and resort to intuition or inspiration, but it is hard to believe in the unaided power of the human mind when faced with the complexities of urbanism.

Education of Engineers

It is probably no exaggeration to say that only today's education of engineers and economists produces men who are capable of conceptualizing these large systems, and manipulating the concepts, so as to predict the results of decisions.

Others, of course, worry whether the rigorous education of the engineer leaves him: (a) unable to imagine the need for change, or (b) psychologically unable to initiate the change, and, as I have indicated before, the engineer is indeed a conservative person. But it will, perhaps, be easier to liberate the engineer from his devotion to economic choice, than it will be to educate others in the intricacies of systems analysis.

How to Put Public Interest First

What, then, must be done if we are to truly place the Public Interest first in engineering design?

First: Oppose the criticisms of the intuitionists, and restate our belief in the power of the engineering method.

Second: Continue research and development of means of conceptualizing and analyzing complex models, particularly of urban systems.

Third: Work with the exponents of the "soft sciences," to learn to develop means of quantitatively treating human aspirations within these complex systems.

Fourth: Loosen our professional ties with an entrepreneural economy, and redevelop our profession's historic service to the government—but to *popular* government rather than to *royal* government.

Fifth: Resist the temptation to power proffered by the fascist ideology, and steadfastly serve the commonwealth.

Sixth: Reform our educational process so that we place additional emphasis on the public interest and the human dimension.

Seventh: Seek managerial independence for the employed engineer in the rendering of his professional decisions.

Eighth: Seek appointment of qualified engineers to Federal and State policy positions.

Such a program would require the efforts of a generation, and some parts of it could be achieved only by organized political effort on the part of the engineering profession. But if we are to make the title of this paper "Public Interest—First Priority in Engineering Design" a statement of professional policy, rather than a question, we must begin.

QUINCY WRIGHT

Technology And Warfare

This selection by the late political scientist Quincy Wright describes the interaction between technology and warfare from the period when firearms were introduced until the early years of World War II, prior to the extensive bombing raids against Germany and Japan and the use of the atomic bomb. The invention of new defensive weapons, Wright states, usually followed closely on the heels of the appearance of new offensive weapons. Most important is the fact that the technology of weaponry has had, in addition to military implications, significant social and political effects, often combining with cultural or economic factors to lead toward aggression.

Throughout the long history of war there has been a cumulative development of military technique. Invention of defensive instruments has usually followed close on the heels of the invention of offensive weapons. This balance of technology has tended to support a balance of power, but the balance has not tended toward increasing stability. Consequently, the political effect of military invention has not been continuous. There have been times when inventions have given the offensive an advantage, and conquerors have been able to overcome the defenses of their neighbors and build huge empires. At other times the course of invention and the art of war have favored the defensive. Local areas have been able to resist oppression, to revolt, and to defend themselves from conquest. Empires have crumbled, local liberties

Quincy Wright, "Technique of Modern War," Chapter XII in *A Study of War.* University of Chicago Press, 1965. Copyright 1965, University of Chicago Press. Reprinted by permission.

have been augmented, and international anarchy has sometimes resulted.

During the last five centuries military invention has proceeded more rapidly than ever before. Important differentials in the making and utilization of such inventions have developed. In general, the inventions have favored the offensive, and there has been a tendency for the size of political units to expand. This tendency was, however, arrested during much of the nineteenth century by inventions favoring the defensive, and many self-determination movements were successful.

Development of Modern Military Technique

"Until within the last few years," wrote Rear Admiral Bradley A. Fiske in 1920, "the most important single change in the circumstances and methods of warfare in recorded history was made by the invention of the gun; but now we see that even greater changes will certainly be caused by the invention of the airplane." Modern civilization began in the fifteenth century with the utilization of the first of these inventions and has witnessed the steady improvement of this utilization through development of accuracy and speed of fire of the gun itself; penetrability and explosiveness of the projectile; steadiness, speed, and security of the vehicle which conveys it over land or sea toward the enemy; and adaptation of military organizations to such utilization.

While the airplane continued this development by providing an even swifter vehicle for carrying the gun, it also introduced the third dimension into warfare. This made possible the use of gravitation to propel explosives, more extensive and accurate scouting, and military action behind the front, over vast areas, and across all barriers of terrain. Both of these inventions, after their use was thoroughly understood, greatly augmented the power of the offensive, though, in the case of the gun, the defense immediately began to catch up, and the general trend of war between equally equipped belligerents was toward a deadlock. A similar tendency is already observable in the case of the airplane.

These two inventions are but the most striking of the numerous applications to war of the technical advances characteristic of modern civilization. In historic civilizations men and animals provided the power for military movement and propulsion. In the modern period wind and sail, coal and the steam engine, petroleum and the internal combustion engine, have successively revolutionized naval, military, and aerial movement, as gunpowder, smokeless powders, and high explosives have successively revolutionized military propulsion. The history of modern military technique falls into four periods, each initiated by certain physical or social inventions and leading to certain military and political consequences: the periods of experimental adaptation of firearms and religious war (1450–1648), of professional armies and dynastic wars (1648–1789), of industrialization and nationalistic wars (1789–1914), and of the airplane and totalitarian war (1914——).

a. Adaptation of firearms (1450–1648). During the period of discoveries and wars of religion medieval armor was being abandoned; pikemen, halberdiers, and heavy cavalry were going. The organization of the Turkish Janizary infantry, well disciplined, equipped with cutlass and longbow, and supported by light cavalry and artillery, was being copied throughout Europe. Heavy artillery had begun to reduce feudal castles in the early fifteenth century and the *Wagenburg* revolutionized field tactics. Hand firearms first used by Spaniards, Hussites, and Swiss in the fifteenth century were adopted by all in the general wars of the early sixteenth century. The experience of the Thirty Years' War ended this period of experimental adaptation of firearms by the mercenary armies, and modern armies began to emerge.

Naval architecture was greatly improved during this period. The clumsy galleons of the Spanish Armada, differing little from those of Columbus a century earlier, and resembling the oar-driven galleys of the Middle Ages, were superseded in the mid-seventeenth century by longer, swifter, and more heavily armed "broadside battleships" which differed little from those of Nelson, nearly two centuries later.

Equipped with the new technique of firearms, Europeans had occupied strategic points in America, Africa, and Asia, readily overcoming the natives whom they found there. The tendency of this new technique was toward political integration inside and expansion outside of Europe. By increasing the relative power of the offensive, firearms made it possible for the more aggressive rulers, especially those of Turkey, Portugal, Spain, France, Britain, Prussia, the Netherlands, and Sweden, to expand their domains in Europe at the expense of feudal princes and to expand overseas at the expense of native chieftains.

b. Professionalization of armies (1648–1789). The seventeenth and eighteenth centuries witnessed the development of the professional army loyal to the king and ready to suppress internal rebellion or to fight foreign wars if paid promptly and if the officers were adequately rewarded by honors and perquisites of victory. Louis XIV and Cromwell contributed greatly to the development of this type of army, which, however, in the eighteenth century tended to be more concerned with safety and booty than with victory. Consequently, military invention emphasized defense and fortification. The art of war prescribed elaborate rules of strategy and siegecraft. Rules also dealt with the treatment of prisoners, with capitulations, with military honors, and with the rights of civilians. The Prussian army with its vigorous discipline, aggressiveness, and new strategic ideas under Frederick the Great to some extent broke through this defensive technique and brought this type of army to the highest point.

The destructiveness of war was limited by the general exemption from the activities of land war of the bourgeois and peasants, who constituted the bulk of the population. The bourgeois were anti-military in attitude and of little influence in the politics of most states. The monarch preferred to leave his own bourgeois and peasants to production, provided they paid taxes, and to recruit his armed forces from the unproductive riffraff, officered by the nobility, whose loyalty could be relied upon. With the existing techniques the army could not easily attack the enemy's middle classes, unless his army was first destroyed and his fortifications taken. In that case such attack was unnecessary because these classes would usually accept whatever peace might be imposed. Lacking in patriotism or nationalism, they were little concerned if the territory on which they lived had a new sovereign, provided they could retain their property. . . .

c. Capitalization of war (1789–1914). The French revolutionary and Napoleonic period developed the idea of the "nation in arms" through revolutionary enthusiasm and the conscription of mass armies. The idea of totalitarian war was developed in the writings of Clausewitz, rationalizing Napoleonic methods. After these wars the issue between professional long-service aristocratically officered armies and conscript short-service democratic armies was debated on the Continent of Europe with a general relapse to the former type during the long peace of Metternich's era. The rise of nationalism, democracy, and industrialism and the mechanization of war in the mid-century re-established the trend toward the nation in arms and totalitarian war.

The use of steam power for land and water military transportation developed in the first half of the nineteenth century and was given its first serious test in the American Civil War. Moltke appreciated the military value of these inventions, and his genius in using railroads for rapid mass mobilization won Bismarck three wars with extraordinary rapidity against Denmark, Austria, and France. The ironclad and heavy naval ordnance were also tested in the American Civil War. The era of military mechanization and of firearms of superior range and accuracy progressed rapidly, adding greatly to military and naval budgets and to the importance of national wealth and industry in war. The new methods were given a further test in the Spanish-American, Boer, and Russo-Japanese wars.

The great nineteenth-century naval inventions—steam power, the screw propeller, the armored vessel, the iron-hulled vessel, heavy ordnance—were at first favorable to British maritime dominance because British superiority was more marked in iron and coal resources and a developed heavy industry than in forests and wooden shipbuilders. But this advantage did not continue. The new battleships were more vulnerable than the wooden ships because ordnance gained in the race with armor, and repair at sea was impossible. Furthermore, the mine, torpedo, submarine, and airplane added new hazards to the surface fleet, especially in the vicinity of the enemy's home bases. Warships, therefore, became more dependent upon well-equipped and secure bases for fueling and repair, and approach to even a greatly inferior enemy became hazardous. With the industrialization of other powers and their development of naval strength, Britain found it increasingly difficult to maintain a three- or even a two-power superiority in the ships themselves, while its distant bases became less secure. . . .

This situation was realized by the Continental powers. They developed their armies and navies with increasing speed after observation of the Russo-Japanese War and after the failure of the Hague conferences to achieve disarmament. They paid particular attention to the potentialities of the improved rifle, machine gun, and artillery as well as to the art of intrenchment. The possibilities of the mine, torpedo, and submarine were developed, especially by France, pointing the way for German utilization of these weapons in World War I. Beginnings were made, especially by France and Germany, in the adaptation of the airplane and dirigible to military purposes. The results anticipated by the Polish banker, Ivan Bloch, in his book published in 1898 occurred. War became deadlocked in the machine-gun-lined trenches and in the mined and submarine-infested seas of World War I. This deadlock was not broken until attrition had ruined all the initial belligerents, and new recruits and resources on the Allied side from the United States made the cause of the Central Powers hopeless.

d. Totalitarianization of war (1914———). The advent of aerial war in the twentieth century ended the relative invulnerability of the British Isles to invasion. The weakening of surface control of the sea by the use of mines, submarines, and airplanes further impaired the position of Great Britain, and that country during the 1920s accepted the thesis that the integrity of the empire depended upon collective security. The possibilities of the airplane and tank, neither of them fully exploited during World War I, supported hope in some quarters and fear in others that the power of the offensive would be increased, the mobility in war would again be possible, and that the deadlock would be broken.

These possibilities encouraged aggression by Japan, Italy, and Germany after 1930. Dissatisfaction with the political results of World War I, resentment at the self-centered economic policies of the democracies, serious deterioration of the middle classes, and the spread of revolutionary ideologies engendered by the costs of war and widespread unemployment flowing from the great depression of 1929 provided motives for aggression; but, if collective security had been better organized and the airplane and tank had not been invented, the prospects would hardly have been sufficiently encouraging to induce action. As it was, the initial success of Japan in Manchuria and the failure of the disarmament conference alarmed the Soviets into rapid rearmament and encouraged Italy and Germany to do likewise, especially in the air. Initial failure of the democracies to support the treaty structure when Germany began to rearm and re-occupied the Rhineland in violation of international obligations encouraged Germany, Italy and Japan to consort together and to continue aggression in weak areas, utilizing aviation with rapid success, while all phases of the national life were organized for total war.

This development of militarism, totalitarianism, and aggressiveness was most completely exhibited in Nazi Germany. Here the reaction from military defeat had been intense, confidence in technical ability to develop mechanized war was great, the economic situation was particularly grave, and the extreme democracy of the Weimar Constitution made government weak. In the background, however, the foundations had been laid by the historic rise of Prussia through war; the methods of the Great Elector, Frederick the Great, and Bismarck; the philosophies of Fichte, Hegel, and Nietzsche; the historical interpretations of Mommsen and Treitschke, and the geopolitics of Ratzel and Haushofer. Similar developments had begun in England under Cromwell and in France under Louis XIV and Napoleon at times when new military techniques, the disciplined use of firearms, superior co-ordination of army and industry, and mass mobilization through nationalistic propaganda appeared to give these governments a strategic initiative. The strength of parliamentarianism and reliance upon the navy stopped the trend in England. In France this trend was stopped by military defeats, the democratic sentiment of the revolution, and the declining population of the nineteenth century. In the United States decentralized institutions, geographic isolation, and the democratic tradition formed a barrier against militarism. In Japan, Italy, and the Soviet Union, however, the circumstances of political ambition, economic frustration, imputations of racial, social, or political inferiority, post-war disorganization, and revolutionary ideas when stimulated by hopes born of new military inventions tended, in varying degrees, toward military totalitarianism similar to that of Germany.

As the development of the gun by the European great powers in the sixteenth and seventeenth centuries extended their imperial control to the overseas countries, followed by the latter's imitation of their techniques and eventual revolt, so the development of the airplane by the totalitarian states in the twentieth century first extended their empires and then compelled the democracies to adopt their techniques. Thus the great powers, whether with a democratic or an autocratic tradition, whether relying on the army or navy, whether European or American or Asiatic, have in a disorganized world felt obliged to follow the lead of that one of their number most advanced in the art of war. . . .

It seems probable, however, that small nationalities can no longer defend themselves from powerful neighbors equipped with a vast superiority of planes. As the invention of artillery made it possible for monarchs to batter down feudal castles and to build nations, so the airplane may destroy the independent sovereignty of nations and create larger regional units. These may be empires resting on conquest or they may be confederations resting on consent.

Characteristics of Modern Military Technique

a. Mechanization. The outstanding characteristic in which modern war has differed from all earlier forms of war has been in the degree of mechanization. The use of long-range striking power (rifles, machine guns, artillery, gases), of power-propelled means of mobility (railroads, motor trucks, battleships, tanks, airships), and of heavy protective covering (armor plate on fortresses, tanks, and warships) has meant that the problem of war manufacture has risen to primary importance. In historic civilizations the soldier provided his own equipment, and it generally lasted as long as the soldier. Some equipment was lost, but even arrows could usually be collected in large numbers from the battlefield. Now a dozen men must be engaged in production and transportation services behind the lines to keep one soldier supplied.

b. Increased size of armies. A second important change has been in the size of the armed forces, both absolutely and in proportion to the population. It might seem that if each soldier needs such a large amount of civilian help there would be fewer soldiers, but this has not proved to be the case. Power transport and electrical communication have made it possible to mobilize and control from the center a much larger proportion of the population than formerly. The men can themselves be transported rapidly by railroad and motor lorry, and canned food can be brought to them. Where formerly 1 per cent of the population was a large number to mobilize, now over 10 per cent can be mobilized, of which a quarter may be at the front at one time. But 10 per cent mobilized reuires most of the remaining adult population to provide them with the essentials for continuing operation. Thus instead of 1 per cent engaging in war and the rest pursuing their peacetime occupations of trade or agriculture, now the entire working population must devote itself to direct or indirect war service.

c. Militarization of population. A third change, consequent upon the second, has been the military organization of the entire nation. The armed forces have ceased to be a self-contained service apart from the general population. The soldiers and sailors must be recruited from those men whose services can be most readily supplied by women, children, and the aged. The experts in transportation and industrial services must be largely exempted in order that they may continue their "civilian" services which, under modern conditions, are no less essential to war. Such a gearing-in of the agricultural, industrial, and professional population to the armed forces requires a military organization of the entire population. Since the perfection of such an organization after the outbreak of war has been impossible, the conditions of war have more and more merged into those of peace. The military organization of the entire population in peace has become necessary as a preparation for war.

Such a militarization of the population must be distinguished from the militia system, illustrated in Switzerland. In this system the duty of military service, though considered a universal burden of citizenship, has involved only a limited training which has not withdrawn persons from normal civilian occupations for long periods. . . .

Both of these defense systems may be distinguished from the professional army system characteristic of the United States and Great Britain and of most European states in the eighteenth century. In this system the army has been voluntarily recruited for long service and has existed with its own organization, discipline, law, and professional standards quite apart from the civilian population. Because of the emphasis upon professional qualifications, its size has not been greatly augmented in time of war, although as emergencies develop voluntary recruitment has often given way to conscription and the press gang.

While all three systems have played their part in the history of most modern states, the development of modern military technique has tended toward the military state.

d. Nationalization of war effort. A fourth change, characteristic of modern military technique, has been the extension of government into the control of economy and public opinion. The military state has tended to become the totalitarian state. Other forces of modern life have, it is true, had a similar tendency. Democracy, under the influence of nationalism, has induced the individual to identify all phases of his life with that of the state, while state socialism, under the influence of depression, has induced the state to intervene in all phases of the life of the individual, but the needs of modern war have led and accelerated the process. Modern war has required propaganda to sustain morale among the civilian population which, contributing directly to the war effort, can no longer expect to be exempt from attack. Modern war has also required an adjustment of the nation's economy to its needs. A free-market system, depending on profits, has proved less adequate than military discipline for reducing private consumption and directing resources and productive energy to war requirements. Since transition from a free economy to a controlled economy would be difficult in the presence of war, preparation for war tends toward such a change in time of peace. Furthermore, autarchy is necessary as a defense against blockade. The controls necessary to confine the nation's economic life to those regions whose resources and markets will be available in time of war must be applied before the war. The modern technique of war has, therefore, led to the autarchic totalitarian state and the elimination both of free economy and of free speech.

e. Total war. A fifth change, characteristic of modern war technique, has been the breakdown of the distinction between the armed forces and the civilians in military operations. The moral identification of the individual with the state has given the national will priority over humanitarian considerations. The civilian's morale and industry support the national will. Thus the population, manufacturing, and transport centers have become military targets. Bombing aircraft and starvation blockades have made it possible to reach these targets over the heads of the army and fortifications; consequently, the principle of military necessity has tended to be interpreted in a way to override the traditional rules of war for the protection of civilian life and property. . . .

Starvation, bombardment, confiscation of property, and terrorization have in World War II been considered applicable against the entire enemy population and territory, except in so far as practical dangers of reprisal and a desire to utilize the population of occupied areas have inhibited. The entire life of the enemy state comes to be an object of attack. The modern doctrine of conquest even extends to the elimination of that population and its property rights in order to open the space it occupied for settlement.

f. Intensification of operations. A sixth characteristic of modern war technique has been a great increase in the intensity of military operations in time and of their extension in space. . . .

The inventions in mechanization and mobility, the organization of the entire population, the increase in the number of important targets for attack, has made it possible to concentrate enormously greater forces at a given point, to supply reserves and to continue attack and resistance at that point for a much longer period, to increase the number of points being attacked simultaneously, to enlarge the theater of the campaign by mutual efforts at outflanking, and to co-ordinate operations on all fronts at all seasons for the entire course of the war. . . .

These six characteristics of modern military technique—increased mechanization and size of armed forces, more general militarization and national-

ization of the people, more comprehensive, intense, and extended operations—collectively tend toward totalitarian military organization of the belligerents and totalitarian military operations during the war. . . .

These changes have been most marked in the characteristics of weapons,
less marked in that of organization and operations, and of little significance
in the fields of policy and strategy. The art of using superior preparedness, a
reputation for ruthlessness, and threats of war for bloodless victory is as old as
history and was expounded by Machiavelli. The vulnerability of civilians to
bombing aircraft may have increased the effectiveness of these methods
against nations which have greater potential power than the threatener.

GIULIO DOUHET

The Command
Of the Air

The airplane was the single most powerful stimulus to the involvement of the entire civilian population in a war. In 1921, the chief of Italy's air arm, Giulio Douhet (1860–1930) wrote a prophetic book entitled (in translation) The Command of the Air. *In it, Douhet expounded his theory of the role of strategic bombing in disorganizing and destroying the capability of an enemy to wage war. Although his ideas initially met violent opposition, Douhet's strategy was eventually accepted by the major powers and was employed by Germany, Great Britain, and the United States in World War II. Douhet believed erroneously that chemical weapons would be extensively used in future wars, and he did not foresee the development of the atomic bomb.*

Aeronautics opened up to men a new field of action, the field of the air. In so doing it of necessity created a new battlefield; for wherever two men meet, conflict is inevitable. In actual fact, aeronautics was widely employed in warfare long before any civilian use was made of it. Still in its infancy at the outbreak of the [first] World War, this new science received then a powerful impetus to military development.

The practical use of the air arm was at first only vaguely understood. This new arm had sprung suddenly into the field of war; and its characteristics, radically different from those of any other arm employed up to that time, were still undefined. Very few possibilities of this new instrument of war were recognized when it first appeared. Many people took the extreme position that it was impossible to fight in the air; others admitted only that it might prove a useful auxiliary to already existing means of war.

At first the speed and freedom of action of the airplane—the air arm chiefly used in the beginning—caused it to be considered primarily an instrument of exploration and reconnaissance. Then gradually the idea of using it as a range-finder for the artillery grew up. Next, its obvious advantages over surface means led to its being used to attack the enemy on and behind his own lines, but no great importance was attached to this function because it was thought that the airplane was incapable of transporting any heavy load of offensive matériel. Then, as the need of counteracting enemy aerial operations was felt, antiaircraft guns and the so-called pursuit planes came into being.

Thus, in order to meet the demands of aerial warfare, it became necessary step by step to increase aerial power. But because the needs which had to be met manifested themselves during a war of large scope, the resulting increase was rapid and hectic, not sound and orderly. And so the illogical concept of utilizing the new aerial weapon solely as an auxiliary to the army and navy prevailed for almost the entire period of the World War. It was only toward the end of the war that the idea emerged, in some of the belligerent nations, that it might be not only feasible but wise to entrust the air force with inde-pendent offensive missions. None of the belligerents fully worked out this idea, however—perhaps because the war ended before the right means for actuating the idea became available....

The form of any war—and it is the form which is of primary interest to men of war—depends upon the technical means of war available. It is well known, for instance, that the introduction of firearms was a powerful influence in changing the forms of war in the past. Yet firearms were only a gradual de-velopment, an improvement upon ancient engines of war—such as the bow and arrow, the ballista, the catapult, et cetera—utilizing the elasticity of solid materials. In our own lifetime we have seen how great an influence the intro-duction of small-caliber, rapid-fire guns—together with barbed wire—has had on land warfare, and how the submarine changed the nature of sea war-fare. We have also assisted in the introduction of two new weapons, the air arm and poison gas. But they are still in their infancy, and are entirely differ-ent from all others in character; and we cannot yet estimate exactly their potential influence on the form of future wars. No doubt that influence will be great, and I have no hesitation in asserting that it will completely upset all forms of war so far known.

These two weapons complement each other. Chemistry, which has already provided us with the most powerful of explosives, will now furnish us with poison gases even more potent, and bacteriology may give us even more formidable ones. To get an idea of the nature of future wars, one need only imagine what power of destruction that nation would possess whose bacteriologists should discover the means of spreading epidemics in the enemy's country and at the same time immunize its own people. Air power makes it possible not only to make high-explosive bombing raids over any sector of the enemy's territory, but also to ravage his whole country by chemical and bacteriological warfare. . . .

Now *it is possible* to go far behind the fortified lines of defense without first breaking through them. It is air power which makes this possible.

The airplane has complete freedom of action and direction; it can fly to and from any point of the compass in the shortest time—in a straight line—by any route deemed expedient. Nothing man can do on the surface of the earth can interfere with a plane in flight, moving freely in the third dimension. All the influences which have conditioned and characterized warfare from the beginning are powerless to affect aerial action.

By virtue of this new weapon, the repercussions of war are no longer limited by the farthest artillery range of surface guns, but can be directly felt for hundreds and hundreds of miles over all the lands and seas of nations at war. No longer can areas exist in which life can be lived in safety and tranquillity, nor can the battlefield any longer be limited to actual combatants. On the contrary, the battlefield will be limited only by the boundaries of the nations at war, and all of their citizens will become combatants, since all of them will be exposed to the aerial offensives of the enemy. There will be no distinction any longer between soldiers and civilians. The defenses on land and sea will no longer serve to protect the country behind them; nor can victory on land or sea protect the people from enemy aerial attacks unless that victory insures the destruction, by actual occupation of the enemy's territory, of all that gives life to his aerial forces.

All of this must inevitably effect a profound change in the form of future wars, because the essential characteristics of those wars will be radically different from those of any previous ones. We may thus be able to understand intuitively how the continuing development of air power, whether in its technical or in its practical aspects, will conversely make for a relative decrease in the effectiveness of surface weapons, in the extent to which these weapons can defend one's country from the enemy.

The brutal but inescapable conclusion we must draw is this: in face of the technical development of aviation today, in case of war the strongest army we can deploy in the Alps and the strongest navy we can dispose on our seas will prove no effective defense against determined efforts of the enemy to bomb our cities. . . .

The Magnitude of Aerial Offensives

Some conception of the magnitude aerial offensives may reach in the future is essential to an evaluation of the command of the air, a conception which the World War can clarify for us in part. . . .

Aerial bombardment can certainly never hope to attain the accuracy of artillery fire; but this is an unimportant point because such accuracy is unnecessary. Except in unusual cases, the targets of artillery fire are designed to withstand just such fire; but the targets of aerial bombardment are ill-prepared to endure such onslaught. Bombing objectives should always be large; small targets are unimportant and do not merit our attention here.

The guiding principle of bombing actions should be this: *the objective must be destroyed completely in one attack, making further attack on the same target unnecessary.* Reaching an objective is an aerial operation which always involves a certain amount of risk and should be undertaken once only. The complete destruction of the objective has moral and material effects, the repercussions of which may be tremendous. To give us some idea of the extent of these repercussions, we need only envision what would go on among the civilian population of congested cities once the enemy announced that he would bomb such centers relentlessly, making no distinction between military and non-military objectives.

In general, aerial offensives will be directed against such targets as peacetime industrial and commercial establishments; important buildings, private and public; transportation arteries and centers; and certain designated areas of civilian population as well. To destroy these targets three kinds of bombs are needed—explosive, incendiary, and poison gas—apportioned as the situation may require. The explosives will demolish the target, the incendiaries set fire to it, and the poison-gas bombs prevent fire fighters from extinguishing the fires. . . .

Such bombing expeditions, however, cannot be undertaken successfully unless they are directed against very large centers of civilian population. In fact, we have no difficulty in imagining what would happen when areas of 500 to 2,000 meters in diameter in the centers of large cities such as London, Paris, or Rome were being unmercifully bombed. With 1,000 bombers of the type described—an actual type in use today, not a hypothetical type in some blueprint of the future—with their necessary maintenance and replacements for daily losses, 100 such operating squadrons can be constituted. Operating 50 of these daily, such an aerial force in the hands of those who know how to use it could destroy 50 such centers every day. This is an offensive power so far superior to any other offensive means known that the power of the latter is negligible in comparison.

As a matter of fact, this same offensive power, the possibility of which was not even dreamed of fifteen years ago, is increasing daily, precisely because the building and development of large, heavy planes goes on all the time. The same thing is true of new explosives, incendiaries, and especially poison gases. What could an army do faced with an offensive power like that, its lines of communication cut, its supply depots burned or blown up, its arsenals and auxiliaries destroyed? What could a navy do when it could no longer take refuge in its own ports, when its bases were burned or blown up, its arsenals and auxiliaries destroyed? How could a country go on living and working under this constant threat, oppressed by the nightmare of imminent destruction and death? How indeed! We should always keep in mind that aerial offensives can be directed not only against objectives of least physical resistance, but against those of least moral resistance as well. For instance, an infantry regiment in a shattered trench may still be capable of some resistance even after losing two-thirds of its effectives; but when the working personnel of a factory sees one of its machine shops destroyed, even with a minimum loss of life, it quickly breaks up and the plant ceases to function.

All this should be kept in mind when we wish to estimate the potential power of aerial offensives possible even today. To have command of the air means to be in a position to wield offensive power so great it defies human imagination. It means to be able to cut an enemy's army and navy off from their bases of operation and nullify their chances of winning the war. It means complete protection of one's own country, the efficient operation of one's army and navy, and peace of mind to live and work in safety. In short, it means to be in a position *to win*. *To be defeated* in the air, on the other hand, is finally to be defeated and to be at the mercy of the enemy, with no chance at all of defending oneself, compelled to accept whatever terms he sees fit to dictate.

This is the meaning of the "command of the air."

ALBERT EINSTEIN

Letter to President Roosevelt

European refugee scientists played a major role in getting the U.S. nuclear weapons program started. In this message, Albert Einstein brings nuclear energy to the attention of President Franklin Roosevelt and urges that his administration start a nuclear development program. Two scientists working on this problem, Leo Szilard and Eugene Wigner, actually wrote the text of this letter in 1939, a month before Germany invaded Poland, and suggested that Einstein send it to the President. The letter stimulated the eventual establishment of the Manhattan Project and the development of the atomic bomb. In this project and through the Office of Scientific Research and Development, many scientists and engineers were mobilized for the war effort.

August 2nd, 1939

F. D. Roosevelt
President of the United States
White House
Washington, D.C.

Sir:

Some recent work by E. Fermi and L. Szilard, which has been communicated to me in manuscript, leads me to expect that the element uranium may be

Reprinted from *Collected Works of Leo Szilard: Scientific Papers, 1972,* Spencer Weart and Gertrude Szilard, eds., by permission of the MIT Press, Cambridge, Massachusetts.

turned into a new and important source of energy in the immediate future. Certain aspects of the situation which has arisen seem to call for watchfulness and, if necessary, quick action on the part of the Administration. I believe therefore that it is my duty to bring to your attention the following facts and recommendations:

In the course of the last four months it has been made probable—through the work of Joliot in France as well as Fermi and Szilard in America—that it may become possible to set up a nuclear chain reaction in a large mass of uranium by which vast amounts of power and large quantities of new radium-like elements would be generated. Now it appears almost certain that this could be achieved in the immediate future.

This new phenomenon would also lead to the construction of bombs, and it is conceivable—though much less certain—that extremely powerful bombs of a new type may thus be constructed. A single bomb of this type, carried by

UPI PHOTO

boat and exploded in a port, might very well destroy the whole port together with some of the surrounding territory. However, such bombs might very well prove to be too heavy for transportation by air.

The United States has only very poor ores of uranium in moderate quantities. There is some good ore in Canada and the former Czechoslovakia, while the most important source of uranium is Belgian Congo.

In view of this situation you may think it desirable to have some permanent contact maintained between the Administration and the group of physicists working on chain reactions in America. One possible way of achieving this might be for you to entrust with this task a person who has your confidence and who could perhaps serve in an inofficial capacity. His task might comprise the following:

a) to approach Government Departments, keep them informed of the further development, and put forward recommendations for Government action, giving particular attention to the problem of securing a supply of uranium ore for the United States.

b) to speed up the experimental work, which is at present being carried on within the limits of the budgets of University laboratories, by providing funds, if such funds be required, through his contacts with private persons who are willing to make contributions for this cause, and perhaps also by obtaining the cooperation of industrial laboratories which have the necessary equipment.

I understand that Germany has actually stopped the sale of uranium from the Czechoslovakian mines which she has taken over. That she should have taken such early action might perhaps be understood on the ground that the son of the German Under-Secretary of State, von Weizsäcker, is attached to the Kaiser-Wilhelm-Institut in Berlin where some of the American work on uranium is now being repeated.

Yours very truly,
[Signed] Albert Einstein

Old Grove Road
Nassau Point
Peconic, Long Island

GENERAL ADVISORY COMMITTEE TO THE ATOMIC ENERGY COMMISSION

Report on the "Super"

In 1949 the Soviet Union exploded its first atomic bomb, giving it the same military capability as the United States. Soon afterwards scientists and many other concerned citizens flooded the government with suggestions about how the United States should react. Most important of these was that of the physicist Edward Teller (a refugee from Nazi Germany), who strongly recommended that the U.S. proceed with the development of the hydrogen bomb, or "super," the possibility of which had been recognized for some time. The General Advisory Committee (GAC) to the Atomic Energy Commission (AEC), made up of scientists who had been involved in the development of the atomic bomb, considered these suggestions. Their report to David Lilienthal, chairman of the AEC, presents the committee's judgments and recommendations to forgo development, not because of any technical incapability but instead on moral grounds. Despite the recommendations of the scientific and engineering advisors, national security concerns prompted President Truman to decide to go forward with the construction of H-bombs.

GENERAL ADVISORY COMMITTEE
to the
U.S. ATOMIC ENERGY COMMISSION
Washington 25, D.C.

October 30, 1949

Dear Mr. Lilienthal:

At the request of the Commission, the seventeenth meeting of the General Advisory Committee was held in Washington on October 29 and 30, 1949 to consider some aspects of the question of whether the Commission was making all appropriate progress in assuring the common defense and security. Dr. Seaborg's absence in Europe prevented his attending this meeting. For purposes of background, the Committee met with the Counsellor of the State Department, with Dr. Henderson of AEC Intelligence, with the Chairman of the Joint Chiefs of Staff, the Chairman of the Military Liaison Committee, the Chairman of the Weapons Systems Evaluation Group, General Norstadt and Admiral Parsons. In addition, as you know, we have had intimate consultations with the Commission itself.

The report which follows falls into three parts. The first describes certain recommendations for action by the Commission directed toward the common defense and security. The second is an account of the nature of the super project and of the super as a weapon, together with certain comments on which the Committee is unanimously agreed. Attached to the report, but not a part of it, are recommendations with regard to action on the super project which reflect the opinions of Committee members.

The Committee plans to hold its eighteenth meeting in the city of Washington on December 1, 2 and 3, 1949. At that time we hope to return to many of the questions which we could not deal with at this meeting.

J. R. Oppenheimer
Chairman

UNITED STATES ATOMIC ENERGY COMMISSION
WASHINGTON, D.C. 20545
HISTORICAL DOCUMENT NUMBER 349

David E. Lilienthal
Chairman
U.S. Atomic Energy Commission
Washington 25, D.C.

PART I

(1) PRODUCTION. With regard to the present scale of production of fissionable material, the General Advisory Committee has a recommendation to make to the Commission. We are not satisfied that the present scale represents either the maximum or the optimum scale. We recognize the statutory and appropriate role of the National Military Establishment in helping to determine that. We believe, however, that before this issue can be settled, it will be desirable to have from the Commission a careful analysis of what the capacities are which are not now being employed. Thus we have in mind that an acceleration of the program on beneficiation of low grade ores could well turn out to be possible. We have in mind that further plants, both separation and reactor, might be built, more rapidly to convert raw material into fissionable material. It would seem that some notion of the costs, yields and time scales for such undertakings would have to precede any realistic evaluation of what we should do. We recommend that the Commission undertake such studies at high priority. We further recommend that projects should not be dismissed because they are expensive but that their expense be estimated.

(2) TACTICAL DELIVERY. The General Advisory Committee recommends to the Commission an intensification of efforts to make atomic weapons available for tactical purposes, and to give attention to the problem of integration of bomb and carrier design in this field.

(3) NEUTRON PRODUCTION. The General Advisory Committee recommends to the Commission the prompt initiation of a project for the production of freely absorbable neutrons. With regard to the scale of this project the figure * * * per day may give a reasonable notion. Unless

obstacles appear, we suggest that the expediting of design be assigned to the Argonne National Laboratory.

With regard to the purposes for which these neutrons may be required, we need to make more explicit statements. The principal purposes are the following:

(a) The production of U-233.
(b) The production of radiological warfare agents.
(c) Supplemental facilities for the test of reactor components.
(d) The conversion of U-235 to plutonium.
(e) A secondary facility for plutonium production.
(f) The production of tritium (1) for boosters, (2) for super bombs.

We view these varied objectives in a quite different light. We have a great interest in the U-233 program both for military and for civil purposes. We strongly favor, subject to favorable outcome of the 1951 Eniwetok tests, the booster program. With regard to radiological warfare, we would not wish to alter the position previously taken by our Committee. With regard to the conversion to plutonium, we would hardly believe that this alone could justify the construction of these reactors, though it may be important should unanticipated difficulties appear in the U-233 and booster programs. With regard to the use of tritium in the super bomb, it is our unanimous hope that this will not prove necessary. It is the opinion of the majority that the super program itself should not be undertaken and that the Commission and its contractors understand that construction of neutron producing reactors is not intended as a step in the super program.

PART II

SUPER BOMBS

The General Advisory Committee has considered at great length the question of whether to pursue with high priority the development of the super bomb. No member of the Committee was willing to endorse this proposal. The reasons for our views leading to this conclusion stem in large part from the technical nature of the super and of the work necessary to establish it as a weapon. We therefore here transmit *an elementary* account of these matters.

The basic principle of design of the super bomb is the ignition of the thermo-nuclear DD reaction by the use of a fission bomb, and of high temperatures, pressure, and neutron densities which accompany it. In over-whelming probability, tritium is required as an intermediary, more easily ignited than the deuterium itself and, in turn, capable of igniting the deuterium. The steps which need to be taken if the super bomb is to become a reality include:

(1) The provision of tritium in amounts perhaps of several ***** per unit.
(2) Further theoretical studies and criticisms aimed at reducing the very great uncertainties still inherent in the behavior of this weapon under extreme conditions of temperature, pressure and flow.
(3) The engineering of designs which may on theoretical grounds appear hopeful, particularly with regard to the ***** problems presented.
(4) Carefully instrumented test programs to determine whether the deuterium-tritium mixture will be ignited by the fission bomb, *****.

It is notable that there appears to be no experimental approach short of actual test which will substantially add to our conviction that a given model will or will not work, and it is also notable that because of the unsymmetric and extremely unfamiliar conditions obtaining, some considerable doubt will surely remain as to the soundness of theoretical anticipation. Thus we are faced with a development which cannot be carried to the point of conviction without the actual construction and demonstration of the essential elements of the weapon in question. This does not mean that further theoretical studies would be without avail. It does mean that they could not be decisive. A final point that needs to be stressed is that many tests may be required before a workable model has been evolved or before it has been established beyond reasonable doubt that no such model can be evolved. Although we are not able to give a specific probability rating for any given model, we believe that an imaginative and concerted attack on the problem has a better than even chance of producing the weapon within five years.

A second characteristic of the super bomb is that once the problem of initiation has been solved, there is no limit to the explosive power of the bomb itself except that imposed by requirements of delivery. This is because one can continue to add deuterium—an essentially cheap material—to make larger and larger explosions, the energy release and radioactive products of which are both proportional to the amount of deuterium itself. Taking into account the probable limitations of carriers likely to be available for the delivery of such a weapon, it has generally been estimated that the weapon would have an explosive effect some hundreds of times that of present fission bombs. This would correspond to a damage area of the order of hundreds of square miles, to thermal radiation effects extending over a comparable area, and to very grave contamination problems which can easily be made more acute, and may possibly be rendered less acute, by surrounding the deuterium with uranium or other material. It needs to be borne in mind that for delivery by ship, submarine or other such carrier, the limitations here outlined no longer apply and that the weapon is from a technical point of view without limitations with regard to the damage that it can inflict.

It is clear that the use of this weapon would bring about the destruction of innumerable human lives; it is not a weapon which can be used exclusively for the destruction of material installations of military or semi-military purposes. Its use therefore carries much further than the atomic bomb itself the policy of exterminating civilian populations. It is of course true that super bombs which are not as big as those here contemplated could be made, provided the initiating mechanism works. In this case, however, there appears to be no chance of their being an economical alternative to the fission weapons themselves. It is clearly impossible with the vagueness of design and the uncertainty as to performance as we have them at present to give anything like a cost estimate of the super. If one uses the strict criteria of damage area per dollar and if one accepts the limitations on air carrier capacity likely to obtain in the years immediately ahead, it appears uncertain to us whether the super will be cheaper or more expensive than the fission bomb.

PART III

Although the members of the Advisory Committee are not unanimous in their proposals as to what should be done with regard to the super bomb, there are certain elements of unanimity among us. We all hope that by one means or another, the development of these weapons can be avoided. We are all reluctant to see the United States take the initiative in precipitating this development. We are all agreed that it would be wrong at the present moment to commit ourselves to an all-out effort toward its development.

We are somewhat divided as to the nature of the commitment not to develop the weapon. The majority feel that this should be an unqualified commitment. Others feel that it should be made conditional on the response of the Soviet government to a proposal to renounce such development. The Committee recommends that enough be declassified about the super bomb so that a public statement of policy can be made at this time. Such a statement might in our opinion point to the use of deuterium as the principal source of energy. It need not discuss initiating mechanisms nor the role which we believe tritium will play. It should explain that the weapon cannot be explored without developing it and proof-firing it. In one form or another, the statement should express our desire not to make this development. It should

J. Robert Oppenheiner testifying before the Joint Congressional Atomic Energy Committee, 1949.

explain the scale and general nature of the destruction which its use would entail. It should make clear that there are no known or foreseen nonmilitary applications of this development. The separate views of the members of the Committee are attached to this report for your use.

J. R. Oppenheimer

October 30, 1949

We have been asked by the Commission whether or not they should immediately initiate an "all-out" effort to develop a weapon whose energy release is 100 to 1000 times greater and whose destructive power in terms of damage is 20 to 100 times greater than those of the present atomic bomb. We recommend strongly against such action.

We base our recommendation on our belief that the extreme dangers to mankind inherent in the proposal wholly outweigh any military advantage that could come from this development. Let it be clearly realized that this is a super weapon; it is in a totally different category from an atomic bomb. The reason for developing such super bombs would be to have the capacity to devastate a vast area with a single bomb. Its use would involve a decision to slaughter a vast number of civilians. We are alarmed as to the possible global effects of the radioactivity generated by the explosion of a few super bombs of conceivable magnitude. If super bombs will work at all, there is no inherent limit in the destructive power that may be attained with them. Therefore, a super bomb might become a weapon of genocide.

The existence of such a weapon in our armory would have far-reaching effects on world opinion: reasonable people the world over would realize that the existence of a weapon of this type whose power of destruction is essentially unlimited represents a threat to the future of the human race which is intolerable. Thus we believe that the psychological effect of the weapon in our hands would be adverse to our interest.

We believe a super bomb should never be produced. Mankind would be far better off not to have a demonstration of the feasibility of such a weapon until the present climate of world opinion changes.

It is by no means certain that the weapon can be developed at all and by no means certain that the Russians will produce one within a decade. To the argument that the Russians may succeed in developing this weapon, we would reply that our undertaking it will not prove a deterrent to them. Should they use the weapon against us, reprisals by our large stock of atomic bombs would be comparably effective to the use of a super.

In determining not to proceed to develop the super bomb, we see a unique opportunity of providing by example some limitations on the totality of war and thus of limiting the fear and arousing the hopes of mankind.

James B. Conant	*L. A. DuBridge*
Hartley Rowe	*Oliver E. Buckley*
Cyril Stanley Smith	*J. R. Oppenheimer*

October 30, 1949

AN OPINION ON THE DEVELOPMENT OF THE "SUPER"

A decision on the proposal that an all-out effort be undertaken for the development of the "Super" cannot in our opinion be separated from considerations of broad national policy. A weapon like the "Super" is only an advantage when its energy release is from 100–1000 times greater than that of ordinary atomic bombs. The area of destruction therefore would run from 150 to approximately 1000 square miles or more.

Necessarily such a weapon goes far beyond any military objective and enters the range of very great natural catastrophes. By its very nature it cannot be confined to a military objective but becomes a weapon which in practical effect is almost one of genocide.

It is clear that the use of such a weapon cannot be justified on any ethical ground which gives a human being a certain individuality and dignity even if he happens to be a resident of an enemy country. It is evident to us that this would be the view of peoples in other countries. Its use would put the United States in a bad moral position relative to the peoples of the world.

Any postwar situation resulting from such a weapon would leave unresolvable enmities for generations. A desirable peace cannot come from such an inhuman application of force. The postwar problems would dwarf the problems which confront us at present.

The application of this weapon with the consequent great release of radioactivity would have results unforeseeable at present, but would certainly render large areas unfit for habitation for long periods of time.

The fact that no limits exist to the destructiveness of this weapon makes its very existence and the knowledge of its construction a danger to humanity as a whole. It is necessarily an evil thing considered in any light.

For these reasons we believe it important for the President of the United States to tell the American public, and the world, that we think it wrong on fundamental ethical principles to initiate a program of development of such a weapon. At the same time it would be appropriate to invite the nations of the world to join us in a solemn pledge not to proceed in the development or construction of weapons of this category. If such a pledge were accepted even without control machinery, it appears highly probable that an advanced stage of development leading to a test by another power could be detected by available physical means. Furthermore, we have in our possession, in our stockpile of atomic bombs, the means for adequate "military" retaliation for the production or use of a "super."

E. Fermi
I. I. Rabi

ANDREI D. SAKHAROV

Nuclear Weapons Development

In 1954, an accident during a nuclear weapons test in the Pacific, in which one man was killed and many others injured, focused attention on the problem of radioactive fallout. It was not only in the West that this threat generated anxieties. In the Soviet Union as well, many individuals who were in a position to influence the course of their country's nuclear test program also became troubled because of the potential hazards of radioactive fallout. In this passage from the memoirs of Andrei Sakharov, a scientist who played a significant role in the development program of the hydrogen bomb in the Soviet Union, his concerns are evidenced.

In the summer of 1950, almost simultaneously with the beginning of work on the thermonuclear weapon, I. E. Tamm and I began work on the problem of a controlled thermonuclear reaction; i.e., on the utilization of the nuclear energy of light elements for purposes of industrial power. In 1950 we formulated the idea of the magnetic thermo-isolation of high-temperature plasma, and completed estimates on the parameters for thermonuclear synthesis installations. This research, which became known abroad through a paper read by I. V. Kurchatov at Harwell in 1956 and through the materials of the First Geneva Conference on the Peaceful Use of Atomic Energy, was recognized as pioneering. In 1961 I proposed, for the same purposes, the heating of deuterium with a beam from a pulse laser. I mention these things here by way of explaining that my contributions were not limited to military problems.

In 1950 our research group became part of a special institute. For the next eighteen years I found myself caught up in the routine of a special world of military designers and inventors, special institutes, committees and learned councils, pilot plants and proving grounds. Every day I saw the huge material, intellectual, and nervous resources of thousands of people being poured into the creation of a means of total destruction, something potentially capable of annihilating all human civilization. I noticed that the control levers were in the hands of people who, though talented in their own way, were cynical. Until the summer of 1953 the chief of the atomic project was Beria, who ruled over millions of slave-prisoners. Almost all the construction was done with their labor. Beginning in the late fifties, one got an increasingly clearer picture of the collective might of the military-industrial complex and of its vigorous, unprincipled leaders, blind to everything except their "job." I was in a rather special position. As a theoretical scientist and inventor, relatively young and (moreover) not a Party member, I was not involved in administrative responsibility and was exempt from Party ideological discipline. My position enabled me to see and know a great deal. It compelled me to feel my own responsibility; and at the same time I could look upon this whole perverted system as something of an outsider. All this prompted me—especially in the ideological atmosphere that came into being after the death of Stalin and the Twentieth Congress of the CPSU [Communist Party of the Soviet Union]— to reflect in general terms on the problems of peace and mankind, and in particular on the problems of a thermonuclear war and its aftermath.

Beginning in 1957 (not without the influence of statements on this subject made throughout the world by such people as Albert Schweitzer, Linus Pauling, and others) I felt myself responsible for the problem of radioactive contamination from nuclear explosions. As is known, the absorption of the radioactive products of nuclear explosions by the billions of people inhabiting the earth leads to an increase in the incidence of several diseases and birth defects, of so-called sub-threshold biological effects—for example, because of damage to DNA molecules, the bearers of heredity. When the radioactive products of an explosion get into the atmosphere, each megaton of the nuclear explosion means thousands of unknown victims. And each series of tests of a nuclear weapon (whether they be conducted by the United States, the USSR, Great Britain, China, or France) involves tens of megatons; i.e., tens of thousands of victims.

In my attempts to explain this problem, I encountered great difficulties— and a reluctance to understand. I wrote memorandums (as a result of one of them I. V. Kurchatov made a trip to Yalta to meet with Khrushchev in an unsuccessful attempt to stop the 1958 tests), and I spoke at conferences. I remember that in the summer of 1961 there was a meeting between atomic scientists and the chairman of the Council of Ministers, Khrushchev. It turned out that we were to prepare for a series of tests that would bolster up the new policy of the USSR on the German question (the Berlin Wall). I wrote a note to Khrushchev, saying: "To resume tests after a three-year moratorium would undermine the talks on banning tests and on disarmament, and would lead to a new round in the armaments race—especially in the sphere of intercontinental missiles and anti-missile defense." I passed it up the line. Khrushchev put the note in his breast pocket and invited all present to dine. At the dinner table he made an off-the-cuff speech that I remember for its frankness, and that did not reflect merely his personal position. He said more or less the following: Sakharov is a good scientist. But leave it to us, who are specialists in this tricky business, to make foreign policy. Only force—only the disorientation of the enemy. We can't say aloud that we are carrying out our policy from a position of strength, but that's the way it must be. I would be a slob, and not chairman of the Council of Ministers, if I listened to the likes of Sakharov. In 1960 we helped to elect Kennedy with our policy. But we don't give a damn about Kennedy if he is tied hand and foot—if he can be overthrown at any moment.

Another and no less dramatic episode occurred in 1962. The Ministry, acting basically from bureaucratic interests, issued instructions to proceed with a routine test explosion that was actually useless from the technical point of view. The explosion was to be powerful, so that the number of anticipated victims was colossal. Realizing the unjustifiable, criminal nature of this plan, I made desperate efforts to stop it. This went on for several weeks—weeks that, for me, were full of tension. On the eve of the test I phoned the minister and threatened to resign. The minister replied: "We're not holding you by the throat." I was able to put a phone call through to Ashkabad, where Khrushchev was stopping on that particular day, and begged him to intervene. The next day I had a talk with one of Khrushchev's close advisors. But by then the time for the test had already been moved up to an earlier hour, and the carrier aircraft had already transported its burden to the designated point for the explosion. The feeling of impotence and fright that seized me on that day has remained in my memory ever since, and it has worked much change in me as I moved toward my present attitude.

HERBERT F. YORK
AND G. ALLEN GREB

Strategic Reconnaissance

The way in which a particular weapons system evolves and becomes part of the U.S. arsenal is largely a mystery to most people, often even to those who work on various aspects of the system's research and development. The major reason for this is the largely fortuitous nature of the process. Typically, no master plan exists for a new weapons system. Instead it emerges as a final product of many isolated technological discoveries which combine and build on one another. The evolution of the U.S. surveillance satellite system, described and analyzed by York and Greb, provides a classic example of how this process works.

From the time the notion of orbiting artificial or man-made satellites was first introduced in U.S. military circles at the end of World War II, the value of using such satellites as military observation posts was realized. In fact, studies made by the RAND Corporation (initially a unit of Douglas Aircraft Company known as Project Rand) during the Truman administration not only demonstrated the feasibility of space reconnaissance but also offered fairly accurate estimates concerning the dimensions and other parameters of the necessary hardware. No program for the development of such a system was initiated during the Truman years, however. This was mainly because the great cost of developing and building the rocket boosters needed to put the satellites in orbit simply could not be justified on the basis of this application alone.

Early in the Eisenhower administration this situation changed radically. A combination of political events and some major advances in technology brought about the initiation of several programs for the development of very large rockets for use as long-range missiles. The political events were the end of the Korean War and the first complete change in personnel at the highest levels of the national security apparatus in a dozen years. These events resulted in a total review of U.S. national security policy, known as the "New Look," and that review in turn led to the proclamation of the Doctrine of Massive Retaliation and an increased emphasis on reliance on military technology rather than manpower.

The major technological advances were the invention of the hydrogen bomb by Edward Teller and Stanislaw Ulam, and the important though less dramatic improvements in rocket propulsion technology (mostly at North American's Rocketdyne Division) and guidance systems (mostly at MIT in Stark Draper's Instrumentation Laboratory). In addition, there was a growing awareness that the Soviets were, or soon would be, exploiting the same technological possibilities, and the notion that we were in a technological arms race with them began to take hold.

Out of this concatenation of events came a quick succession of authorizations for no less than five high priority programs to develop large long-range rockets: the Atlas in 1954; the Titan, Thor, and Jupiter in 1955; and the Polaris in 1956. The first of these authorizations came out of a review of the situation by a very special group, the Strategic Missile Evaluation Committee (SMEC, also sometimes known as the Teapot Committee). The principal figures in that review were Trevor Gardner, John von Neumann, George Kistiakowsky, Simon Ramo, and Jerome Wiesner. Each of the other authorizations flowed from a more complex set of circumstances which are not of direct interest to this discussion.

It was obvious at the time that the first four of the rockets listed above would be suitable for use as the main booster in a multistage rocket system capable of launching satellites of the size (1,000 pounds and up) described by RAND as being adequate for reconnaissance from space. And the first three rockets were, in fact, eventually used for such purposes.

On March 1, 1954, just a month after von Neumann's Strategic Missile Evaluation Committee (SMEC) had submitted its report to the Air Force urging the development of the first ICBM (Atlas), RAND issued a report on "Project FEED BACK," the then current name for the reconnaissance satellite proposal. This report, edited by James E. Lipp and Robert M. Salter, described all of the major subsystems in some detail, confirmed the validity and feasibility of the concept, and estimated that the development of a complete satellite system based on the existing state of the art would require seven years and cost $165 million.

As a result of a combination of events—the publication of RAND's "Project FEED BACK" report, the initiation of a "highest priority" program to develop the Atlas missile; the growing attention being given to the importance of reconnaissance by such influential groups as the Air Staff, the Technological Capabilities Panel, and the Air Force Science Advisory Board—the Air Force decided to go ahead with the program. On March 16, 1955, they issued "General Operational Requirement No. 80," a formal requirements document calling for the development of a reconnaissance satellite. Lockheed, RCA and Martin participated in the design competition that followed.

Lockheed's missiles and space division had already been interested in this project for some time. In order to strengthen its hand in the competition, Lockheed recruited a number of key people from RAND. The RAND people themselves saw the move as providing the opportunity to work on the actual design and construction of the systems they had been studying for some time, an opportunity which RAND itself by its nature could never provide.

Among those who joined Lockheed and who had long been involved in the RAND satellite studies were Louis Ridenour, L. Eugene Root (who shortly after became President of Lockheed Missiles and Space Company), Robert Salter, and James Lipp. Salter had been the author or a co-author of almost every major RAND report on the subject. Accordingly, he was able to formulate for Lockheed a proposal for a satellite system which in effect incorporated almost everything the Air Force knew and thought about the subject.

Not surprisingly, then, the Air Force considered Lockheed's proposal the most satisfactory of those submitted and on Oct. 29, 1956 that firm was awarded a letter contract making it the prime contractor on what became known formally as Weapons System 117L. The Air Force Ballistic Missile Division (AFBMD) under General Schriever's command was given the responsibility for managing the project.

The Ramo-Wooldridge organization and the von Neumann committee, which were involved in all other AFBMD programs, were left out of the management loop in this case. At about the time the contract was awarded and "the money really started coming in," the analysts who had been working on the theoretical aspects of the program were shunted aside and a program management expert, John H. Carter, formerly a colonel on Shriever's staff at the BMD, became the program director.

In addition to a reconnaissance payload, the original 1956 contract also included the development of a system for conducting round-the-clock over-head surveillance of all large rocket launches by means of detection of the large amounts of infrared radiation emitted by such rockets during take-off. This particular subsystem became known as Midas and, after a somewhat fitful development program, it is today widely presumed to be in service as one of the major components of the American missile early warning system.

Agena Spacecraft

Since the Atlas alone could not put a significant payload in orbit, a major component of the Lockheed program involved the development of a new additional rocket stage designed to sit atop the Atlas to provide the final push for getting the payload in orbit. This booster-satellite stage is known as the Agena. The first version was 19 feet long, 5 feet in diameter, and was powered by a Bell Aerosystems rocket engine capable of generating 15,000 pounds of thrust. Fully fueled, it weighed 8,500 pounds but its orbital weight (without propellants) was only about 1,700 pounds.

The Agena A was the first large rocket whose engines were designed to be stopped and restarted in space, an absolute essential for providing any sort of capability to maneuver. This innovation, in turn, has allowed Lockheed to boast of myriad "firsts" achieved by various models of Agena spacecraft since 1959. Agenas "were first to achieve a circular orbit, to achieve a polar orbit, to be stabilized in all three axes in orbit, to be controlled in orbit by ground command, to return a man-made object from space, to propel themselves from one orbit to another, to propel spacecraft on successful Mars and Venus flyby missions, to achieve a rendezvous and docking by spacecraft in orbit, and to provide propulsion power in space for another spacecraft."...

The Agena stage and its descendants, and the Thor, Atlas and Titan boosters, have continued to provide the means for launching practically all U.S. military satellites and most large civilian satellites as well. The only major exception is the Saturn V system used in connection with the Apollo program; but even that huge rocket got the thrust required for its first stage from engines whose development was originally started under the aegis of the Air Force missile and space programs and only later transferred to NASA....

The foregoing is written largely in terms of what certain leading technologists and public officials thought, recommended, and did. It is also instructive to consider the development programs of the U-2 and the reconnaissance satellites purely in terms of the status of the requisite technology. When viewed in this way, one finds that the developments under examination were approved at precisely the time the underlying technology matured and became available, and that the technology reached that state of readiness for reasons having nothing to do with reconnaissance objectives.

The example of the reconnaissance satellite is the clearest case in point. The development of the Atlas rocket was authorized in 1954 when U.S. intelligence indicated the Soviets were moving along similar lines, and when it became evident that the technologies for building the necessary thermonuclear warhead, propulsion system, guidance system, and space frame were all either in hand or clearly in sight.

When it became more widely evident a year later that such a rocket could and would really be built, the Air Force then issued its requirement for a reconnaissance satellite of the type and weight that could be launched into orbit using such a first stage booster plus a special upper stage based on very similar technology. As pointed out above, the idea of a reconnaissance satellite had been present for nearly a decade, but a program to build one had not been authorized previously because the very expensive and difficult booster development could not be justified for that application alone. . . .

In summary, then, . . . the reconnaissance satellites appeared precisely when the general forward movement of technology made it possible for them to do so. Does this in turn mean that all the persons named above were, as individuals, unimportant? Were they just puppets in a world totally governed by technological imperatives?

We think the most fruitful way to look at the matter is this: the general Cold War situation plus the high level of technological exuberance that has characterized American defense thinking since World War II created implicit niches for these people. The individuals who filled the niches had both the opportunity and the intellectual ability and energy to do so.

This situation is not different from most success stories. The opportunity that makes success possible comes along in large part by chance, but making something of the opportunity does depend on the personal characteristics of the person involved. Each of the individuals in this story influenced the details of the events that took place according to the dictates of his own individual technological and political personality; but each of them could have been replaced by any one of a small number of not too dissimilar individuals. The overall result would very likely have been the same.

What was important, then, is not that these particular men were active in American technology at that particular time, but that there was a pool of such people available—perhaps just barely large enough, perhaps adequate several times over. Without such a pool, a technological imperative could not per se have produced the results described.

It was not only the state of the art of American technology that mattered; the state of mind was equally important as well. And that state of mind was the collective creation of the principal actors and their colleagues.

In the same vein, how is one to assess the role of President Eisenhower?

Neither the U-2 nor the reconnaissance satellite project originated with the President or his immediate political advisers. Rather, they were urged on him by technical advisers who were convinced both of their efficacy and feasibility. Given the Cold War situation, and particularly the growing fear of a possibly decisive surprise attack, it is difficult to believe that any man who might have been elected President at that time could have turned down the proposal to build a reconnaissance satellite. In this case, the role of the President as a decision-maker seems to have been less than the role of his chief technological advisers. . . .

Nearly all Americans in the 1950s implicitly but firmly believed that U.S. goals were moral, that Soviet goals were evil, and that this difference was obvious even to the Soviet leadership itself. Again, any president would have shared in this national mood. But, because of his wartime experience and reputation, Eisenhower in particular demonstrated an unusual degree of certainty and confidence when confronted with problems of a military nature.

REYNOLD M. WIK

The Government
And Agricultural
Technology

Since the founding of the United States, the state and federal governments have encouraged applied science and technology. Over the years, at the urging of government officials or ordinary citizens, Presidents and Congresses have initiated and supported legislation to improve social or economic conditions in many areas through science and technology. Historian Reynold Wik, in this selection, describes how this process has contributed to American agriculture. Wik stresses the fact that government support has not only improved crop yields and developed important farm by-products, but has also lightened the burdens of farmers.

In 1942 Secretary of Agriculture Clinton R. Anderson summarized the revolution that science and technology had wrought in farming in the United States. No longer did the American farmer have to rely on sheer muscle. He now had tractors to pull heavy loads, dig holes, and grind feed and thus cultivate vast areas of land with a minimum of labor. Poultry scientists had shown him how to double egg production. Farmers now had strains of alfalfa, wheat, and oats that resisted plant diseases, thanks to the agronomists. Researchers

Reynold M. Wik, "Science and American Agriculture," from David D. Van Tassel and Michael G. Hall, eds., *Science and Society in the United States* (Homewood, Ill.: The Dorsey Press, 1966), pp. 81–104. © 1966 by The Dorsey Press.

had provided him with chemicals that killed weeds and fertilized the soil, and insecticides that sanitized farm buildings. Improved breeds of sheep, cattle, and hogs produced more wool, beef, and bacon. Airplanes hauled perishable produce to market, while new refrigerating processes kept it fresh and wholesome. Hybrid fruit trees and hybrid corn brought higher yields. Science could even convert corncobs to industrial uses. Agricultural research had revealed the mysteries of the good earth, and increased knowledge could make possible the abundant life for every family. In a sanguine mood, the Secretary promised that "more was yet to come."

Eighteen years later the *United States Census Reports* of 1960 showed that farm families possessed these modern conveniences: Of the farm families having electricity, 98 percent owned refrigerators, 90 percent used electric irons, 72 percent had vacuum cleaners, and 93 percent owned motor-driven washing machines. In addition, the farm population operated 4,685,000 tractors, 3,000,000 trucks and 4,260,000 automobiles. In fact, for seven million farm workers in the United States there were now seven million farm tractors and trucks.

Yields per acre of crop land doubled between 1930 and 1960:

	1930	1960
Wheat (bushels per acre)	14	24
Corn " "	26	54
Cotton (pounds per acre)	171	440

Output per manhour more than quadrupled. . . .

These impressive achievements had a long background in American agricultural history. Traditionally, Americans approached their problems pragmatically. Throughout the entire colonization venture following the settlement in Jamestown in 1607, they employed trial and error methods to sustain life, to defend themselves, and to improve their status. In the process of adjustment they increased their knowledge, because they could not escape the observations of their own experience. . . .

Early Government Aid to Agriculture and Industry

The student of scientific agriculture will notice that public funds were frequently used to foster improved conditions on the American farm. It would be misleading to assume that science took root in early America because our forefathers kept the government out of the affairs of the individual and rejected all aspects of the welfare state. On the contrary, the nation's early history reflected considerable economic pluralism rather than unadulterated laissez-faire capitalism.

The colonial governments subsidized a variety of projects. New Jersey, Pennsylvania and Massachusetts granted bounties for the establishment of iron foundries, while Virginia aided the textile manufacturers. Town officials of East Bridgewater, Massachusetts, hired several Scotsmen to come to America in 1786 to build Arkwright spinning machines. In the following year, $14,000 was spent on a similar project in Beverly. Richard Morris, in *Government and Labor in Early America,* has pointed out that entrepreneurs in the seventeenth century were not free to carry on business as they saw fit, nor were workers always free to choose their vocations. At times, craftsmen in Virginia were forbidden to leave their work to take up farming. Men were required to work specified days each week in the harvest fields or on public works. Likewise, legislation frequently set wage and price ceilings and regulated the fees charged by lawyers and doctors. Physicians caring for the poor or treating men in the army were paid by the state. It appears that laissez-faire economics was far from the minds of the founding fathers.

Because of the mixed economy, it seemed natural that government should encourage work in science and technology. During the Revolutionary War, military arms were produced in government armories as well as private firms. Congress made the Springfield Armory a national institution in 1794. Here Thomas Blanchard designed copying lathes for manufacturing standardized parts for rifles. The federal government also rescued Eli Whitney from financial ruin by offering him contracts for arms manufacture. By 1809 he had received $134,000 in Federal funds. Samuel F. B. Morse accepted $30,000 from the federal treasury to help build a telegraph line from Baltimore to Washington. . . .

Since public funds were used for a variety of other purposes, it seemed logical to assume that government should finance agricultural projects as well. George Washington doubtless reflected the sentiment of his day when he stated in his last message to Congress on December 7, 1796, that agriculture was of prime importance to the national welfare. He advocated the establishment of a Federal Board of Agriculture and insisted that:

> In proportion as nations advance in population and other circumstances of maturity, this truth becomes more apparent, and renders the cultivation of the soil more and more an object of public patronage. Institutions for promoting it grow up, supported by the public purse; and to what object can it be dedicated with greater propriety.

Although the federal government hesitated to dole out money directly for agriculture, the several states did divert funds in this direction. New Hampshire, in 1810, subsidized several agricultural societies, while Massachusetts paid $115,800 from the state treasury to support similar organizations over a

15-year period. Massachusetts also paid farmers $27,000 in bounties from 1838 to 1840 to encourage the planting of wheat. In Maine $200,000 went as bounty to farmers for growing wheat and corn from 1837 to 1839.

The first specific congressional appropriation for agriculture came in 1839, when the Patent Office was given $1,000 for the collection of farm statistics. Henry L. Ellsworth, the Patent Commissioner, urged Congress to increase these appropriations. Apparently he had Justus von Liebig's influential work, *Chemistry in Its Application to Agriculture and Physiology,* because he insisted that chemistry could be a boon to the farmer. "If the application of the sciences be yet further made to husbandry," he said, "what vast improvements may be anticipated." He pointed out that chemistry had provided a way for western farmers to market oil, because now pork fat could be converted into stearine for candles. Chemistry also showed how 10 gallons of oil could be extracted from 100 bushels of corn meal. The Patent Office officials looked to the science of agriculture to make folklore yield to scientific analysis. Potato diseases could be eradicated, and the Hessian fly, so destructive to wheat, could be controlled. Experience had proved that every grain, vegetable, and fruit could be improved by scientific cultivation. Indeed, genius had stooped from its lofty height to lessen the farmer's burden.

When Dr. Daniel Lee joined the Patent Office to supervise the scientific work of the agency, he begged Congress to increase appropriations for agriculture to prevent universal impoverishment of the soil. "The prayers of generations had gone unheard," he asserted. "How could the appropriation of $1,000 restore the fertility of 100 million acres of land? . . . Agricultural schools should be established to teach young people the principles of soil-building science." Thus the struggle to secure federal funds for agricultural education was intensified, and the battle was won when the Land-Grant College Act was passed in 1862.

Farm journals joined in the crusade for more federal aid to agriculture. *De Bow's Review,* in 1858, maintained that the nation had good soil, labor-saving machinery, and industrious workers, but the farmer lacked theoretical knowledge, inventive skills, and a system of science to make adequate progress. Above all:

> No species of human pursuit is more depending and more indebted to Chemistry than agriculture. Chemistry does not only give instruction to the farmer on everything there is, but teaches him what is wanting, and how it can be got. . . . If but a single series of such investigations would be undertaken on the part of the Federal or State Government, we do not for a moment doubt that the results would be looked upon with astonishment and hailed with delight by legislators, statesmen and personal agriculturists.

In the face of mounting pressures, Congress abandoned its position that agriculture needed no direct financial aid and, in 1854, granted $35,000 for agricultural experimental work on a two-acre plot near Missouri Avenue in Washington, D.C. Glover Townsend, an entomologist, joined the Patent Office staff; a chemist and botanist were employed to do research; and arrangements were made to have the Smithsonian Institution publish meteorological statistics. Thus the same Congress that passed the Kansas-Nebraska Bill, which was destined to split the nation politically into two factions and lead to violence in "Bleeding Kansas," initiated the first major financial aid to agriculture. Henceforth the farmer was to receive some of the same consideration which had been bestowed on the businessman by the founding fathers of the Constitution and the administration of George Washington. . . .

Scientific Study of Agriculture

When Abraham Lincoln, on May 15, 1862, signed a bill establishing the Department of Agriculture, and on July 2, signed the Land-Grant College bill, the foundations were laid for scientific studies in agriculture for the next century. Isaac Newton, the first Commissioner of Agriculture, stated that his objectives would be to collect, arrange, and publish agricultural information; distribute seeds; answer farmers' queries; institute chemical investigations; investigate cotton culture; seek to introduce silkworms from China; and promote the culture of flax and hemp as substitutes for cotton. He also proposed a chemical laboratory for analysis of soils, grains, fruits, plants, and vegetables and urged the establishment of professorships of botany and entomology. He wished to use science for the benefit of agriculture, but as A. Hunter Dupree has pointed out, "he had no specific ideas as to what problems could be solved or how they should be attacked." During the first year, the Department consisted of a horticulturist, a chemist, an entomologist, a statistician, and 24 other staff members. Some experimental work was begun in a garden between Fourth and Sixth streets, and a larger area between Independence and Constitution avenues was used for an experimental farm when the Union army no longer needed it for a cattle yard.

Early scientific achievements included C. M. Wetherill's study of sugars and his report on the *Chemical Analysis of Grapes*. Later, Henri Erni reported on fermentation studies and made analyses of wines, coal, soils, asphalt, and guano. . . .

Inadequate financing plagued the Department of Agriculture for the first 20 years of its existence. Appropriations for the Department were only $199,770 in 1864 and $199,500 in 1880. However, the Department had a renaissance in the 1880's and 1890's, which continued into the twentieth century. The Hatch Act of 1887 established the Office of Experiment Stations in the Department.

The federal government provided money for distribution to the state experiment stations for carrying out basic research. The Adams Act of 1906 provided additional funds for experiment stations engaged in research "bearing directly on the agricultural industry of the United States."

Increased prestige came to the Department in 1889, when Congress elevated it to Cabinet status. When the Army Signal Corps was transferred to the Department in 1891 and the Weather Bureau was established, the number employed by the Department reached 1,577. Besides, various commissions which had acted independently were now absorbed by the Department. For example, the United States Entomological Commission attempted to cope with the grasshopper plagues which swept the Midwest during the 1870's. In 1880 this agency became part of the Department of Agriculture, where its appropriations grew from $7,000 in 1880 to $42,000 in 1885. This division studied insects and attempted to prevent the introduction of pests from Europe and to control the spread of the boll weevil in the cotton states. After 1906 the Division became a Bureau in name, and under more aggressive leadership, field laboratories were set up in infested areas rather than waiting for farm problems to be brought to Washington. By securing grants for money for research in a general area instead of allocating it for specific purposes, the Department of Agriculture gained flexibility to shift programs as conditions warranted.

The Department developed rapidly under the leadership of James Wilson of Iowa, who served as Secretary from 1897 to 1913. During these years he transformed his agency into an outstanding research, regulatory, educational, and custodial institution. In 1900 the Arlington Farm of 400 acres in Virginia was acquired for experimental work. The Bureau of Plant Industry used this land until 1941, when it was transferred to the War Department as a site for the Pentagon. Additional research facilities were secured for the Agricultural Research Center at Beltsville, Maryland.

During the depression of the 1930's, the Agricultural Adjustment Act established in the Department of Agriculture the Agricultural Adjustment Administration, which assumed responsibility for initiating programs for soil conservation, rural electrification, and research in the branches of economics and science. The AAA initiated, for instance, the Forest Products Laboratory near Madison, Wisconsin, as one of four regional laboratories dedicated to the work of developing industrial uses for farm products.

Although the Congress never established a national university, the federal government endeavored to meet the practical needs of rural Americans by passage of the Land-Grant College Act of 1862, which endowed state colleges with 11 million acres of land, equal in size to twice the area of Vermont. The legislation introduced the principle of federal grants-in-aid to the states.

As a result of the Act, 68 institutions, sometimes referred to as "peoples colleges" or "cow colleges," were established. They provided education at a minimum cost. From these schools came large numbers dedicated to the ideal of a more scientific agriculture. . . .

Today, . . . these institutions enroll one fifth of the total college population, grant 40 percent of the doctoral degrees in all subjects and all of the doctoral degrees in agriculture. From these institutions flow scientific advances which have bettered the lot of all Americans.

Work of Individual Scientists in the Department of Agriculture

Some indication of the profound significance of science in American agriculture can be seen in the work of individual scientists in the Department of Agriculture. In 1883 livestock diseases received little attention from governmental agencies. At the time, hog cholera caused deaths amounting to the value of $30 million annually. Cattlemen feared the Texas fever and pleuropneumonia, anthrax and blackleg. The causes of these diseases were unknown or in dispute, and veterinary science had not been able to provide adequate treatment. Constant demands for assistance resulted in the act of Congress which established the Bureau of Animal Industry in 1884 and appropriated $150,000 to put the law into effect.

Farmers had long suspected that the cattle tick transmitted Texas fever. Theobald Smith, the Bureau's first pathologist, began experimenting with tick fever in 1888. He discovered that the cause of the disease was a protozoan parasite that multiplied in the blood of infected animals. Working with Cooper Curtice and F. L. Kilborne, Smith announced, in 1892, confirmation of the suspicion that infection could be carried from one animal to another by the tick. This was a momentous discovery—the first demonstration that disease-producing microorganisms could be transmitted by an insect carrier. The obvious cure was to get rid of the ticks, and this was achieved by a long-range eradication program. The original research cost $65,000, but the saving to cattlemen today amounts to $65 million a year. In addition, this discovery led others to scrutinize the mosquito as a possible disease carrier, thus paving the way for the understanding and control of such serious maladies as malaria, typhus, bubonic plague, and Rocky Mountain spotted fever.

Again, when Marion Dorset joined the Department in 1894, he knew that hog cholera created losses of $65 million annually. . . . Dorset found . . . that the disease was caused by an ultramicroscopic virus rather than a bacterium. He also proved that hogs that recovered from the disease were immune for life. While conducting experiments in Iowa in 1903, his team of scientists found that the blood from immune hogs gave only temporary immunity to other hogs. Then Dorset began using two injections. The first was a serum from the blood of a pig which had survived the cholera. The second was an injection of the virus from an infected hog. Today, this double injection method provides protection against this dreaded disease.

Another scientist in the Department, S. Henry Ayers, contributed to the health of the American people by convincing them that... pasteurized milk was safer than raw milk.

Still another Department scientist, Dr. Harvey W. Wiley, conducted research that contributed to better health. He spent 21 years investigating the adulteration of food products on the market. His analyses showed that cottonseed oil was commonly used as an adulterant in making butter. Poisonous dyes were often used to color food, while foreign matter contaminated much of the foodstuffs. His findings contributed to the passage of the Pure Food and Drug Act of 1906.

In 1890 Professor S. M. Babcock of the University of Wisconsin came to the aid of housewives by devising a test for butterfat in milk and cream. . . . The Babcock test gave dairy farmers a better way to measure the value of their cows. This led to official testing of dairy herds by the members of the Dairy Herd Improvement Association, in which over 40,000 dairy farmers keep production records on more than a half-million cows. The test provided a method to check the legal requirements for butterfat in every bottle of milk sold in the market, to the benefit of the consumer.

In the field of plant industry, it is known that 30,000 different diseases attack plants in the United States with a cost of $3 billion a year. Agricultural pathology gained new impetus from Louis Pasteur's discovery that bacteria produced some animal diseases. Following Pasteur's lead Professor T. J. Burrill of Illinois, in 1878, proved bacteria to be the cause of "fire blight," a devastating disease which swept through apple and pear orchards. The discovery led to a whole new theory of plant disease. Dr. Erwin F. Smith and his associates of the United States Department of Agriculture carried the work forward. Smith summarized his findings in *Bacterial Disease,* published in 1920. Today, there are 170 known kinds of bacteria which attack plant life. In addition, more than 200 plant viruses have been isolated since 1900. Among these are "curly top," which injures sugar beets, tomatoes, and beans, and tristeza, which attacks orange groves.

To cope with these types of diseases, William A. Orton, a plant pathologist in the Department, concentrated on the wilt which damaged cotton crops. He used experimental plots to breed a new variety of cotton which would not wilt. This became a landmark in agricultural research because it pioneered in breeding plants which were resistant to disease. . . .

Since the quest for improved plants never ceases, the Department of Agriculture, in 1898, set up an agency to introduce new plants into the United

States from other countries. As a result, there have been more than 275,000 plant introductions from all parts of the world. . . . Today these introductions continue, with tests of adaptability being made at state experimental stations. Seeds thought to be essential for the needs of crop breeders in the future are preserved in the National Seed Laboratory in Fort Collins, Colorado.

In this tradition, Mark Alfred Carleton conducted exploration of wheat varieties in Russia from 1894 to 1918. Through his efforts, hard red winter and Durum wheats became widely accepted in the United States. During subsequent years, the Department of Agriculture has preserved the wheat germ plasm, the hereditary material in the wheat kernel. This collection, now numbering 16,000 accessions of wheat varieties, is used for research in pedigree selection, hybridization, and crossbreeding. . . .

Present-Day Research

Today the focal point for scientific studies conducted by the Department of Agriculture is the Agricultural Research Center near Beltsville, Maryland. Here 5,000 scientists are engaged in 3,000 research projects costing approximately $135 million a year. The Center includes 1,160 buildings, 35 greenhouses, 10,500 acres of land, 3,000 head of livestock, and 10,000 in poultry. The National Fungus Collection contains 665,000 specimens of fungi and plant disease organisms, and another collection includes 60,000 lots of parasites which affect animal life.

Since the Center employs so many specialists, it is easy for them to consult one another on mutual problems dealing with the life sciences, physical sciences, and social sciences. Here an agronomist who studies weeds may readily confer with botanists, hydrologists, and soil scientists, while an investigation of anaplasmosis in cattle would bring together physiologists, pathologists, entomologists, and chemists. In recent years, work has been extended to basic research in investigating life processes, the changes in living cells, and the physiology of virology. Sixteen of the laboratories carrying on this work are at Beltsville, while others are located in Peoria, New Orleans, Washington, and Albany, and Berkeley in California and Wyndmoor near Philadelphia.

A visit to the Albany laboratory would reveal 200 professional people making various types of analyses of food. One project measures the effect of temperature on color, texture, flavor, and vitamin content of frozen foods. A new technique called gas-liquid-partition chromatography is so sensitive that the expert can make an analysis of flavors and aromas which are too subtle to be detected by the human nose. The pharmacologists run safety checks on foods to spot toxic and allergenic substances present in some farm crops. Tests are made on growth-promoting or growth-inhibiting substances to aid the feeders of livestock and poultry.

In recent years, engineering research conducted by land-grant universities and private corporations has greatly improved farm machinery.... To provide more comfort for the operator, tractors can be equipped with power steering, foam rubber seats, umbrellas, air-conditioned cabs with windshields and radios....

In addition, mechanical innovations have led to the manufacture of self-propelled grain combines, corn pickers, sugar beet harvesters, and cotton pickers.... Various hydraulic and electrical equipment makes the tractor versatile enough to perform virtually every job encountered on the farm....

Similarly, the miracles of science become evident in all phases of farm life. Although the field of genetics is only about 70 years old, it has already influenced agricultural practices. One of the most spectacular developments has been the hybridization of corn, a discovery which provided a key to all scientific breeding. In addition, plant hormones are used to regulate plant growth, either to accelerate growth or retard it. For instance, if only a speck, one-millionth of an ounce of dichlorophenoxyacetic acid is put on one side of the stem of a bean seedling, the cells along the treated side grow faster than those on the untreated side, thus bending the plant in a direction away from the treated surface. Also, by using chemicals, chrysanthemums can be made to grow on a prescribed schedule.

Developmental research has created thousands of new products such as cellulose, rayon, plastics, casein wool fiber, and other synthetic articles. Marketing research has replaced the corner store with the supermarket, while engineering techniques have provided 200 different uses of electrical power on the modern farm.

CARROLL W. PURSELL, JR.

The Government and Industrial Technology

In the twentieth century, and particularly after World War II, the federal gov-
ernment became increasingly involved with the encouragement of scientific and
technological development. The government not only established such massive
agencies to promote new technologies as the Atomic Energy Commission and
the National Aeronautics and Space Administration, but also poured billions of
dollars into educational institutions to train scientists, engineers, and physi-
cians. Historian Carroll Pursell, in this selection, describes this development,
which culminated in the establishment by President Eisenhower of the office of
a White House science advisor and of a President's Science Advisory Com-
mittee.

As American industry moved out of the free-swinging nineteenth century and began to face the problems of modern industrialism which it had itself created, it turned to government science for help. Many industries, especially the newer ones based on electricity and chemicals, were already beginning to support their own industrial research laboratories, but there were certain functions which were either clearly unprofitable or impossible for industry to provide for itself.

The government's constitutional responsibility concerning weights and measures had for many years been handled by the Coast and Geodetic Survey

Carroll W. Pursell, "Science and Government Agencies," from David D. Van Tassel and Michael G. Hall, eds., *Science and Society in the United States* (Homewood, Ill.: The Dorsey Press, 1966), pp. 232–249. © 1966 by The Dorsey Press.

on the theory that surveying had a close connection with standard measures. With the rising need for such exotic new standards as the ohm, henry, and watt, this rationale was clearly inadequate. An American economy which was becoming truly national, increasingly technological, and incredibly complex stood in urgent need of all kinds of standards from screw threads to drug purity. When Henry S. Pritchett became head of the Coast Survey in 1897, he began a careful, and eventually successful, campaign to have a special bureau set up to handle national standards. After study of foreign laboratories and consultations with interested executive bureaus, a bill was presented to Congress and was passed in 1901 with the endorsement of such organizations as the National Academy of Sciences, the American Association for the Advancement of Science, and the American Institute of Electrical Engineers.

The organic act of the new National Bureau of Standards authorized it not only to care for the standards themselves but to enter into research to determine such things as physical constants and properties of materials. The Bureau was permitted to conduct research and provide standards for private firms and professional societies, and a visiting committee of civilian scientists was established to insure that the nature and quality of work undertaken would conform to the best thinking of the scientific community. Eventually, the Bureau established standards for, and tested materials purchased by, the federal government, thus providing measures of quality which could be used by civilian consumers as well.

During these same years, the nation's mineral industry, already important as the center of industrial prosperity, came more and more to the attention of conservationists concerned with nonrenewable resources as well as those who sought to prevent the dramatic and tragic mine disasters which took an enormous toll of workers. The Geological Survey had been interested in the industry since [the 1860s], and it now received some funds for specific investigations of mine safety, testing of coals, and the like. Just as the Bureau of Reclamation had spun off from the Survey with a cluster of responsibilities more or less related, so now it appeared reasonable that a Bureau of Mines should do the same. Congress set up such a Bureau in 1910, giving it an emphasis on mine safety. However, Joseph A. Holmes, the first bureau chief, raised prevention of waste to equal importance.

The newly born aviation industry received early support from government-sponsored research. Although the Smithsonian Institution's role as a research center was handicapped by growing curatorial duties and its relatively paltry endowment, the work of Secretary Samuel Pierpont Langley in aeronautics gave the government an early interest in that field. After Langley's death, Secretary C. D. Walcott helped organize a Smithsonian advisory committee on aeronautics and, after the start of World War I in France, succeeded in getting it formalized as a government body. The National Advisory Committee for Aeronautics (NACA), established in 1915, presided over aeronautical research in the United States for over 40 years, and its unusual structure made it a model for later scientific agencies within the government. . . .

Under Herbert Hoover the Department of Commerce during the 1920s attempted to relate itself to industry in somewhat the same way that the Department of Agriculture serviced agriculture. It was a reasonable analogy, which has persisted to the present, and Secretary Hoover pursued it with alacrity and imagination. Industrial standards, radio, aviation, and even conservation became objects of particular departmental concern, and all of them required the aid of science at some level. In keeping with his preference for voluntary self-help, Hoover also fought a losing battle to create a National Research Fund, financed by private industry, administered by the quasi-governmental National Research Council, and applied to the pursuit of basic research. The high position of Hoover as both Secretary and President, combined with his unique membership in the National Academy of Sciences, led most scientists to expect much from his tenure and some to be disappointed with his performance.

When Hoover left office, the nation was facing the bitter reality of the Great Depression. Because of the delays involved in federal budgeting, it was not until 1932 that the budgets of the various science agencies felt the pinch. When the cutbacks came, however, they were in many cases substantial. So successful had been scientists in convincing the public that science and the common welfare were connected, that there were even demands for a moratorium on research. It seemed quixotic to many that money should be spent to increase farm crop yield while it was necessary to burn surplus crops already harvested. The first move of both President Hoover and President Franklin D. Roosevelt was to cut the normal expenditures of government. Among these, science budgets suffered severely, the Bureau of Standards receiving cuts of 50 percent. The first responsibility of government science, therefore, was just to stay alive, and in this it was successful. By 1936, budgets were back to normal and, in most cases, were even beginning to grow again. . . .

During World War I, the government had brought scientific men to Washington and put many of them into uniform. When the war was over, this whole effort could be rapidly abandoned because no basic alignments had been changed. In World War II, the government supported vast amounts of research *in situ,* leaving scientists on campuses and in industrial research laboratories. In the previous decade, serious doubts had been raised as to whether the government could devise ways in which to stimulate directly civilian science without ruining its unique structure. Under the pressure of war, ways were quickly found.

The main scientific effort of the government during the war was funneled through the National Defense Research Committee, set up in June 1940 and a year later absorbed into the Office of Scientific Research and Development (OSRD). Once again the traditional Academy-Research Council was bypassed by reformers. A small group of scientists, including James Conant, president of Harvard University, Karl T. Compton, president of the Massachusetts Institute of Technology, and Vannevar Bush, president of the Carnegie Institution of Washington, decided, on the basis of World War I experience and subsequent developments, to bypass all existing agencies and establish an independent agency, patterned on the National Advisory Committee for Aeronautics which Bush had lately chaired. The resulting National Defense Research Committee was made up of part-time administrators capable of spending government money and authorized both to initiate research projects without being asked and to farm those projects out to the best available contractor. There was some small precedent in the government for each of these innovations, but each flew directly in the face of the great bulk of previous experience. It was the initial and critical decision of the National Defense Research Committee to concentrate on that research in weaponry which would have a most decisive and immediate impact on the war effort, and, secondly, to do this in a manner which would be least likely to disturb prewar arrangements in science and technology. . . .

After months of debate, an Office of Production Research and Development was organized within the War Production Board to give industry the kind of scientific aid that the Office of Scientific Research and Development was providing for the military. All of these agencies faced the same basic problem: the need to rationalize quickly the vast edifice of American science which had been built up over two centuries of salutary neglect, and to concentrate its efforts toward winning a war that none had wanted and for which few had prepared.

Neither the new nor the old agencies attempted to provide any real overall supervision of America's scientific war effort. Even the powerful and effective Office of Scientific Research and Development operated to win the cooperation of the several estates of science rather than to dictate their various roles, and limited itself strictly to problems of medicine and weaponry. For some critics, it seemed that the disparate and individualistic nature of American scientific effort—an essential quality which the government tended to accept rather than challenge—was inherently too weak and inefficient to meet the challenge of world war, and perhaps after that, renewed depression. Led by Senator Harley M. Kilgore of West Virginia, these critics sought to make fundamental changes in the social structure of American science.

This division of opinion as to whether the government could or should impose social controls upon scientific and technical research, which began under the pressure of depression in the 1930s and persists into our own day, under-lay the debate over technological mobilization during World War II and much of the subsequent discussion over national science policy. Only on this one point could everyone agree: that the necessary and deep involvement of sci-ence in the war effort had brought about basic changes in the relationship between science and the federal government. The latter had, from its earliest days, made use of science to promote the common welfare, but now had also to rely upon science to provide for the common defense.

As the war drew to a close, there was an almost unanimous consensus among leading scientists that, in the postwar years, the government would not only have to support a sizable scientific effort of its own but would also have to give massive support to science in industry and the universities. The nature and extent of this support was a matter of considerable debate. Through three years of hearings by the Kilgore committee of the Senate, sentiment had been building for a program of subvention, centering around a proposed National Science Foundation. The proposals of the Kilgore committee were considered to be somewhat radical by most of the powerful science administrators who centered around Vannevar Bush. In order to provide an alternative to this, Bush prevailed upon President Roosevelt, in November of 1944, to ask for answers to four questions involving postwar science policy. These questions were, of course, referred to Bush and offered him an opportunity to present a conservative plan for the federal support of science.

Bush's report, made public several days before that of Kilgore, also en-visioned a central National Research Foundation. This foundation was to con-tinue much of the good work of the Office of Scientific Research and De-velopment, including independent supervision of medical and weapons re-search. The essential differences between the plans of Bush and Kilgore were that the former would support only physical and not social science, would create a foundation virtually independent of presidential control, would make no attempt to distribute funds on anything like a geographical basis, and would allow grantees to retain patent rights where applicable. Kilgore's plan would have preserved governmental control of research sponsored by govern-ment funds, while Bush's plan sought to limit the role of the government to merely that of paying the bill.

The debate over the relative merits of these two plans dragged on in the Congress from 1945 until 1950. In 1948 a bill was passed creating a National Science Foundation free of political responsibility, but it was vetoed by Presi-dent Truman. The essential question to be decided was whether the govern-ment should be able to use science as a tool for social conservation and re-construction, or whether science should remain the exclusive property of the expert and essentially beyond the democratic control of society as a whole.

It was not so much a matter of whether science would be free: it was more a matter of whether it should be responsible to the government or continue to be responsible only to big business and big education. It was universally agreed that the people should pay for it in either case.

The National Science Foundation, which was finally established in 1950, was more a victory for the conservative than liberal faction but, on the whole, was quite different from what either Bush or Kilgore had anticipated. Had it been set up in 1945 it would have been of towering importance in the structure of all American science. But between 1945 and 1950, the perpetual realities of bureaucratic life and the new element of the East-West cold war sapped its incipient primacy. In actual fact, the Foundation became only another pillar in the edifice of government science rather than the capstone of the whole structure.

In the first place, several vital areas of research needed immediate attention and could not wait five years while the American people decided on the fundamental direction of the postwar nation. Less than a month after Bush and Kilgore had recommended their respective foundations, the United States dramatically brought the war to an end with the use of the atomic bomb. Neither Bush nor Kilgore had mentioned atomic research in his report: the former because he was bound by secrecy and the latter because he had never heard of it. The genie thus so precipitously loosed had just as quickly to be bottled up again in some kind of government agency. With the setting up in 1946 of an independent Atomic Energy Commission, the still unborn Foundation lost the most lucrative and dramatic field in all of science.

Medical research, which had been handled by the Medical Research Committee of the Office of Scientific Research and Development, had its own humanitarian urgency and the alert National Institutes of Health quickly took over a large number of wartime research contracts from the Office when the latter went out of business. The powerful medical profession, more used to doing business with the Public Health Service, had never been enthusiastic about submerging the identity of medical research within some giant multipurpose science agency such as the proposed foundation. . . .

By the time the National Science Foundation was finally established in 1950, control of medical, atomic, and military research (before Sputnik the three greatest scientific efforts of the government) had already been wrested from it. The mission of the new agency became, partly through default, the support of "basic research," a concern for science education, and a weak mandate to provide the government with a national science policy. Even these were not its exclusive responsibility. The postwar Office of Naval Research had early made a place for itself in the hearts of academic scientists by generously supporting a wide spectrum of basic research. The old Office of

Education, never a very vigorous agency, was still the logical place for educational responsibilities, and it was apparent to all that very few people concerned, and certainly not civilian scientists or government scientific bureaus, would permit any agency to make a single scientific policy for the whole nation.

By the end of the Truman Era, the postwar federal scientific establishment was virtually complete. The most important research tasks had been provided for in some manner, and if there was no overall policy, at least the most important jobs got done. The domestic issues which had dictated the lines of controversy in the late 1940s were, if not irrelevant, at least pushed into the background by the increasing tempo of cold war, already become hot in Korea. . . .

As the cold war expanded through the years, it raised fundamental questions for American science. In the first place, the prospect of peace breaking out, though unlikely, posed a real threat to the continuation of government subvention on such a high scale. It has never been established that the support of science for its own sake is something the government could or should do solely as a form of *noblesse oblige.* In the second place, cold war motivation has placed a strain on many of the most cherished myths of the scientific profession. It has been decisively demonstrated, for example, that unfree science (either Soviet or American) can flourish and produce astonishing results. Scientists have also demonstrated rather clearly that the "scientific method," except to the extent that it shares in the common virtues of honesty and fairmindedness, provides no panaceas for the solution of complex political problems. . . .

The government's response to the launching of Sputnik in October of 1957 placed some strain upon the accommodation reached in the postwar debates. The metamorphosis of th National Advisory Committee for Aeronautics into the first truly cold war science agency, the National Aeronautics and Space Administration (NASA), created the latest in that long line of predominant agencies which, by virtue of the importance of their mission and scale of their support, have been able to warp the government's whole scientific effort. . . .

A more conscious and direct attempt to coordinate the vast structure of government science has also been a part of the post-Sputnik response. In 1957 President Eisenhower appointed James R. Killian as his personal science advisor and head of the President's Science Advisory Committee. Congressional talk of a Department of Science, going back at least as far as the Allison Commission of 1884, was revived, and the Office of Science and Technology was created as a separate entity within the Executive Office of the President, so that it might be answerable to Congress. . . .

This burst of administrative and organizational ingenuity is symptomatic of a fundamental aspect of the history of governmental involvement in scientific activities. The government's need for science has frequently stimulated it to new organizational experiments. The basic problem has been that, much as the government needs science, science has, by and large, offered its services only on its own terms. Those terms have been support without control or, in other terms, power without responsibility. By and large, the government has refused to grant such terms, and the two have compromised in such a manner as to blur the traditional boundaries between the private and public sectors of American life.

The fundamental trend over the past century has been the rise of scientific research in the universities, private foundations, and industrial establishments of the nation, and the spinning of a web of mutual interest between these institutions and the federal government. The trend has been hastened by the steadily increasing need of the government for scientific data to aid it in discharging its growing responsibilities. As these scientific needs have increased, they have, in turn, exerted a growing influence on that scientific activity taking part outside the government. Both to protect their own interests and to influence federal science policy, civilian scientists have sought to create institutions through which they could affect policy but which would not allow politics to affect them. At the same time, government scientists have sought to create institutions through which they could benefit from the support of their civilian colleagues against their political masters. And finally, those vested with political responsibility sought to create institutions through which they could best accomplish their scientific purposes without abdicating their responsibility to the people as a whole. It is typical of the American political genius that all interested parties have settled for something less than they desired. The institutional maze produced by the interplay of history and necessity confounds the enemies as well as the friends of federal science.

JOHN G. BURKE

Bursting Boilers and The Federal Power

In addition to the intended results, technology produces unforeseen and some-times very undesirable effects. In the early decades of the nineteenth century, Americans began to experience some unwanted effects of new technology. As described in this selection by historian John G. Burke, steam boiler explosions, primarily in marine service, provoked public outcry over the loss of life and injuries. Urged over decades to take positive action to prevent the accidents, the federal government in 1852 finally established its first regulatory agency, the Steamboat Inspection Service, which was the prototype of many others that followed. In addition to aiding agriculture and industry in developing new tech-nologies, the government has also assumed a policing role, in attempting to prevent the occurrence of undesirable social effects.

When the United States Food and Drug Administration removes thou-sands of tins of tuna from supermarket shelves to prevent possible food poisoning, when the Civil Aeronautics Board restricts the speed of cer-tain jets until modifications are completed, or when the Interstate Commerce Commission institutes safety checks of interstate motor carriers, the federal government is expressing its power to regulate dangerous processes or prod-ucts in interstate commerce. Although particular interests may take issue with a regulatory agency about restrictions placed upon certain products or seek to alleviate what they consider to be unjust directives, few citizens would argue that government regulation of this type constitutes a serious invasion of private property rights.

Though federal regulatory agencies may contribute to the general welfare, they are not expressly sanctioned by any provisions of the U.S. Constitution. In fact, their genesis was due to a marked change in the attitude of many early nineteenth-century Americans who insisted that the federal government exercise its power in a positive way in an area that was non-existent when the Constitution was enacted. At the time, commercial, manufacturing, and business interests were willing to seek the aid of government in such matters as patent rights, land grants, or protective tariffs, but they opposed any action that might smack of governmental interference or control of their internal affairs. The government might act benevolently but never restrictively.

The innovation responsible for the changed attitude toward government regulation was the steam engine. The introduction of steam power was transforming American culture, and while Thoreau despised the belching locomotives that fouled his nest at Walden, the majority of Americans were delighted with the improved modes of transportation and the other benefits accompanying the expanding use of steam. However, while Americans rejoiced over this awesome power that was harnessed in the service of man, tragic events that were apparently concomitant to its use alarmed them—the growing frequency of disastrous boiler explosions, primarily in marine service. At the time, there was not even a governmental agency that could institute a proper investigation of the accidents. Legal definitions of the responsibility or negligence of manufacturers or owners of potentially dangerous equipment were in an embryonic state. The belief existed that the enlightened self-interest of an entrepreneur sufficed to guarantee the public safety. This theory militated against the enactment of any legislation restricting the actions of the manufacturers or users of steam equipment. . . .

The regulatory power of the federal government, then, was not expanded in any authoritarian manner. Rather, it evolved in response to novel conditions emanating from the new machine age, which was clearly seen by that community whose educations or careers encompassed the new technology. In eventually reacting to this danger, Congress passed the first positive regulatory legislation and created the first agency empowered to supervise and direct the internal affairs of a sector of private enterprise in detail. Further, certain congressmen used this precedent later in efforts to protect the public in other areas, notably in proposing legislation that in time created the Interstate Commerce Commission. Marine boiler explosions, then, provoked a crisis in the safe application of steam power, which led to a marked change in

American political attitudes. The change, however, was not abrupt but evolved between 1816 and 1852.

*　　*　　*

Oliver Evans in the United States and Richard Trevithick in England introduced the relatively high-pressure non-condensing steam engine almost simultaneously at the turn of the nineteenth century. This development led to the vast extension in the use of steam power. The high-pressure engines competed in efficiency with the low-pressure type, while their compactness made them more suitable for land and water vehicular transport. But, simultaneously, the scope of the problem faced even by Watt was increased, that is, the construction of boilers that would safely contain the dangerous expansive force of steam. Evans thoroughly respected the potential destructive force of steam. He relied chiefly on safety valves with ample relieving capacity but encouraged sound boiler design by publishing the first formula for computing the thickness of wrought iron to be used in boilers of various diameters carrying different working pressures.

Despite Evans' prudence, hindsight makes it clear that the rash of boiler explosions from 1816 onward was almost inevitable. Evans' design rules were not heeded. Shell thickness and diameter depended upon available material, which was often of inferior quality. In fabrication, no provision was made for the weakening of the shell occasioned by the rivet holes. The danger inherent in the employment of wrought-iron shells with cast-iron heads affixed because of the different coefficients of expansion was not recognized, and the design of internal stays was often inadequate. The openings in the safety valves were not properly proportioned to give sufficient relieving capacity. Gauge cocks and floats intended to ensure adequate water levels were inaccurate and subject to malfunction by fouling with sediment or rust.

In addition, there were also problems connected with boiler operation and maintenance. The rolling and pitching of steamboats caused alternate expansion and contraction of the internal flues as they were covered and uncovered by the water, a condition that contributed to their weakening. The boiler feedwater for steamboats was pumped directly from the surroundings without treatment or filtration, which accelerated corrosion of the shell and fittings. The sediment was frequently allowed to accumulate, thus requiring a hotter fire to develop the required steam pressure, which led, in turn, to a rapid weakening of the shell. Feed pumps were shut down at intermediate stops without damping the fires, which aggravated the danger of low water and excessive steam pressure. With the rapid increase in the number of steam engines, there was a concomitant shortage of competent engineers who understood the necessary safety precautions. Sometimes masters employed mere stokers who had only a rudimentary grasp of the operation of steam equipment. Increased competition also led to attempts to gain prestige by arriving first at the destination. The usual practice during a race was to overload or tie down the safety valve, so that excessive steam pressure would not be relieved.

* * *

The first major boiler disasters occurred on steamboats, and, in fact, the majority of explosions throughout the first half of the nineteenth century took place on board ship. By mid-1817, four explosions had taken five lives in the eastern waters, and twenty-five people had been killed in three accidents on the Ohio and Mississippi rivers. The city council of Philadelphia appears to have been the first legislative body in the United States to take cognizance of the disasters and attempt an investigation. A joint committee was appointed to determine the causes of the accidents and recommend measures that would prevent similar occurrences on steamboats serving Philadelphia. The question was referred to a group of practical engineers who recommended that all boilers should be subjected to an initial hydraulic proof test at twice the intended working pressure and additional monthly proof tests to be conducted by a competent inspector. Also, appreciating the fact that marine engineers were known to overload the safety valve levers, they advocated

COURTESY THE MARINERS MUSEUM

The explosion of the steamboat Helen McGregor *on the Mississippi River, February 24, 1830. More than fifty people were killed and an equal number injured.*

placing the valve in a locked box. The report of the joint committee incorporated these recommendations, but it stated that the subject of regulation was outside the competence of municipalities. Any municipal enactment would be inadequate for complete regulation. The matter was referred, therefore, to the state legislature, and there it rested....

* * *

From 1818 to 1824 in the United States, the casualty figures in boiler disasters rose, about forty-seven lives being lost in fifteen explosions. In May 1824 the "Aetna," built in 1816 to Evans' specifications, burst one of her three wrought-iron boilers in New York harbor, killing about thirteen persons and causing many injuries. Some experts attributed the accident to a stoppage of feedwater due to incrustations in the inlet pipes, while others believed that the rupture in the shell had started from an old fracture in a riveted joint. The accident had two consequences. Because the majority of steamboats plying New York waters operated at relatively low pressures with copper boilers, the public became convinced that wrought-iron boilers were unsafe. This prejudice forced New York boat builders who were gradually recognizing the superiority of wrought iron to revert to the use of copper even in high-pressure boilers. Some owners recognized the danger of this step, but the outcry was too insistent. One is reported to have said: "We have concluded therefore to give them [the public] a copper boiler, the strongest of its class, and have made up our minds that they have a perfect right to be scalded by copper boilers if they insist upon it." His forecast was correct, for within the next decade, the explosion of copper boilers employing moderate steam pressures became common in eastern waters.

The second consequence of the "Aetna" disaster was that it caught the attention of Congress. A resolution was introduced in the House of Representatives in May 1824 calling for an inquiry into the expediency of enacting legislation barring the issuance of a certificate of navigation to any boat operating at high steam pressures. Although a bill was reported out of committee, it was not passed due to lack of time for mature consideration.

In the same year, the Franklin Institute was founded in Philadelphia for the study and promotion of the mechanical arts and applied science. The institute soon issued its *Journal,* and, from the start, much space was devoted to the subject of boiler explosions. The necessity of regulatory legislation dealing with the construction and operation of boilers was discussed, but there was a diversity of opinion as to what should be done. Within a few years, it became apparent that only a complete and careful investigation of the causes of explosions would give sufficient knowledge for suggesting satisfactory regulatory legislation. In June 1830, therefore, the Institute empowered a committee of its members to conduct such an investigation and later authorized it to perform any necessary experiments.

The statement of the purpose of the committee reflects clearly the nature of the problem created by the frequent explosions. The public, it said, would continue to use steamboats, but if there were no regulations, the needless waste of property and life would continue. The committee believed that these were avoidable consequences; the accidents resulted from defective boilers, improper design, or carelessness. The causes, the committee thought, could be removed by salutary regulations, and it affirmed: "That there must be a power in the community lodged somewhere, to protect the people at large against any evil of serious and frequent recurrence, is self-evident. But that such power is to be used with extreme caution, and only when the evil is great, and the remedy certain of success, seems to be equally indisputable."

Here is a statement by a responsible group of technically oriented citizens that public safety should not be endangered by private negligence. It demonstrates the recognition that private enterprise was considered sacrosanct, but it calls for a reassessment of societal values in the light of events. It proposes restrictions while still professing unwillingness to fetter private industry. It illustrates a change in attitude that was taking place with respect to the role of government in the affairs of industry, a change that was necessitated by technological innovation. The committee noted that boiler regulation proposals had been before Congress twice without any final action. Congressional committees, it said, appeared unwilling to institute inquiries and elicit evidence from practical men, and therefore they could hardly determine facts based upon twenty years of experience with the use of steam in boats. Since Congress was apparently avoiding action, the committee asserted, it was of paramount importance that a competent body whose motives were above suspicion should shoulder the burden. Thus, the Franklin Institute committee began a six-year investigation of boiler explosions.

From 1825 to 1830, there had been forty-two explosions killing about 273 persons, and in 1830 a particularly serious one aboard the "Helen McGregor" near Memphis, which killed 50 or 60 persons, again disturbed Congress. The House requested the Secretary of the Treasury, Samuel D. Ingham of Pennsylvania, to investigate the boiler accidents and submit a report.... Ingham committed government funds to the [Franklin] Institute to defray the cost of apparatus necessary for the experiments. This was the first research grant of a technological nature made by the federal government....

In his State of the Union message in December 1833, President Jackson noted that the distressing accidents on steamboats were increasing. He suggested that the disasters often resulted from criminal negligence by masters of the boats and operators of the engines. He urged Congress to pass precautionary and penal legislation to reduce the accidents. . . . [But a] reported Senate bill failed to pass.

*　　*　　*

A program of experiments [was] carried out by the Franklin Institute from 1831 to 1836. . . .

Taken as a whole, the Franklin Institute reports demonstrate remarkable experimental technique as well as a thorough methodological approach. They exposed errors and myths in popular theories on the nature of steam and the causes of explosions. They laid down sound guidelines on the choice of materials, on the design and construction of boilers, and on the design and arrangement of appurtenances added for their operation and safety. Further, the reports included sufficient information to emphasize the necessity for good maintenance procedures and frequent proof tests, pointing out that the strength of boilers diminished as the length of service increased.

*　　*　　*

The Franklin Institute report on steam boiler explosions was presented to the House through the secretary of the treasury in March 1836, and the report on boiler materials was available in 1837. The Franklin Institute committee also made detailed recommendations on provisions that any regulatory legislation should incorporate. It proposed that inspectors be appointed to test all boilers hydraulically every six months; it prohibited the licensing of ships using boilers whose design had proved to be unsafe; and it recommended penalties in cases of explosions resulting from improper maintenance, from the incompetence or negligence of the master or engineer, or from racing. It placed responsibility for injury to life or property on owners who neglected to have the required inspections made, and it recommended that engineers meet certain standards of experience, knowledge, and character. The committee had no doubt of the right of Congress to legislate on these matters.

Congress did not act immediately. In December 1836 the House appointed a committee to investigate the explosions, but there was no action until after President Van Buren urged the passage of legislation in December 1837. That year witnessed a succession of marine disasters. Not all were attributable to boiler explosions, although the loss of 140 persons in a new ship, the "Pulaski," out of Charleston, was widely publicized. The Senate responded quickly to Van Buren's appeal, passing a measure on January 24, 1838. The House moved less rapidly. An explosion aboard the "Moselle" at Cincinnati in April 1838, which killed 151 persons, caused several Congressmen to request suspension of the rules so that the bill could be brought to the floor, but in the face of more pressing business the motion was defeated. The legislation was almost caught in the logjam in the House at the end of the session, but on June 16 the bill was brought to the floor. Debate centered principally upon whether the interstate commerce clause in the Constitution empowered Congress to pass such legislation. Its proponents argued affirmatively, and the bill was finally approved and became law on July 7, 1838.

The law incorporated several sections relating to the prevention of collisions, the control of fires, the inspection of hulls, and the carrying of lifeboats. It provided for the immediate appointment by each federal judge of a competent boiler inspector having no financial interest in their manufacture. The inspector was to examine every steamboat boiler in his area semiannually, ascertain its age and soundness, and certify it with a recommended working pressure. For this service the owner paid $5.00—his sole remuneration—and a license to navigate was contingent upon the receipt of this certificate. The law specified no inspection criteria. It enjoined the owners to employ a sufficient number of competent and experienced engineers, holding the owners responsible for loss of life or property damage in the event of a boiler explosion for their failure to do so. Further, any steamboat employee whose negligence resulted in the loss of life was to be considered guilty of manslaughter, and upon conviction could be sentenced to not more than ten years imprisonment. Finally, it provided that in suits against owners for damage to persons or property, the fact of the bursting of the boilers should be considered prima facie evidence of negligence until the defendant proved otherwise. . . .

The elimination of inspection criteria and the qualification of engineers rendered the measure ineffectual. . . .

The disappointment of the informed public concerning the law was voiced immediately in letters solicited by the secretary of the treasury, contained in a report that he submitted to Congress in December 1838. There were predictions that the system of appointment and inspection would encourage corruption and graft. There were complaints about the omission of inspection criteria and a provision for the licensing of engineers. One correspondent pointed out that it was impossible legally to determine the experience and skill of an engineer, so that the section of the law that provided penalties for owners who failed to employ experienced and skillful engineers was worthless. One critic who believed that business interests had undue influence upon the government wrote: "We are mostly ruled by corporations and joint-stock companies. . . . If half the citizens of this country should get blown up, and it should be likely to affect injuriously the trade and commerce of the other half by bringing to justice the guilty, no elective officer would risk his popularity by executing the law.". . .

*   *   *

Experience proved that the 1838 law was not preventing explosions or loss of life. In the period 1841–48, there were some seventy marine explosions that killed about 625 persons. In December 1848 the commissioner of patents, to whom Congress now turned for data, estimated that in the period 1816–48 a total of 233 steamboat explosions had occurred in which 2,563 persons had been killed and 2,097 injured, with property losses in excess of $3 million.

In addition to the former complaints about the lack of proof tests and licenses for engineers, the commissioner's report included testimony that the inspection methods were a mockery. Unqualified inspectors were being appointed by district judges through the agency of highly placed friends. The inspectors regarded the position as a lifetime office. Few even looked at the boilers but merely collected their fees. The inspector at New York City complained that his strict inspection caused many boats to go elsewhere for inspections. He cited the case of the "Niagara," plying between New York City and Albany, whose master declined to take out a certificate from his office because it recommended a working pressure of only 25 p.s.i. on the boiler. A few months later the boiler of the "Niagara," which had been certified in northern New York, exploded while carrying a pressure of 44 p.s.i. and killed two persons.

Only eighteen prosecutions had been made in ten years under the manslaughter section of the 1838 law. In these cases there had been nine convictions, but the penalties had, for the most part, been fines which were remitted. It was difficult to assemble witnesses for a trial, and juries could not be persuaded to convict a man for manslaughter for an act of negligence, to which it seemed impossible to attach this degree of guilt. Also, the commissioner's report pointed out that damages were given in cases of bodily injury but that none were awarded for loss of life in negligence suits. It appeared that exemplary damages might be effective in curbing rashness and negligence.

The toll of life in 1850 was 277 dead from explosions, and in 1851 it rose to 407. By this time Great Britain had joined France in regulatory action, which the Congress noted. As a consequence of legislation passed in 1846 and 1851, a rejuvenated Board of Trade was authorized to inspect steamboats semiannually, to issue or deny certificates of adequacy, and to investigate and report on accidents. The time had come for the Congress to take forceful action, and in 1852 it did.

John Davis, Whig senator from Massachusetts, who had favored stricter legislation in 1838, was the driving force behind the 1852 law. In prefacing his remarks on the general provisions of the bill, he said: "A very extensive correspondence has been carried on with all parts of the country... there have been laid before the committee a great multitude of memorials, doings of chambers of commerce, of boards of trade, of conventions, of bodies of

engineers; and to a considerable extent of all persons interested, in one form or another, in steamers . . . in one thing . . . they are all . . . agreed—that is, that the present system is erroneous and needs correction."

Thus again, the informed public submitted recommendations on the detailed content of the measure. An outstanding proponent who helped shape the bill was Alfred Guthrie, a practical engineer from Illinois. With personal funds, Guthrie had inspected some two hundred steamboats in the Mississippi valley to ascertain the causes of boiler explosions. Early in the session, Senator Shields of Illinois succeeded in having Guthrie's report printed, distributed, and included in the Senate documents. Guthrie's recommendations were substantially those made by the Franklin institute in 1836. His reward was the post as first supervisor of the regulatory agency which the law created.

After the bill reached the Senate floor, dozens of amendments were proposed, meticulously scrutinized, and disposed of. The measure had been, remarked one senator, "examined and elaborated . . . more patiently, thoroughly, and faithfully than any other bill before in the Senate of the United States." As a result, in place of the 1838 law which embodied thirteen sections and covered barely three pages, there was passed such stringent and restrictive legislation that forty-three sections and fourteen pages were necessary. . . .

A new feature of the law, which was most indicative of the future, was the establishment of boards of inspectors empowered to investigate infractions or accidents, with the right to summon witnesses, to compel their attendance, and to examine them under oath. Above the local inspectors were nine supervisors appointed by the President. Their duties included the compilation of evidence for the prosecution of those failing to comply with the regulations and the preparation of reports to the secretary of the treasury on the effectiveness of the regulations. Nor did these detailed regulations serve to lift the burden of presumptive negligence from the shoulders of owners in cases of explosion. The explosion of boilers was not made prima facie evidence as in the 1838 law, but owners still bore a legal responsibility. This was made clear in several court decisions which held that proof of strict compliance with the 1852 law was not a sufficient defense to the allegations of loss by an explosion caused by negligence.

The final Senate debate and the vote on this bill shows how, in thirty years, the public attitude and, in turn, the attitudes of its elected representatives had changed toward the problem of unrestricted private enterprise, mainly as a result of the boiler explosions. The opponents of the bill still argued that the self-interest of the steamboat companies was the best insurance of the safety of the traveling public. But their major argument against passage was the threat to private property rights which they considered the measure entailed. . . . This expression of a belief that Congress should in no circumstances interfere with private enterprise was now supported by only a small minority. One proponent of the bill replied: "I consider that the only question involved in the bill is this: Whether we shall permit a legalized, unquestioned, and peculiar class in the community to go on committing murder at will, or whether we shall make such enactments as will compel them to pay some attention to the value of life." It was, then, a question of the sanctity of private property rights as against the duty of government to act in the public weal. On this question the Senate voted overwhelmingly that the latter course should prevail.

Though not completely successful, the act of 1852 had the desired corrective effects. During the next eight years prior to the outbreak of the Civil War, the loss of life on steamboats from all types of accidents dropped to 65 per cent of the total in the corresponding period preceding its passage. A decade after the law became effective, John C. Merriam, editor and proprietor of the *American Engineer,* wrote: "Since the passage of this law steamboat explosions on the Atlantic have become almost unknown, and have greatly decreased in the west. With competent inspectors, this law is invaluable, and we hope to hail the day when a similar act is passed in every legislature, touching locomotive and stationary boilers."

There was, of course, hostility and opposition to the law immediately after its passage, particularly among the owners and masters of steamboats. It checked the steady rise in the construction of new boats, which had been characteristic of the earlier years. The effect, however, was chastening rather than emasculating. Associations for the prevention of steam boiler explosions were formed; later, insurance companies were organized to insure steam equipment that was manufactured and operated with the utmost regard for safety. In time, through the agency of the American Society of Mechanical Engineers, uniform boiler codes were promulgated and adopted by states and municipalities.

Thus, the reaction of the informed public, expressed by Congress, to boiler explosions caused the initiation of positive regulation of a sector of private enterprise through a governmental agency. The legislation reflected a definite change of attitude concerning the responsibility of the government to interfere in those affairs of private enterprise where the welfare and safety of the general public was concerned. . . .

Bursting steamboat boilers, then, should be viewed not merely as unfortunate and perhaps inevitable consequences of the early age of steam, as occurrences which plagued nineteenth-century engineers and which finally, to a large degree, they were successful in preventing. They should be seen also as creating a dilemma as to how far the lives and property of the general public might be endangered by unrestricted private enterprise. The solution was an important step toward the inauguration of the regulatory and investigative agencies in the federal government.

SANFORD A. LAKOFF

Knowledge, Power, And Social Purpose

The strong government support of science and technology that was noted by Pursell has continued through the 1970s. In 1978, for example, approximately $26.3 billion was appropriated for research and development, of which slightly less than half was devoted to national defense. Some people, pointing to the unprecedented achievements that have been made in nuclear energy, space, and medicine over the past few decades, view the government's encouragement and active promotion of science and technology as having been entirely beneficial. Others agree with Jerome Ravetz, whose ideas were presented in an earlier selection, in the belief that massive government support has corrupted science. In this selection, political scientist Sanford Lakoff asserts that government involvement in science and technology in an advanced society is absolutely necessary. As knowledge becomes a source of power, he concludes, it is only government that can make certain that the power is used for broad social purposes.

One major element in the advance of industrial society is of course the change from a dependence on mere applications of scientific knowledge to a situation where basic improvements in scientific understanding are sought in the hope that these will stimulate innovation. In this sense science has indeed become, as Vannevar Bush described it, an "endless frontier" replacing the incentives and opportunities afforded by open spaces and expanding markets. Whereas the old frontier functioned to stimulate investment, to keep up demand for labor, and to dampen social protest, the new indefinite frontier opened by science offers opportunity for the development of new knowledge which can be used to prolong life, curb population growth, increase the supply of food and other consumer articles, speed communication throughout the world and provide weapons of unprecedented destructiveness.

The role of government in the advancement of science has been recognized as crucial. Although privately sponsored industrial research accounts for much of the scientific and technical progress achieved in the last twenty years and much more of the marketable goods, private sources cannot be expected to provide the trained scientific personnel, the facilities, and the projects which are necessary for a full range of scientific activity and which may or may not have immediate dividends. Understanding the relations of government and science has therefore become more important than ever.

Some scientific progress may be of great utility to the developing society, especially in the effort to increase food supply, improve public health, and bring population increase under control. On the whole, however, science of a highly sophisticated order is more of a requisite to the advanced society. It is in the advanced society, for example, that the volume of information required in ordinary business transactions is so high and the rate of flow of such a frequency that computerized systems of storage and retrieval become indispensable. It is in the advanced society where separate avenues of research are so well developed that they overlap and reenforce each other, making possible collaborative advances in industrial products, medicine, and even fundamental knowledge. It is in the advanced society that metropolitan congestion has become so acute that only the imaginative application of engineering and the ecological arts can prevent inconvenience and even paralysis. It is in the advanced society that the industrial waste and pollution of natural resources become a problem requiring a comprehensive effort to understand, to plan, and carry into execution a systematic effort to conserve the natural endowment. It is the advanced society, finally, that finds itself called upon to use its productive and scientific superiority to assist other nations struggling to modernize. Self-interest, no less than idealism, demands that this call be heeded by the commitment of scientific resources as well as by the shipment of relief supplies which, in the long run, do not provide lasting help.

In America the early stages of government involvement with science were almost exclusively centered about military needs. Lately they have been concerned with the broader needs for science as well as with the exploration of outer space. Increasingly they are becoming related to the needs of the economy and the state of social health and welfare. Throughout the phases of this effort, government funding has altered the relationship between public and nonpublic bodies. It is not only that government provides the wherewithal.

To some extent government must also determine or influence the pattern of activity by deciding how and where to spend the money. Its spending in effect induces movements of both personnel and capital. Its own research activities provide both jobs and training and vital services. In the past two decades government support of science has provided the impetus not only for a breakdown of the traditional separation between public and private but for the development of new intermediate institutions—such as the nonprofit corporation and the university-affiliated research center. It has become necessary for government to solicit advice from those in the scientific professions concerning the allocation of its funds, decisions to approve or disapprove projects for sponsorship, and a broad spectrum of policy issues in which science plays a role, such as proposals in defense policy.

But these developments are only a beginning. They provide only a shadowy outline of the scientific society that is gradually emerging from the industrial society. We can only guess at the shape this new society will take. From what we know now, it would appear that the changeover need not involve the sort of radical transformation in the social structure that many early theorists foresaw. Some changes, however, that are in some sense fundamental, do seem in prospect. It is hard to imagine, for example, that individuals, institutions, or societies will be as independent and as isolated in the future as they have been in the past. Larger units of organization seem inevitable in many areas of social life. Patterns of work and enterprise may well have to be adjusted to an economic system in which none of the components can be permitted great autonomy without endangering the entire system. Much will depend, of course, upon how humanity responds to the complexities and challenges that knowledge brings. Will the nations of the world succeed in securing themselves against the unprecedented dangers posed by modern weapons of mass destruction? Will particular societies appreciate in time the need to make social adjustments to accommodate scientific and technical change?

Questions like these are in the first instance questions of practicality. But they are also, in concrete terms, questions of human values. The answers to them inevitably require an effort to bend both knowledge and power to some conception of social purpose. In the largest sense, this is precisely the challenge that modern science holds for society. The social theorists of earlier times were right to predict that scientific knowledge would become a source of enormous power to society. They also saw that, after the science of nature had begun to make its great contributions, an effort would have to be made to establish a science of man as a complementary source of power. It is by no means certain that this effort can succeed and it may be argued that if it does succeed, it will pose far more of a threat than natural science has ever posed.

But there is no escaping the fundamental fact that as knowledge of any kind becomes a source of power, questions of purpose will arise. The test of a great society is surely not simply that it amasses knowledge and turns that knowledge into power, but rather that it does so with a careful and constant attention to the purposes which learning may serve. The test will not be met successfully by reliance on a technocratic elite. In a self-governing society, only the citizenry, guided but not governed by professional specialists, has the right to decide what shall be done with the resources and the human ability which are its inheritance.

PART FIVE

Retrospect And Prospect

Introduction

In the previous sections we have read some of the criticisms and defenses of modern technology and have learned about some of the effects of innovation. We have also explored the conditions underlying technological change and have surveyed the principal sources of innovation. In this section, we shall study more closely the process of invention and innovation, investigate some of the ethical issues involved in the process of technological change, and review some current attempts to monitor new technologies.

We know that technological innovation produces social and cultural change, but it is surprising how little we really know about inventive activity. How and under what conditions invention or innovation occurs is the subject of selections by White, Fairbairn, Schmookler, Ferguson, and Jewkes. Lynn White describes eleven inventions that occurred in antiquity and the Middle Ages. He concludes from his study that we learn nothing by attributing invention to a spark of genius. Invention, he thinks, results from the application of the human intellect to a particular problem existing in the environment. Sir William Fairbairn's account appears to bear out White's analysis. In this instance, the invention of a riveting machine was provoked by a strike of workers, who were performing the riveting by hand. Jacob Schmookler also agrees with White's thinking in its main points. Inventors become intensely preoccupied with the problem at hand, but he maintains that the desire for economic gain stimulates their inventive activity.

Eugene Ferguson delves more deeply into the inventive process, which he attributes to "visual thinking." This, Ferguson believes, is a peculiar mode of thought, allied to artistic talent, in which the inventor imagines the completed product or process in his mind's eye, after which he carries out the image into the actual design. Ferguson cites historical examples to prove his thesis, and concludes that modern engineering schools are actually blunting inventive activity by their emphasis on a mathematical approach to solving problems.

John Jewkes and his collaborators also submit the assumptions about invention and inventive activity to a searching analysis. They wonder whether our current attempts to institutionalize invention in industrial research laboratories may not be misguided, since this may actually paralyze inventive talent, and they question the extent to which scientific research results in techno-logical innovation. Recall that Derek Price in his selection on the increase of scientific activity believed that its rate of growth would stabilize within a cen-tury. If the misgivings voiced by Ferguson and Jewkes have any validity, then the pace of innovation and resulting social change may also slacken. It seems clear, in any case, that what innovations the future may bring are largely unpredictable.

We should also inquire as to whether the changes in society and among individuals brought about by technology have been beneficial. Friedrich Engels, writing a century ago, asserted that the technology of industrializa-tion had enslaved the workers. People could only be free, he preached, if the workers revolted and seized the means of production, thus creating a socialist state. The ideal type of state socialism envisaged by Engels has not come into existence. The major reason it has not, Clarence Ayres believes, is that in-dustrial technology gave workers security and abundance, which he stresses are the preconditions of freedom and equality. David Lilienthal thinks that technology can also enhance democracy. The massive development of the Tennessee Valley was not mandated in all of its details by Washington ex-perts, he writes. Instead, it was the participation of local citizens that guided the development of the valley's resources and ensured the reclamation of agricultural lands.

But, as the selection by Arthur Morgan demonstrates, technology in the hands of arrogant and powerful bureaucrats can be, and has been, a de-humanizing force. The insensitivity of the U.S. Corps of Engineers in plan-ning the site of the Garrison Dam, he charges, destroyed the civilization of the Arikara Indians, who had lived peacefully on the land for centuries. James Wallace, too, warns that just because a project is technically feasible, it should not necessarily be implemented. Increasing scientific knowledge and techno-logical capability, he writes, should not be permitted to set the moral and ethical standards for human beings.

Kenneth Keniston believes that we are at a point of crisis insofar as tech-nology and ethics are concerned. Each generation, he says, knows more about nature and possesses greater technical capability than the previous one, with the result that each creates new definitions of what is ethical. What Keniston fears, however, is the emergence of a "technological man," who being ethically neutral may employ technology for either ethical or unethical purposes without truly knowing the difference. Thus far, he concludes, ad-vancing technology has neither created nor extended an ethic which has as its central focus the preservation and enhancement of human life on earth. Until such an ethic is developed, he warns, the future of humanity remains uncertain.

It is just possible that current technological change has encouraged us to take the first faltering steps toward developing the ethical stance that Keniston thinks is necessary. There is a strong commitment to aid the peoples of the Third World, even though some attempts have been abortive or have failed. In our own country many people have voluntarily chosen to live simpler lives. Further, it has been through the efforts and activity of citizens' groups, some led or advised by scientists and engineers, that the federal government in the last decade has established the Environmental Protection Agency, the Consumer Products Safety Commission, and the Occupational Health and Safety Administration. The goal of all of these agencies is to ensure that technological advance does not threaten life and safety.

Congress, in creating an Office of Technology Assessment in 1972, recognized the need to monitor inventions and innovations prior to their development and widespread application. There has been much debate about the process of technology assessment. To begin with, because the effects of innovation are largely unpredictable, many critics think that assessment is a waste of time. Peter Drucker is among these, but he goes further in his criticism by pointing out that assessment may actually lead to the shelving of new technologies that we badly need. Monitoring technology is fine, he concludes, but it should be performed after the impact of an innovation is recognized and when it can be fully evaluated.

According to Langdon Winner, technology assessment is of no particular value, because it is just another method of government regulation. It is, he asserts, just a new façade to cover the usual trade-offs made by politicians, in which the powerless will suffer as usual. Winner does not conceal his cynicism about our present political process. Modern technology, he writes, has created institutions, politics included, in which the public has no voice. The only way to put things right, he concludes, is to disassemble the present technological structure and create new technologies that will ensure freedom, equality, and democracy.

Although admitting that there are problems involved in political decision-making regarding technological matters, Dorothy Nelkin sees this as a challenge to democracy. She believes public participation is necessary but recognizes that it must be informed by technical experts.

In his review of the Office of Technology Assessment and of other agencies monitoring new technologies, Harvey Brooks admits the deficiencies noted by Drucker and Winner. Nevertheless, he thinks that these are not sufficient reasons for abandoning the assessment system, and he rejects the idea that technology determines the political process. He is, however, wary of public participation in technological decision-making, because he fears that people can be manipulated too easily by advocates of one point of view or another. Brooks thinks that some benefits have emerged from technology assessment, but concedes that it may not be the final answer for controlling technological innovation.

There are no simple explanations, then, of how technological change comes about, and there are no simple solutions to the problems engendered by such change. Experience and education, though, can give us the capability of recognizing when either new or old technologies are not being used to enhance human life the world over. We have seen that the democratic process has eliminated or alleviated problems caused by technological change, and that active and informed public participation in this process can make a difference. Technology can serve human purposes, and the task ahead is to be certain that it does.

LYNN WHITE, JR.

The Act of Invention

Historian Lynn White describes eleven instances of technological invention that occurred in antiquity and in the Middle Ages, and discusses the problems of the nature, the motivations, the conditioning circumstances, and the effects of these acts. He also notes that many inventions made in China and other parts of Asia diffused to Europe, where they were more effectively exploited. He asserts that we learn nothing if we ascribe the act of invention to genius. Looking at the range of cases, White believes that quite possibly a cluster of different factors enters into technological creativity, which emerges as a result of the application of intellect to existing environmental conditions.

The rapidly growing literature on the nature of technological innovation and its relation to other activities is still largely rubbish because so few of the relevant concrete facts have thus far been ascertained. It is an inverted pyramid of generalities, the apex of which is very nearly a void.... The best that we can do at present is to work hard to find the facts and then to think cautiously about the facts which have been found.

In view of our ignorance, then, it would seem wise to discuss the problems of the nature, the motivations, the conditioning circumstances, and the effects of the act of invention far less in terms of generality than in terms of specific instances about which something seems to be known.

The beginning of wisdom may be to admit that even when we know some facts in the history of technology, these facts are not always fully intelligible—that is, capable of "explanation"—simply because we lack adequate contextual information. The Chumash Indians of the coast of Santa Barbara County built plank boats which were unique in the pre-Columbian New World: their activity was such that the Spanish explorers of California named a Chumash village "La Carpintería." A map will show that this tribe had a particular inducement to venture upon the sea: they were enticed by the largest group of offshore islands along the Pacific Coast south of Canada. But why did the tribes of South Alaska and British Columbia, of Araucanian Chile, or of the highly accidented Eastern coast of the United States never respond to their geography by building plank boats? Geography would seem to be only one element in explanation. To be sure, the Chumash coast is one of the few Pacific areas where there are seepages of tarlike petroleum good for calking plank boats. But if one has never seen such a boat, this use is not obvious.

Can a plank-built East Asian boat have drifted on the great arc of currents in the North Pacific to the Santa Barbara region? It is entirely possible; but such boats would have been held together by pegs, whereas the Chumash boats were lashed, like the dhows of the Arabian Sea or the early Norse ships. Diffusion seems improbable.

Since a group can conceive nothing which is not first conceived by a person, we are left with the hypothesis of a genius: a Chumash Indian who at some unknown date achieved a breakaway from log dugout and reed balsa to the plank boat. But the idea of "genius" is itself an ideological artifact of the age of the Renaissance when painters, sculptors, and architects were trying to raise their social status above that of craftsmen. Does the notion of genius "explain" Chumash plank boats? On the contrary, it would seem to be no more than a traditionally acceptable way of labeling the great Chumash innovation as unintelligible. All we can do is to observe the fact of it and hope that eventually we may grasp the meaning of it.

A symbol of the rudimentary nature of our thinking about technology, its development, and its human implications, is the fact that while the *Encyclopaedia Britannica* has an elaborate article on "Alphabet," thus far it contains no discussion of its own organizational presupposition, alphabetization. Alphabetization is the basic invention for the storage and recovery of information; it is fully comparable in significance to the Dewey decimal system and to the new electronic devices for these purposes. Modern big business, big government, big scholarship are inconceivable without alphabetization. One hears that the chief reason why the Chinese Communist regime has decided to Romanize Chinese writing is the inefficiency of trying to classify everything from telephone books to tax registers in terms of 214 radicals of ideographs. . . .

Sterling Dow of Harvard University . . . believes that the earliest evidence of alphabetization is found in Greek materials of the third century B.C. In other words, there was a thousand-year gap between the invention of the alphabet as a set of phonetic symbols and the realization that these symbols, and their sequence in individual written words, could be divorced from their phonetic function and used for an entirely different purpose: an arbitrary but very useful convention for storage and retrieval of verbal materials. That we have neglected thus completely the effort to understand so fundamental an invention should give us humility whenever we try to think about the larger aspects of technology.

Coinage was one of the most significant and rapidly diffused innovations of late Antiquity. The dating of it has recently become more conservative than formerly: the earliest extant coins were sealed into the foundation of the temple of Artemis at Ephesus about 600 B.C., and the invention of coins—that is, lumps of metal the value of which is officially certified—was presumably made in Lydia not more than a decade earlier.

Here we seem to know something, at least until the next archaeological spades turn up new testimony. But what do we know with any certainty about the impact of coinage? We are compelled to tread the slippery path of *post hoc ergo propter hoc.** There was a great acceleration of commerce in the Aegean, and it is hard to escape the conviction that this movement, which is the economic presupposition of the Periclean Age, was lubricated by the invention of coinage.

If we dare to go this far, we may venture further. Why did the atomic theory of the nature of matter appear so suddenly among the philosophers of the Ionian cities? Their notion that all things are composed of different arrangements of identical atoms of some "element," whether water, fire, ether, or something else, was an intellectual novelty of the first order, yet its sources have not been obvious. The psychological roots of atomism would seem to be found in the saying of Heraclitus of Ephesus that "all things may be reduced to fire, and fire to all things, just as all goods may be turned into gold and gold into all goods." He thought that he was just using a metaphor, but the metaphor had been possible for only a century before he used it.

Here we are faced with a problem of critical method. Apples had been dropping from trees for a considerable period before Newton discovered gravity: we must distinguish cause from occasion. But the appearance of coinage is a phenomenon of a different order from the fall of an apple. The unprecedented element in the general life of sixth-century Ionia, the chief stimulus to the prosperity which provided leisure for the atomistic philosophers, was the invention of coinage: the age of barter was ended. Probably no Ionian was conscious of any connection between this unique new technical instrument and the brainstorms of the local intellectuals. But that a causal

*Editors' note: After this, therefore because of this.

relationship did exist can scarcely be doubted, even though it cannot be "proved" but only perceived.

Fortunately, however, there are instances of technological devices of which the origins, development, and effects outside the area of technology are quite clear. A case in point is the pennon.

The stirrup is first found in India in the second century B.C. as the big-toe stirrup. For climatic reasons its diffusion to the north was blocked, but it spread wherever India had contact with barefoot aristocracies, from the Philippines and Timor on the east to Ethiopia on the west. The nuclear idea of the stirrup was carried to China on the great Indic culture wave which also spread Buddhism to East Asia, and by the fifth century the shod Chinese were using a foot stirrup.

The stirrup made possible, although it did not require, a new method of fighting with the lance. The unstirruped rider delivered the blow with the strength of his arm. But stirrups, combined with a saddle equipped with pommel and cantle, welded rider to horse. Now the warrior could lay his lance at rest between his upper arm and body: the blow was delivered not by the arm but by the force of a charging stallion. The stirrup thus substituted horse-power for man-power in battle.

The increase in violence was tremendous. So long as the blow was given by the arm, it was almost impossible to impale one's foe. But in the new style of mounted shock combat, a good hit might put the lance entirely through his body and thus disarm the attacker. This would be dangerous if the victim had friends about. Clearly, a baffle must be provided behind the blade to prevent penetration by the shaft of the lance and thus permit retraction.

Some of the Central Asian peoples attached horse tails behind the blades of lances—this was probably being done by the Bulgars before they invaded Europe. Othes nailed a piece of cloth, or pennon, to the shaft behind the blade. When the stirrup reached Western Europe about 730 A.D., an effort was made to meet the problem by adapting to military purposes the old Roman boar spear which had a metal crosspiece behind the blade precisely because boars, bears, and leopards had been found to be so ferocious that they would charge up a spear not so equipped.

This was not, however, a satisfactory solution. The new violence of warfare demanded heavier armor. The metal crosspiece of the lance would sometimes get caught in the victim's armor and prevent recovery of the lance. By the early tenth century, Europe was using the Central Asian cloth pennon, since even if it got entangled in armor it would rip and enable the victor to retract his weapon.

Until our dismal age of camouflage, fighting men have always decorated their equipment. The pennons on lances quickly took on color and design. A lance was too long to be taken into a tent conveniently, so a knight usually set it upright outside his tent, and if one were looking for him, one looked first for the flutter of his familiar pennon. Knights riding held their lances erect, and since their increasingly massive armor made recognition difficult, each came to be identified by his pennon. It would seem that it was from the pennon that distinctive "connoissances" were transferred to shield and surcoat. And with the crystallization of the feudal structure, these heraldic devices became hereditary, the symbols of status in European society.

In battle, vassals rallied to the pennon of their liege lord. Since the king was, in theory if not always in practice, the culmination of the feudal hierarchy, his pennon took on a particular aura of emotion: it was the focus of secular loyalty. Gradually a distinction was made between the king's two bodies, his person and his "body politic," the state. But a colored cloth on the shaft of a spear remained the primary symbol of allegiance to either body, and so remains even in polities which have abandoned monarchy. The grimly functional rags first nailed to lance shafts by Asian nomads have had a great destiny. But it is no more remarkable than that of the cross, a hideous implement in the Greco-Roman technology of torture, which was to become the chief symbol of the world's most widespread religion.

In tracing the history of the pennon, and of many other technological items, there is a temptation to convey a sense of inevitability. However, a novel technique merely offers opportunity; it does not command. As has been mentioned, the big-toe stirrup reached Ethiopia. It was still in common use there in the nineteenth century, but at the present time Muslim and European influences have replaced it with the foot stirrup. However, travelers tell me that the Ethiopian gentleman, whose horse is equipped with foot stirrups, rides with only his big toes resting in the stirrups.

Indeed, in contemplating the history of technology, and its implications for our understanding of ourselves, one is as frequently astonished by blindness to innovation as by insights of invention. The Hellenistic discovery of the helix was one of the greatest of technological inspirations. Very quickly it was applied not only to gearing but also to the pumping of water by the so-called Archimedes screw. Somewhat later the holding screw appears in both Roman and Germanic metalwork. The helix was taken for granted thenceforth in Western technology. Yet, despite the sophistication of the Chinese in many technical matters, no form of helix was known in East Asia before modern times: it reached India but did not pass the Himalayas. Indeed, I have not been able to locate any such device in the Far East before the early seventeenth century when Archimedes screws, presumably introduced by the Portuguese, were used in Japanese mines.

Next to the wheel, the crank is probably the most important single element in machine design, yet until the fifteenth century the history of the crank is a dismal record of inadequate vision of its potentialities. It first appears in China under the Han dynasty, applied to rotary fans for winnowing hulled rice, but its later applications in the Far East were not conspicuous. In the West the crank seems to have developed independently and to have emerged from the hand quern. The earliest querns were fairly heavy, with a handle, or handles, inserted laterally in the upper stone, and the motion was reciprocating. Gradually the stones grew lighter and thinner, so that it was harder to insert the peg-handle horizontally: its angle creeps upward until eventually it stands vertically on top. All the querns found at the Saalburg had horizontal handles, and it is increasingly clear that the vertical peg is post-Roman.

Seated before a quern with a single vertical handle, a person of the twentieth century would give it a continuous rotary motion. It is far from clear that one of the very early Middle Ages would have done so. Crank motion was a kinetic invention more difficult than we can easily conceive. Yet at some point before the time of Louis the Pious the sense of the appropriate motion changed; for out of the rotary quern came a new machine, the rotary grindstone, which (as the Latin term for it, *mola fabri,* shows) is the upper stone of a quern turned on edge and adapted to sharpening. Thus, in Europe at least, crank motion was invented before the crank, and the crank does not appear before the early ninth century. As for the Near East, I find not even the simplest application of the crank until al-Jazarī's book on automata of 1206 A.D.

Once the simple crank was available, its develoment into the compound crank and connecting rod might have been expected quite quickly. Yet there is no sign of a compound crank until 1335, when the Italian physician of the Queen of France, Guido da Vigevano, in a set of astonishing technological sketches, illustrates three of them. By the fourteenth century, Europe was using crankshafts with two simple cranks, one at each end. Guido was interested in the problem of self-moving vehicles: paddlewheel boats and fighting towers propelled by windmills or from the inside. For such constricted situations as the inside of a boat or a tower it apparently occurred to him to consolidate the two cranks at the ends of the crankshaft into a compound crank in its middle. It was an inspiration of the first order, yet nothing came of it. Evidently the Queen's physician, despite his technological interests, was socially too far removed from workmen to influence the technology of his time. The compound crank's effective appearance was delayed for another three generations.... Thereafter the idea spread like wildfire, and European applied mechanics was revolutionized.

How can we understand the lateness of the discovery, whether in China or Europe, of even the simple crank, and then the long delay in its wide application and elaboration? Continuous rotary motion is typical of inorganic matter, whereas reciprocating motion is the sole movement found in living things. The crank connects these two kinds of motion; therefore we who are organic find that crank motion does not come easily to us. The great physicist and philosopher Ernst Mach noticed that infants find crank motion hard to learn. Despite the rotary grindstone, even today razors are whetted rather than ground: we find rotary motion a bar to the greatest sensitivity. Perhaps as early as the tenth century the hurdy-gurdy was played with a cranked resined wheel vibrating the strings. But by the thirteenth century the hurdy-gurdy was ceasing to be an instrument for serious music. It yielded to the reciprocating fiddle bow, an introduction of the tenth century from Java which became the foundation of modern European musical development. To use a crank, our tendons and muscles must relate themselves to the motion of galaxies and electrons. From this inhuman adventure our race long recoiled.

A sequence originally connected with the crank may serve to illustrate another type of problem in the act of technological innovation: the fact that a simple idea transferred out of its first context may have a vast expansion. The earliest appearance of the crank, as has been mentioned, is found on Han-dynasty rotary fans to winnow husked rice. The identical apparatus appears in the eighteenth century in the Palatinate, in upper Austria and the Sieben-bürgen, and in Sweden. I have not seen the exact channel of this diffusion traced, but it is clearly part of the general Jesuit-inspired *chinoiserie* of Europe in that age. I strongly suspect, but cannot demonstrate, that all subsequent rotary blowers, whether in furnaces, dehydrators, wind tunnels, air-conditioning systems, or the simple electric fan, are descended from this Han machine, which seems, in China itself, to have produced no progeny.

Doubtless when scholarship in the history of technology becomes firmer, another curious device will illustrate the same point. To judge by its wide distribution, the fire piston is an old invention in Malaya. Thomas Kuhn of Princeton, who has made careful studies of the history of our knowledge of adiabatic heat, assures me that when the fire piston appeared in late eighteenth-century Europe not only for laboratory demonstrations but as a commercial product to light fires, there is no hint in the purely scientific publications that its inspiration was Malayan. But the scientists, curiously, also make no mention of the commercial fire pistons then available. So many Europeans, especially Portuguese and Netherlanders, had been trading, fighting, ruling, and evangelizing in the East Indies for so long a time before the fire piston is found in Europe that it is hard to believe that the Malayan fire piston was not observed and reported. The realization of its potential in Europe was considerable, culminating in the diesel engine.

Why are such nuclear ideas sometimes not exploited in new and wider applications? What sorts of barriers prevent their diffusion? Why, at times, does what appeared to be a successful technological item fall into disuse? The history of the faggoted forging method of producing sword blades may assist our thinking about such questions.

In late Roman times, north of the Alps, Celtic, Slavic, and Germanic metallurgists began to manufacture swords with laminations made by welding together bundles of rods of different qualities of iron and steel, hammering the resulting strip thin, folding it over, welding it all together again, and so on. In this way a fairly long blade was produced which had the cutting qualities of steel but the toughness of iron. Although such swords were used at times by barbarian auxiliaries in the Roman army, the Roman legions never adopted them. Yet as soon as the Western Empire crumbled, the short Roman stabbing sword vanished and the laminated slashing blade alone held the field of battle. Can this conservatism in military equipment have been one reason for the failure of the Empire to stop the Germanic invasions? The Germans had adopted the new type of blade with enthusiasm, and by Carolingian times were manufacturing it in quantities in the Rhineland for export to Scandinavia and to Islam, where it was much prized. Yet, although such blades were produced marginally as late as the twelfth century, for practical purposes they ceased to be used in Europe in the tenth century. Does the disappearance of such sophisticated swords indicate a decline in medieval metallurgical methods?

We should be cautious in crediting the failure of the Romans to adopt the laminated blade to pure stupidity. The legions seem normally to have fought in very close formation, shield to shield. In such a situation, only a stabbing sword could be effective. The Germans at times used a "shield wall" formation, but it was probably a bit more open than the Roman and permitted use of a slashing sword. If the Romans had accepted the new weapon, their entire drill and discipline would have been subject to revision. Unfortunately, we lack studies of the development of Byzantine weapons sufficiently detailed to let us judge whether, or to what extent, the vigorously surviving Eastern Roman Empire adapted itself to the new military technology.

The famous named swords of Germanic myth, early medieval epic, and Wagnerian opera were laminated blades. They were produced by the vast patience and skill of smiths who themselves became legendary. Why did they cease to be made in any number after the tenth century? The answer is found in the rapid increase in the weight of European armor as a result of the consistent Frankish elaboration of the type of mounted shock combat made possible by the stirrup. After the turn of the millennium a sword in Europe had to be very nearly a club with sharp edges: the best of the earlier blades was ineffective against such defenses. The faggoted method of forging blades survived and reached its technical culmination in Japan, where, thanks possibly to the fact that archery remained socially appropriate to an aristocrat, mounted shock combat was less emphasized than in Europe and armor remained lighter.

Let us now turn to a different problem connected with the act of invention. How do methods develop by the transfer of ideas from one device to another? The origins of the cannon ball and the cannon may prove instructive.

Hellenistic and Roman artillery were activated by the torsion of cords. This was reasonably satisfactory for summer campaigns in the Mediterranean basin, but north of the Alps and in other damper climates the cords tended to lose their resilience. In 1004 A.D. a radically different type of artillery appeared in China with the name *huo p'ao*. It consisted of a large sling-beam pivoted on a frame and actuated by men pulling in unison on ropes attached to the short end of the beam away from the sling. It first appears outside China in a Spanish Christian illumination of the early twelfth century, and from this one might assume diffusion through Islam. But its second appearance is in the northern Crusader army attacking Lisbon in 1147, where a battery of them were operated by shifts of one hundred men for each. It would seem that the Muslim defenders were quite unfamiliar with the new engine of destruction and soon capitulated. This invention, therefore, appears to have reached the West from China not through Islam but directly across Central Asia. Such a path of diffusion is the more credible because by the end of the same century the magnetic needle likewise arrived in the West by the northern route, not as an instrument of navigation but as a means of ascertaining the meridian, and Western Islam got the compass from Italy. When the new artillery arrived in the West it had lost its name. Because of structural analogy, it took on a new name borrowed from a medieval instrument of torture, the ducking stool, or *trebuchetum*.

Whatever its merits, the disadvantages of the *huo p'ao* were the amount of man-power required to operate it and the fact that since the gang pulling the ropes would never pull with exactly the same speed and force, missiles could not be aimed with great accuracy. The problem was solved by substituting a huge counterweight at the short end of the sling-beam for the ropes pulled by men. With this device a change in the weight of the caisson of stones or earth, or else a shift of the weight's position in relation to the pivot, would modify the range of the projectile and then keep it uniform, permitting concentration of fire on one spot in the fortifications to be breached. Between 1187 and 1192 an Arabic treatise written in Syria for Saladin mentions not only Arab, Turkish, and Frankish forms of the primitive trebuchet, but also credits to Iran the invention of the trebuchet with swinging caisson. This ascription, however, may be in error; for from about 1220 onward Oriental sources frequently call this engine *magribī*—that is, "Western." Moreover, while the counterweight artillery has not yet been documented for Europe before 1199, it quickly displaced the older forms of artillery in the West, whereas this new and more effective type of siege machinery became dominant in the Mameluke army only in the second half of the thirteenth century. Thus the trebuchet with counterweights would appear to be a European improvement on the *huo p'ao*. Europe's debt to China was repaid in 1272 when, if we may believe Marco Polo, he and a German technician, helped by a Nestorian Christian, delighted the Great Khan by building trebuchets which speedily reduced a besieged city.

But the very fact that the power of a trebuchet could be so nicely regulated impelled Western military engineers to seek even greater exactitude in artillery attack. They quickly saw that until the weight of projectiles and their friction with the air could be kept uniform, artillery aim would still be variable. As a result, as early as 1244 stones for trebuchets were being cut in the royal arsenals of England calibrated to exact specifications established by an engineer; in other words, the cannon ball before the cannon.

The germinal idea of the cannon is found in the metal tubes from which, at least by the late ninth century, the Byzantines had been shooting Greek fire.* It may be that even so early they were also shooting rockets of Greek fire, propelled by the expansion of gases, from bazooka-like metal tubes. When, shortly before 673, the Greek-speaking Syrian refugee engineer Callinicus invented Greek fire, he started the technicians not only of Byzantium but also of Islam, China, and eventually the West in search of ever more combustible mixtures. As chemical methods improved, the saltpeter often used in these compounds became purer, and combustion tended toward explosion. In the thirteenth century one finds, from the Yellow Sea to the Atlantic, incendiary bombs, rockets, firecrackers, and fireballs shot from tubes like Roman candles. The flame and roar of all this have made it marvelously difficult to ascertain just when gunpowder artillery, shooting hard missiles from metal tubes, appeared. The first secure evidence is a famous English illumination of 1327 showing a vase-shaped cannon discharging a giant arrow. Moreover, our next certain reference to a gun, at Rouen in 1338, shows how long it took for technicians to realize that the metal tube, gunpowder, and the calibrated trebuchet missile could be combined. However, iron shot appear at Lucca in 1341; in 1346 in England there were two calibers of lead shot; and balls appear at Toulouse in 1347.

The earliest evidence of cannon in China is an extant example of 1332. It is not necessary to assume the miracle of an almost simultaneous independent

---

*Editors' note: Greek fire was a combustible chemical mixture.

Chinese invention of the cannon; enough Europeans were wandering the Yuan realm to have carried it eastward. And it is very strange that the Chinese did not develop the cannon further. Neither India nor Japan knew cannon until the sixteenth century when they arrived from Europe. As for Islam, despite several claims to the contrary, the first certain use of gunpowder artillery by Muslims comes from Cairo in 1366 and Alexandria in 1376; by 1389 it was common in both Egypt and Syria. Thus there was roughly a forty-year lag in Islam's adoption of the European cannon.

Gunpowder artillery, then, was a complex invention which synthesized and elaborated elements drawn from diverse and sometimes distant sources. Its impact upon Europe was equally complex. Its influences upon other areas of technology such as fortification, metallurgy, and the chemical industries are axiomatic, although they demand much more exact analysis than they have received. The increased expense of war affected tax structures and governmental methods; the new mode of fighting helped to modify social and political relationships. All this has been self-evident for so long a time that perhaps we should begin to ask ourselves whether the obvious is also the true.

For example, it has often been maintained that a large part of the new physics of the seventeenth century sprang from concern with military ballistics. Yet there was continuity between the thought of Galileo or Newton and the fundamental challenge to the Aristotelian theory of impetus which appeared in Franciscus de Marchia's lectures at the University of Paris in the winter of 1319–1320, seven years before our first evidence of gunpowder artillery. Moreover, the physicists both of the fourteenth and of the seventeenth centuries were to some extent building upon the criticisms of Aristotle's theory of motion propounded by Philoponus of Alexandria in the age of Justinian, a time when I can detect no new technological stimulus to physical speculation. While most scientists have been aware of current technological problems, and have often talked in terms of them, both science and technology seem to have enjoyed a certain autonomy in their development.

It may well be that continued examination will show that many of the political, economic, and social as well as intellectual developments in Europe which have traditionally been credited to gunpowder artillery were in fact taking place for quite different reasons. But we know of one instance in which the introduction of firearms revolutionized an entire society: Japan.

Metallurgical skills were remarkably high in Japan when, in 1543, the Portuguese brought both small arms and cannon to Kyushu. Japanese craftsmen quickly learned from the gunsmiths of European ships how to produce such weapons, and within two or three years were turning them out in great quantity. Military tactics and castle construction were rapidly revised. Nobunaga and his successor, Hideyoshi, seized the new technology of warfare and utilized it to unify all Japan under the shogunate. In Japan, in contrast to Europe, there is no ambiguity about the consequences of the arrival of firearms. But from this fact we must be careful not to argue that the European situation is equally clear if only we would see it so.

In examining the origins of gunpowder artillery, we have seen that its roots are multiple, but that all of them (save the European name *trebuchet*) lie in the soil of military technology. It would appear that each area of technology has a certain self-contained quality; borrowings across craft lines are now as frequent as might be expected. Yet they do occur, if exceptionally. A case in point is the fusee.

In the early fifteenth century, clock-makers tried to develop a portable mechanical timepiece by substituting a spring drive for the weight which powered stationary clocks. But this involved entirely new problems of power control. The weight on a clock exerted equal force at all times, whereas a spring exerts less force in proportion as it uncoils. A new escapement was therefore needed which would exactly compensate for this gradual diminution of power in the drive.

Two solutions were found, the stackfreed and the fusee, the latter being the

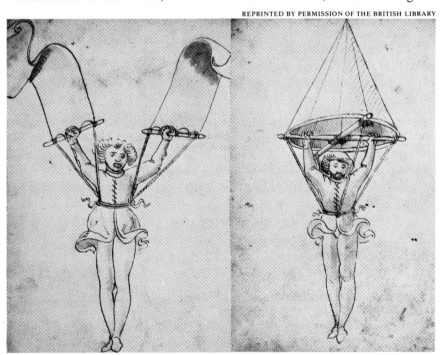

Ahead of its time: evolution of the parachute. The idea for a parachute dates from the fifteenth century, but there was no need for it until the hot air balloon was invented three hundred years later.

more satisfactory. Indeed, a leading historian of horology has said of the fusee, "Perhaps no problem in mechanics has ever been solved so simply and so perfectly." Its first appearance was about 1449. The fusee equalizes the changing force of the mainspring by means of a brake of gut or fine chain which is gradually wound spirally around a conical axle, the force of the brake being dependent upon the leverage of the radius of the cone at any given point and moment. It is a device of great mechanical elegance. Yet the idea did not originate with the clock-makers: they borrowed it from the military engineers. In Konrad Keyser's monumental treatise on the technology of warfare, *Bellifortis,* completed in 1405, we find such a conical axle in an apparatus for spanning a heavy crossbow. With very medieval humor, this machine was called "the virgin," presumably because it offered least resistance when the bow was slack and most when it was taut.

In terms of eleven specific technological acts, or sequences of acts, we have been pondering an abstraction, the act of technical innovation. It is quite possible that there is no such thing to ponder. The analysis of the nature of creativity is one of the chief intellectual commitments of our age. Just as the old unitary concept of "intelligence" is giving way to the notion that the individual's mental capacity consists of a large cluster of various and varying factors mutually affecting each other, so "creativity" may well be a lot of things and not one thing.

Thirteenth-century Europe invented the sonnet as a poetic form and the functional button as a means of making civilized life more nearly possible in boreal climes. Since most of us are educated in terms of traditional humanistic presuppositions, we value the sonnet but think that a button is just a button. It is doubtful whether the chilly Northerner who invented the button could have invented the sonnet then being produced by his contemporaries in Sicily. It is equally doubtful whether the type of talent required to invent the rhythmic and phonic relationships of the sonnet pattern is the type of talent needed to perceive the spatial relationships of button and buttonhole. For the button is not obvious until one has seen it, and perhaps not even then. The Chinese never adopted it: they got no further than to adapt the tie cords of their costumes into elaborate loops to fit over cord-twisted knobs. When the Portuguese brought the button to Japan, the Japanese were delighted with it and took over not only the object itself but also its Portuguese name. Humanistic values, which have been cultivated historically by very specialized groups in quite exceptional circumstances, do not encompass sufficiently the observable human values. The billion or more mothers who, since the thirteenth century, have buttoned their children snugly against the winter weather might perceive as much spirituality in the button as in the sonnet and feel more personal gratitude to the inventor of the former than of the latter. And the historian, concerned not only with art forms but with population, public health, and what S. C. Gilfillan long ago identified as "the coldward course" of culture, must not slight either of these very different manifestations of what would seem to be very different types of creativity.

There is, indeed, no reason to believe that technological creativity is unitary. The unknown Syrian who, in the first century B.C., first blew glass was doing something vastly different from his contemporary who was building the first water-powered mill. For all we now know, the kinds of ability required for these two great innovations are as different as those of Picasso and Einstein would seem to be. . . .

SIR WILLIAM FAIRBAIRN

The Invention of
The Riveting Machine

One of the more important inventions in the construction industry in the nine-teenth century was the riveting machine. Machine-riveted plates or beams were far superior to hand-riveted work, and time and labor were also saved by the machines. This selection describes the circumstances surrounding the inven-tion. Provoked into a fit of pique by what he considered to be an unreasonable strike by boilermakers at his factory in Manchester, England, Sir William Fair-bairn and his assistant engineer developed a successful machine within two days. The basic stimulus was, of course, economic; yet Fairbairn apparently had not planned the invention until the strike occurred.

M r. Fairbairn introduced an invention which has been of the greatest utility in engineering manufacture—namely, the riveting machine. He gives the following account of its origin:

> I have before alluded to a circumstance which occurred at this time, namely, the stoppage of a part of the works at Manchester by a strike of the boiler-makers. For some time previously we had been busily engaged in the construc-tion of boilers, and nothing could have been more injurious than the stoppage of the works at such a time. I remonstrated with the men, but without effect; and perceiving no chance of coming to terms in any reasonable time, I deter-mined to do without them, and effect by machinery what we had heretofore been in the habit of executing by manual labor.

William Pole, editor, *The Life of Sir William Fairbairn, Bart.* London, Longman's Green and Co., 1877.

In arranging this Mr. Fairbairn took into his counsels his assistant-engineer, Mr. Robert Smith. Two plans were proposed, one to act on the rivet by a lever (on the principle of the ordinary punching machine), the other to compress it by a screw. Mr. Smith was in favour of the latter plan, and wished to make drawings of a new machine on that principle; but Mr. Fairbairn says:—

> I replied that the screw would be too slow; and before any further steps were taken, I insisted on making a trial with the punching-machines which were in daily use.
>
> This was done on the following day, and Mr. Smith produced as fine a specimen of riveted work as I have seen either before or since. This was the origin and history of the riveting machine, which so much improves the quality and reduces the price of labour in this important branch of mechanical construction.
>
> Previous to the experiment made with the punching machine, which was accomplished by the simple introduction of two steel dies corresponding with the ends of the rivet, it was argued that compressed rivets would never be tight, that they would become loose and spoil the work; and many other objections were brought against the project by persons interested in the maintenance of the old process. To these, and also to the threats that were held out by the workmen, I turned a deaf ear; and after the first trial I was fully convinced that the principle was sound, and that we had nothing to fear from one or the other. Having convinced myself of the practicability of this new invention, a patent was taken out for it; and as Mr. Smith was the person first to accomplish the task, it was taken out in his name, but at my expense, and he was given an interest in it.

The patent, in the name of Robert Smith, is dated February 16, 1837 (No. 7,302), and entitled "Certain Improvements in the means of connecting Metallic Plates for the Construction of Boilers and other purposes." It gives a full description and drawings of the riveting machine, and it claims "the manner of connecting metallic plates for the construction of boilers and other purposes, by riveting them together by compression obtained by the aid of machinery."

Mr. Fairbairn continues his account:—

> The new machine effected a complete revolution in boiler-making and riveting, and has substituted the rapid and noiseless work of compression for the eternal din of the hammer; besides making the work infinitely superior in quality and strength.
>
> The introduction of the riveting machine gave great facilities for the despatch of business. It fixed, with two men and a boy, as many rivets in one hour as could be done with three men and a boy in a day of twelve hours on the old plan; and such was the expedition and superior quality of the work, that in less than twelve months the machine-made boilers were preferred to those made by hand, in every part of the country where they were known. This success was not attained without opposition; and, as happens in all similar cases, I had not

only to contend against modifications and improvements, but I had to combat prejudice and opposition from quarters where it was least expected. The patent, however, expired some years since, and the machine is now in general use; and I have reason to be satisfied that it has not only answered the purpose intended, but has been of use to the public in the development of a new and important principle in the constructive arts.

JACOB SCHMOOKLER

Economic Sources Of Inventive Activity

The late Jacob Schmookler, a distinguished American economic historian, was convinced that almost all inventions were economically motivated. Inventive activity, he thought, requires intense preoccupation with the object as well as a certain dissatisfaction with existing conditions. But the incentive to invention, he argued, was the expected profit, that is, the excess of expected returns to the inventor over his expected costs.

The fundamental conclusion of this paper is that technological progress is intimately dependent on economic phenomena. The evidence suggests that society may indeed affect the allocation of inventive resources through the market mechanism somewhat as it affects the allocation of economic resources generally. If this is true, then technological progress is not an independent cause of socio-economic change, and an interpretation of history as largely the attempt of mankind to catch up to new technology is a distorted one. Cultural lags undoubtedly exist in social history. The automobile—to use an obvious example—rendered obsolete many pre-existing social arrangements and behavior patterns. But the reverse is also true. New goods and new techniques are unlikely to appear, and to enter the life of society without a pre-existing—albeit possibly only latent—demand. Even a long-standing demand may have been intensified shortly before a technique to satisfy it is

Jacob Schmookler, "Economic Sources of Inventive Activity." *Journal of Economic History,* Vol. XXII (March 1962). Copyright 1962 by The Economic History Association. Reprinted by permission.

invented. In addition to cultural lag, there exists technological lag—a chronic tendency of technology to lag behind demand.

The problem . . . may be viewed in still another way. If one charts over time a firm's research and development expenditures leading to a given product (for example, a given model of railway car) and the volume of sales of the resulting product, one would expect in the nature of the case to find—if the product were marketed at all—that the research expenditures were made largely before the product was marketed. What would one find if total research and development expenditures of all firms on a given *class* of product, such as railroad passenger cars, were compared over a period of time with sales of all products in the class? . . . Inventive effort, it would appear, usually varies directly with the output of the class of goods the inventive effort is intended to improve, with invention tending to lag slightly behind output. The explanation of this pattern, in my judgment, is that variations in invention are a consequence of economic conditions with which output is also positively correlated. While the discussion and evidence relate to classes of products already in existence, the conclusion may well apply with almost equal force to basic, industry-establishing inventions. . . .

To the degree that inventions are made either by producers or consumers of a commodity, more money will be available for invention when the industry's sales are high than when they are low. Increased sales imply that both the producing firms and their employees will be in a better position than before to bear the expenses of invention. Large purchases of a product also suggest that its buyers are better able to finance invention. Finally, the current business practice of setting research budgets at a fixed percentage of sales tends to assure, in recent years, the relation we have observed.

Certain aspects of the psychology of invention are noteworthy in this context. Effective inventive activity apparently requires an intense, almost obsessive preoccupation with its object. Buyers, for example, are more likely to show such a preoccupation when they are shopping for a product, with standards and expectations about performance and cost beginning to form in their minds. The number of potential inventors among the buyers or producers of a product, then, might vary with sales and employment.

Another psychological requirement for invention, which is also involved in new purchases, is a certain dissatisfaction with existing possessions. Every new purchase indicates such dissatisfaction. New buildings are erected because old ones are crowded or inadequate. New machines are purchased because old ones are breaking down or cannot handle an increased volume or produce unsatisfactory products. The same conditions which lead some men to buy existing goods may lead others to try to improve them. Such dissatisfaction with existing products may stem from several sources whose influence usually varies directly with sales. First, in periods of general economic expansion and rising sales, when proportions of many kinds are changing, products which were satisfactory under earlier conditions may be unsatisfactory under the new. To meet the changed requirements of customers, the producers or the customers themselves may invent new or modified products. Second, increased sales often mean that some people have begun to use a product for the first time. Since the circumstances and preferences of new customers may differ from those of old ones, either the producers or the new customers themselves may invent modifications in the product to suit the altered requirements.

Finally, newly purchased goods embodying recent improvements are also likely to have the "bugs" which seem almost the inevitable accompaniment of progress. These emerge after use of a product has begun, and provide the occasion for corrective invention by the user or the producer. Such invention will tend to vary directly with, but lag behind, sales of the product involved.

To summarize, the evidence strongly suggests that the output of a commodity and invention relating to it vary together, with invention tending to lag. The relation cannot be explained by the hypothesis that the variations in invention cause those in output. Rather, it seems probable that expected profits from invention, the ability to finance it, the number of potential inventors, and the dissatisfaction which invariably motivates it—are all likely to be positively associated with sales. That major and minor inventions in an established industry tend to go together is a further indication that inventive effort is responsive to economic pressures and opportunities.

One may suggest further that even basic inventions which establish new industries are often, and perhaps usually, induced by economic forces like those which appear to operate in established industries. Even when a scientific discovery underlies an invention, the discovery may contain the seeds of many potential inventions, and economic factors may then determine which potential applications are selected for exploitation. These economic factors may take the form either of new, or newly intensified, latent demands, or of a greater intensity of the latent demands satisfied by the inventions actually made, as compared to that of the demands which remain unsatisfied. For example, while the invention of wireless telegraphy cannot be understood without reference to prior scientific discoveries in electromagnetism, it cannot be understood with reference to those discoveries alone. Other inventions based on the same discoveries could presumably have been made, and a substantial and probably increased demand for rapid and low-cost communication techniques probably influenced Marconi's efforts.

The essential point is that the incentive to make an invention, like the incentive to produce any other good, is affected by the excess of expected returns over expected costs. Scientific progress may reduce expected costs and so increase the probability that a given invention will be sought and made. However, every invention represents a fixed cost, and the expected benefits from it vary with circumstances. Those circumstances, arising from changes in the prospective market for a commodity or a process, depend not on scientific discovery, but on socio-economic change—urbanization, declining family size, changing status of women, changes in relative factor costs, increases in population and per capita income, etc. Antecedent scientific discoveries are sometimes necessary, but seldom sufficient, conditions for invention. The historical shifts in inventive attention appear to reflect the interplay of advancing knowledge, which opens up new inventive opportunities for exploitation, and the unfolding economic needs and opportunities arising out of a changing social order. The evidence adduced in this article suggests that the influence of the latter has been substantial, at least in established industries.

EUGENE S. FERGUSON

Nonverbal Thought
In Technology

Eugene S. Ferguson, a historian of technology, believes that invention and opti-
mum engineering design result from nonverbal thought, what he terms "visual
thinking." Because of this, technological innovation is allied to art much more
than it is to science. Engineering schools today, Ferguson concludes, stress a
mathematical approach to problems through numerical systems analysis, which
is a sterile approach to the solution of major problems.

This scientific age too readily assumes that whatever knowledge may be incorporated in the artifacts of technology must be derived from science. This assumption is a bit of modern folklore that ignores the many nonscientific decisions, both large and small, made by technologists as they design the world we inhabit. Many objects of daily use have clearly been influenced by science, but their form and function, their dimensions and appearance, were determined by technologists—craftsmen, designers, inventors, and engineers—using nonscientific modes of thought. Carving knives, comfortable chairs, lighting fixtures, and motorcycles are as they are because over the years their designers and makers have established shape, style, and texture.

Many features and qualities of the objects that a technologist thinks about cannot be reduced to unambiguous verbal descriptions; they are dealt with in his mind by a visual, nonverbal process. His mind's eye is a well-developed

Eugene S. Ferguson, "The Mind's Eye: Nonverbal Thought in Technology," *Science,* Vol. 197, pp. 827–835, August 26, 1977. Copyright 1977 by the American Association for the Advancement of Science.

organ that not only reviews the contents of his visual memory but also forms such new or modified images as his thoughts require. As he thinks about a machine, reasoning his way through successive steps in a dynamic process, he can turn it over in his mind. The designer and the inventor, who bring elements together in new combinations, are each able to assemble and manipulate in their minds devices that as yet do not exist.

If we are to understand the development of Western technology, we must appreciate this important, if unnoticed, mode of thought. It has been nonverbal thinking, by and large, that has fixed the outlines and filled in the details of our material surroundings for, in their innumerable choices and decisions, technologists have determined the kind of world we live in, in a physical sense. Pyramids, cathedrals, and rockets exist not because of geometry, theory of structures, or thermodynamics, but because they were first a picture — literally a vision — in the minds of those who built them. . . .

The Nature of Design

There may well be only one acceptable arrangement or configuration of a complex technological device, such as a motorcycle, but that arrangement is neither self-evident nor scientifically predictable. The early designers of motorcycles could not ask science to tell them where to put engine, battery, fuel tank, and spark coil; they had to make their choices on other grounds. In time, wrong choices would be revealed, but not by scientific analysis. Making wrong choices is the same kind of game as making right choices; there is often no a priori reason to do one thing rather than another, particularly if neither had been done before. No bell rings when the optimum design comes to mind. Nor has the plight of designers changed fundamentally in the twentieth century. They must still weigh the imponderable and sound the unfathomable. All of our technology has a significant intellectual component that is both nonscientific and nonliterary.

The creative shaping process of a technologist's mind can be seen in nearly every man-made object that exists. The sweep of a suspension bridge, for example, is much more than an exercise in geometry. The distinctive features of three great suspension bridges in New York — the Brooklyn, George Washington, and Verrazano Narrows — reflect more strongly the conceptualization of their designers and the times of their construction than they do the physical requirements of their respective sites. Different builders of large power boilers use many common elements in their designs, but certain characteristics of internal "style" distinguish the boilers of one maker from those of another. The opportunities for a designer to impress his particular way of nonverbal thinking upon a machine or a structure are literally innumerable. This open-ended process can be seen in the design of a familiar, compact machine such as a diesel engine.

The designer of a diesel engine is a technologist who must continually use his intuitive sense of rightness and fitness. What will be the shape of the combustion chamber? Can I use square corners to gain volume, or must I use a fillet to gain strength? Where shall I place the valves? Should it have a long or short piston? Such questions have a range of answers that are supplied by experience, by physical requirements, by limitations of available space, and not least by a sense of form. As the designer draws lines on paper, he translates a picture held in his mind into a drawing that will produce a similar picture in another mind and will eventually become a three-dimensional engine in metal. Some decisions, such as wall thickness, pin diameter, and passage area may depend upon scientific calculations, but the non-scientific component of design remains primary. It rests largely on the nonverbal thought and nonverbal reasoning of the designer, who thinks with pictures. . . .

Object Teaching

The utility of visual images in elementary schooling was recognized in the seventeenth century by Joannes Comenius, whose little picture book, *Orbis Sensualium Pictus* (1658), matched objects to words and introduced the idea of object teaching. If a word were associated with an object or a picture, Comenius reasoned, it would be more readily learned and better understood. Although he surely realized that abstract concepts, such as fortitude and prudence, are more difficult than material objects to associate with pictures, he assumed that most minds think with pictures as well as words. . . .

Needless to say, the principles of object teaching have not been universally adopted. Beyond kindergarten, as Rudolph Arnheim has pointed out, "the senses lose educational status," and the school child is fitted into the verbal world of school teachers who are generally unaware of the significance of nonverbal components of thought.

Even in an engineering school, a course in "visual thinking" is regarded as an aberration rather than as a discipline that should be incorporated into an engineer's repertoire of skills. A problem such as building a container that will keep an egg from breaking when dropped out of a third-story window is labeled "Rube Goldberg," and the course in which it occurs is picked up as news in the *New York Times*. Accustomed to maintaining control of his class, an engineering instructor finds it difficult to pose any problem that requires mere nonverbal thought, that does not have a single unique solution, and that cannot be solved in rigorous mathematical steps.

Art and Technology

In Renaissance engineering, art (as opposed to science) was the guiding discipline, and it might be supposed that the symbiotic relationship between technology and art would be noted by the authors of the illustrated books. On the contrary, a number of enthusiastic technologists of the Renaissance insisted that the mechanic arts had been brought to their current perfection by the power of mathematics, universally applied. Ramelli, for example, devoted eight large folio pages in the preface of his machine book of 1588 to proving that all mechanic arts rest upon mathematics. Yet when we turn to the body of the work, we find no evidence whatever of his use of geometry or arithmetic or any other branch of mathematics or formal mechanics in the designing of his machines. On close study, the drawings do reveal his intimate knowledge of mechanical principles as derived from experience in workshop and field. They exhibit also a restless originality that is manifested not only in totally new devices, such as Ramelli's rotary pump, but also in numerous detailed variations of common elements that appear again and again, such as fastenings and shaft bearings. The thoughts expressed in his many drawings of machines were clearly nonverbal. Despite his discourse on the "excellence" of mathematics and its universal utility, Ramelli's intellectual resources were not derived from the science of his day.

The association of technology with science has become so close that it is difficult to realize that the Renaissance engineer, trained as an artist and retaining the artist's habits of nonverbal thought, had significant counterparts as recently as the nineteenth century. Benjamin Henry Latrobe, a prominent consulting engineer and architect, was an accomplished watercolorist. Robert Fulton of steamboat fame and Samuel Morse, American inventor of the electrical telegraph, were both artists before they turned to careers in technology. Around the time of the 1876 Centennial Exhibition, John Rogers, who was trained in a machine shop, became the "people's sculptor," producing for American homes many thousands of copies of his nostalgic groups of ordinary people.

The organization of American technology in the first half of the nineteenth century tended naturally to follow the pattern set by the world of art. . . . American societies for the promotion of the mechanic arts were influenced heavily by the London Society of Arts, which was concerned with both fine arts and the mechanic arts. The Franklin Institute, organized in 1824, awarded prizes at its annual exhibitions and sponsored classes in mechanical and architectural drawing. The American Institute of the City of New York held annual industrial fairs, and, like the London Society of Arts, maintained a permanent exhibition of machines and models for the observation and study of mechanics and inventors.

Visual Thinking

In 1880, when Francis Galton, founder of the science of eugenics, was interviewing scientists in his study of the human intellect, he was astonished to find that most scientists' thought processes were quite different from his own. Galton thought in visual images, he said, while the majority of scientists reported that they thought in words, with seldom any suggestion of an image. In deference to his subjects, most of whom were of "very high repute," he was willing to assume that the "visualizing faculty" was probably inferior to "the higher intellectual operations." Although scientists might at one time have had the visualizing faculty, he added, it had probably been lost by disuse in minds that "think hard."

In the early years of the twentieth century, the American philosopher William James remarked that a favorite topic of discussion among philosophers was "whether thought is possible without language." There was no question in James's mind: he recognized the possibility of visual and even tactile imagery. Albert Einstein claimed that he rarely thought in words at all. His visual and "muscular" images had to be "laboriously" translated into conventional language and symbols. . . .

Whatever may be the mode of thought of scientists, it is certain that technologists who make choices and decisions regarding the design of objects employ a species of nonverbal thought. Over the years, a number of prominent technologists have revealed their mode of thought as they explained how they solved technical problems.

Oliver Evans invented the automatic flour mill, in which bucket elevators and screw conveyors were coordinated to eliminate the need for manually lifting or carrying the grain or flour. He first put the system together in his head, explaining later that "I have in my bed viewed the whole operation with much mental anxiety."

When James Brindley, engineer of the Duke of Bridgewater's pioneering English canal, was faced with a particularly difficult problem of design, he would go to bed for one, two, or three days. When he arose he was ready to carry out the designs that existed only in his mind, for he made neither plans nor models.

In a letter to his partner, Matthew Boulton, James Watt referred to the "straight-line" mechanism, which became his favorite invention, when he wrote, "I have got a glimpse of a method of causing the piston-rod to move up and down perpendicularly, by only fixing it to a piece of iron upon the beam."

Marc Isambard Brunel, who designed semiautomatic machines to make ships' pulley blocks in sequential operations, remarked upon the ease with which he expressed his ideas in his drawings; he considered drawing techniques to be the true *"alphabet of the engineer."*...

Francis Galton, speaking of "inventive mechanicians," observed that "they invent their machines as they walk, and see them in height, breadth, and depth as real objects, and they can also see them in action." How to cultivate the "visualizing faculty" without prejudice to the practice of abstract thought in words or symbols was, to Galton, one of the "many pressing desiderata in the yet unformed science of education."...

Conclusion

Much of the creative thought of the designers of our technological world is nonverbal, not easily reducible to words; its language is an object or a picture or a visual image in the mind. It is out of this kind of thinking that the clock, printing press, and snowmobile have arisen. Technologists, converting their nonverbal knowledge into objects directly (as when an artisan fashioned an American ax) or into drawings that have enabled others to build what was in their minds, have chosen the shape and many of the qualities of our man-made surroundings. This intellectual component of technology, which is non-literary and nonscientific, has been generally unnoticed because its origins lie in art and not in science.

As the scientific component of knowledge in technology has increased markedly in the nineteenth and twentieth centuries, the tendency has been to lose sight of the crucial part played by nonverbal knowledge in making the "big" decisions of form, arrangement, and texture that determine the parameters within which a system will operate.

Two results of the abandonment of nonverbal knowledge in engineering colleges can be predicted; indeed, one is already evident. The movement toward a four-year technician's degree reflects a demand for persons who can deal with the complexities of real machines and materials and who have the nonverbal reasoning ability that used to be common among graduates of engineering colleges. In the longer run, engineers in charge of projects will lose their flexibility of approach to solving problems as they adhere to the doctrine that every problem must be treated as an exercise in numerical systems analysis. The technician, lower in status than the systems engineer, will have the ability but not the authority to make the "big" decisions, while the systems engineer in charge will be unaware that his nonverbal imagination and sense of fitness have been atrophied by the rules of a systematic but intellectually impoverished engineering approach.

JOHN JEWKES, DAVID SAWERS, AND RICHARD STILLERMAN

The Sources Of Invention

John Jewkes, an emeritus professor at Oxford, and his co-authors contend that we are still ignorant about the nature of technological innovation and are basing decisions on a number of unproved assumptions. For example, we cannot be certain that technological change is proceeding more rapidly now than in the past, or that there is any necessary connection between scientific knowledge and inventive activity. They warn that the present trend to depend upon large industrial research laboratories for innovation may be a miscalculation, because inventive activity may be stifled by over-organization. Chance, they assert, remains an important factor in invention, and the individual inventor, who was so important in the past, should not be relegated to the scrap heap.

Technical progress, it is commonly assumed, is now going on more rapidly than heretofore. But there is no statistical proof of it. It would, in any case, be necessary to distinguish between net and gross. Much new technical improvement is called for, to alleviate the evils of earlier technical progress; we must go faster to stay where we are: if, in the future, for example, ways are invented of making the jet engine quiet, then the skies will merely be as silent as they were before the invention of the jet engine. But, putting that

Selections are reprinted from *The Sources of Invention* by John Jewkes, David Sawers, and Richard Stillerman, with the permission of W. W. Norton & Company, Inc., and Macmillan, London and Basingstoke. Copyright © 1969, 1958 by John Jewkes, David Sawers, and Richard Stillerman.

on one side, why should it be assumed that the scientific advances or the inventions of the nineteenth century were any less revolutionary, possessed less powers to change ways and improve standards of living, than those of recent days? To every generation the events new to it naturally loom larger in the mind than those of the past to which it has become accustomed: but there is no objective measurement by which it can be established that the jet engine was more significant than the steam engine, the discoveries in atomic energy than the discoveries of the molecular constitution of matter, the development of pre-stressed concrete than that of the gothic arch.

There is no evidence which establishes definitely that technical or economic progress receives greater contributions from the few and rare large advances in knowledge than from the many and frequent smaller improvements. Economically, it might for a period well pay a community to starve its scientific and major technical work and to devote resources to the most thorough and systematic gathering together and exploitation of all the immediate and tiny practical improvements in ways of manufacture and design.

It is not known whether there is any necessary connection between the growth of scientific knowledge and the growth of technology and invention or, if there is a connections, what are its laws. If science and technology have different motives and criteria of success, it is a rash assumption that the one immediately and proportionately stimulates the other. It is not inconceivable that for long periods scientific advance may lie wholly in fields which have no immediate, or even ultimate, utility in the narrow sense. It is not inevitable that the country with the outstanding scientific successes will be the richest country. Indeed it is often suggested that, even where scientific advance has ultimately contributed to technology, the lag has been so great that it automatically rules out the possibility either of prediction or of calculated investment to produce results. If, for instance, the growth of technology in the nineteenth century ultimately arose out of the great progress in mathematics in the seventeenth century, the gap in time is so great as to destroy completely any value that that fact might otherwise possess as a basis for foresight; no one would invest now in any branch of scientific research on the assumption that it might be useful in A.D. 2300. Any community, therefore, which deliberately invests in pure science solely as a way of producing returns in technology and invention is not merely setting out on a course which threatens the ultimate values of science itself but is also engaging in a blind gamble. We do not know whether there is an optimum rate of invention and technical advance or, if such an optimum is accepted as a conceptual device, how it would be defined or determined. Technical changes frequently cause social upset, which may conceivably be great enough to frustrate the powers for economic advance inherent in improved techniques. That, apparently, is the argument sometimes made for monopoly: that by rendering more deliberate the absorption of new ideas into industrial methods, the real economic advantages can, over the longer period, be made the greater.

But even if there were no doubts on this score, if the community sets out to encourage innovation it is by no means clear how best it could be brought about, what institutional framework will most effectively stimulate and encourage the men with powers of originality, most swiftly distinguish between the channels open to progress and the blind alleys, and most thoroughly glean the economic harvest from innovation. . . .

There is, indeed, a deeply rooted belief that sufficient is never made of invention and inventive genius, that in one way or another the community is casual and wasteful of ideas. The grounds for these suspicions vary from time to time. Not so long ago it was common talk, although with very little evidence, that private enterprise would lead to the deliberate suppression of important inventions. This myth has now largely fallen out of fashion, and it is now more commonly asserted that there are some much less deliberate and sinister influences at work, usually described as "resistance to change," which stand between the community and the bounties which the innovator might well confer upon it. Whatever the truth here, it is a fair comment that industrial societies have shown little originality or ingenuity in creating institutions to ensure that all new ideas will be swept into the net and that nothing will be lost. They have not been very prolific in the invention of institutions to encourage invention. . . .

What is an invention? Technical progress is an indivisible moving stream from which it seems impossible, except in an arbitrary fashion, to isolate one fragment for independent examination. Every item seems in the last analysis to be linked with every other item, so that nothing can be thought about or explained unless everything is taken into account. The windscreen wiper, the zip fastener, the jet engine, the cyclotron, nylon: all these have been described as inventions. Yet any definition that includes them all would seem to include also every technical or product variation that has ever occurred. Is the invention the idea; or the first conception of a way of using the idea; or the actual working utilisation of the idea; or the compounding together of two existing ideas; or the effective fusion of two ideas for a useful purpose? And if an invention cannot be defined, what becomes of the attempts to classify inventions? Thus the distinction between a "cost reducing" invention and an invention which consists of a "new product" seems theoretically valid. But in practice, every device for reducing cost is a new product. Every new product is a method of reducing cost in one form or another. A jet-engine aircraft for crossing the Atlantic in place of a sailing-boat is a new product but it is a new product only because it reduces the cost of the journey either in money or time or personal hazard or exertion. It is, indeed, not surprising that in the long history of patent litigation the efforts of the courts to define "invention" have produced such contradictions and confusions.

What is an *important* invention? Anyone seeking to generalise about inventions from case histories must confine himself to the salient novelties. There is no way of measuring the totality of invention; only its sporadic dominating features can be picked out. Yet there are no economic principles to which to appeal to determine whether one invention is more "important" than another. On what grounds could it be asserted that the ball-point pen is a more important invention than the cotton picker, unsplinterable glass than colour photography, the safety razor than the steam turbine?

Who is *the* inventor in any particular case? Which of the long line of thinkers and manipulators, each of whom has added something to the final appearance of a useful thing, should take the palm? When Carothers heated chemicals together in a test tube and discovered nylon, was it the original inventor of fire or the inventor of the bunsen burner or Carothers himself who gave this particular ball the biggest push? Where, as often happens, men working independently of each other appear to have reached the same ideas at about the same time, what tests of priority can be applied? And, an even more difficult question which the persons actually concerned might often be at trouble to answer, how can the real originator of an idea be picked out from among a group of men who have been working closely together for a period? . . .

The part played in invention by individuals cannot be looked upon as of little or no account. It is the practice of some writers to present a fuzzy picture of invention as "a social process"; to suggest that, if one inventor had not done what he did when he did, someone else would have done it; that inventions have come to ripeness at a predestined point in time influenced little, if at all, by human will, courage and pertinacity; that if Watt and the other great names had never been heard of, the world would have been much the same. But this attitude—that nothing can be understood unless all is understood, that by piling one unresolved enigma upon another some all-comprehending solution is made the more likely—involves the error of "seeing depth in mere darkness," as Sir Isaiah Berlin once put it.

It is just as important to be clear about what is here being accepted as about what is being challenged. It is not to be doubted that any and every invention, if the story of it be probed long and deeply enough, can be traced back almost indefinitely. The history of an invention has almost unlimited dimensions in space and time. Every inventor, however original he may appear to have been, is laying bricks upon a building which has long been in the course of construction from innumerable and mainly unknown hands. . . .

Even if the great inventors of the past century and a half had never been born, but some powerful, dynamic, impersonal push had ultimately produced the same results, the delay in itself would be significant. Timing here is of the essence of the matter. If invention is a continuing process, any loss of time is a permanent loss, the whole course of technical progress is set back permanently, the time lost is never made up. But, in fact, individuals have made a permanent impress on the shape of things. It is true that the rapid advances in the past 150 years seem to give meaning to such phrases as "the march of science," "the march of technology" and to create the impression of movements pressed along by forces independent of the will, the decisions, the struggles of individuals. In that period one genius trod closely upon the heels of another. But to describe an epoch especially rich in outstanding inventors as one in which individual contribution can no longer be identified is a curious inversion of facts, and as great a misapprehension as it would be to talk of the "march of painting" in the Netherlands or "the march of drama" in the England of the sixteenth and seventeenth centuries.

The third presupposition is that a useful working distinction can be made between "invention" and "development." Just as a distinction is made between science and technology, so technology itself can be divided into these two parts. Invention is something which comes before development. The essence of invention is the first confidence that something should work, and the first rough tests that it will, in fact, work.

When Watt conceived of his engine and then, years later, made the first satisfactory model; when Cartwright with the help of mechanics put his ideas of a power-loom into a shape which operated; Whitney built his first primitive cotton gin; Perkin produced the first aniline dye; Goodyear his first batch of vulcanised rubber; Gilchrist-Thomas lined a steel converter with a basic material; Bell transmitted speech over a short wire; Cross produced viscose on a small scale; Diesel found a way of producing ignition in a cylinder by compression; Baekeland, ignoring traditional ideas, manufactured bakelite; the chemists at Tootal Broadhurst Lee imparted to a piece of cotton cloth crease-resisting properties; Lawrence constructed his first rickety cyclotron; Whittle ran his first turbo-jet; Poulsen magnetised a wire and recorded sound on it; Carothers drew through an improvised spinneret a fibre now known as nylon; Fleming watched the action of penicillin on bacteria; Farnsworth conceived of his image dissector tube for television and constructed his first crude model to demonstrate the soundness of his notions; Whinfield first saw the fibre subsequently named Terylene; Kroll in his private laboratory produced tiny quantities of ductile titanium; the chemists at Imperial Chemical Industries cast sight on a white plastic material, now known as Polythene; Carlson, working with the principles of photoconductivity and electrostatics, devised xerography, a novel system of copying documents: in all these cases an invention, subsequently proving of great importance, had been made.

Development is a term which is loosely used in general discussion to cover a wide range of activities and purposes, but all these activities seem to satisfy three conditions. One, development is the stage at which known technical methods are applied to a new problem which, in wider or narrower terms, has been defined by the original invention. Of course, it may happen that in the course of development a blockage occurs, existing technology may provide no answers, and then, what is strictly another invention, is called for to set the ball rolling once more. Two, and consequentially, development is the stage at which the task to be performed is more precisely defined, the aim more exactly set, the search more specific, the chances of final success more susceptible to measurement than is true at the stage of invention. Invention is the stage at which the scent is first picked up, development the stage at which the hunt is in full cry. All the money in the world could not have produced nylon or the jet engine or crease-resisting fabrics or the cyclotron in 1900. At the time of writing it is possible to say that all the money in the world may not produce a cure for most forms of cancer, or lead to the discovery of economical methods of storing electricity on a large scale. Three, development is the phase in which commercial considerations can be, and indeed must be, more systematically examined, the limits of feasibility imposed by·the market are narrowed down. As one moves from invention to development the technical considerations give way gradually to the market considerations. . . .

The history of nylon is one good illustration: in 1935, after seven years' work of varying fortune and many disappointments, work which might have led anywhere or nowhere, W. H. Carothers, in the laboratories of the du Pont Company, produced the first nylon fibre and du Pont undertook to translate it into a marketable product. By 1939 large-scale production of nylon hosiery had commenced. Thus, in a matter of four years of development, du Pont had reached its appointed goal. Estimates put the total cost of the early stages of research and development at about $6 million; at that time 230 technical experts were engaged in the work. What precisely was involved in the development undertaken after Carothers's initial discovery?

First it was necessary to find ways of producing on a large scale the intermediate constituents of nylon which, up to that time, had been made only on a small scale. The two important materials were adipic acid and hexamethylenediamine. Adipic acid had been manufactured in Germany for some time but there had been no commercial exploitation of it in the United States. The

German processes were not readily adaptable to the plants of du Pont and it became imperative to develop a new catalytic technique for this purpose. Hexamethylenediamine posed even greater difficulties; it was merely a laboratory curiosity and had never been manufactured on a commercial scale before. Success here required the discovery of new catalysts and the proper handling of heat transfer problems. Next, a great deal of work had to be done at the stage where the materials react to form the long chain molecules of the nylon polymer. The first polymers were made in glass equipment in Carothers's laboratory, but glass equipment was completely unsuitable for commercial manufacture and metal equipment had to be designed. Methods of controlling the degree of polymerisation had to be evolved, since a failure to stop the reaction at precisely the right time resulted in the production of different and far less useful polymers than nylon. The technologists had to learn how to make one batch of the product exactly like another.

At the next stage of manufacture the flakes of the polymer had to be melted and some means found to transfer the molten mass to the spinning machines. Only pumping gave the filaments adequate uniformity, but unfortunately there were no existing pumps suitable for the task. A new type of pump was required embodying new alloys capable of withstanding the heat of the molten polymer. At the next stage of spinning, the machinery had to be specially designed for the task, since nylon could not be spun in the same manner as cotton, wool, viscose or cellulose acetate. The winding and the cold drawing processes also confronted the developers with problems which were novel and for which specially designed machines were required. Thus at each one of these stages—the mass production of what had formerly been made only on a small scale, the maintenance of unusual degrees of purity, the flexible controlling of the chemical processes and the devising of mechanical aids for handling materials with novel properties, the developers were confronted by one hurdle after another. It was only when the process reached the stage of knitting and weaving the existing and familiar techniques could be called in to help. But at every stage workers knew what they were looking for, and, with varying degrees of certainty, they knew it could be found. . . .

*   *   *

There is nothing in the history of technology in the past century and a half to suggest that infallible methods of invention have been discovered or are, in fact, discoverable. It may be true that in these days the search for new ideas and techniques is pursued with more system, greater energy and, although this is more doubtful, greater economy. Yet chance still remains an important factor in invention and the intuition, will and obstinacy of individuals spurred on by the desire for knowledge, renown or personal gain the great driving forces in technical progress. As with most other human activities, the monotony and sheer physical labour in research can be relieved by the use of expensive equipment and tasks can thereby be attempted which would otherwise be wholly impossible. But it does not appear that new mysteries will only be solved and new applications of natural forces made possible by ever increasing expenditure. In many fields of knowledge, discovery is still a matter of scouting about on the surface of things where imagination and acute observation, supported only by simple technical aids, are likely to bring rich rewards.

The theory that technical innovation arises directly out of, and only out of, advance in pure science does not provide a full and faithful story of modern invention. As in the past three centuries, there is still a to-and-fro stimulus between the two; each has a momentum and a potential of its own. The case for scientific enquiry is not a utilitarian one. It may be that the flow of inventions is just as likely to be increased by stimulating the fuller exploitation of the myriads of technical possibilities inherent in the existing stock of scientific knowledge as by increasing that stock.

The history of invention shows no sharp break in continuity. The inventive drive continues to be found in people who, because of their temperament and outlook, are not easy to organise. The sharp contrasts sometimes drawn between the present and the last century seem to be the product of distortions, in the one direction, of what was happening in the nineteenth century and distortions, in the opposite direction, of what has been happening in this. The impression thereby created of the sharp passage from one epoch to another can be reinforced by comparing extremes—for example, an important discovery in a large modern industrial research laboratory contrasted with the struggles... of some impoverished and obscure individual inventor of a hundred years ago. Extremes today are undoubtedly wider apart than they were. But the broad band of middle cases provides little support for spectacular interpretations. In both periods there were inventors of scientific outlook, inventions of the intuitive and empirical type, men who worked in teams, men who worked essentially alone.

Of course, the immediate future may have something entirely different in store for us. It may be that we have now passed into a stage of very rapid transition (dating from about 1950 or perhaps the day when the first atom bomb blew off) which amounts to a violent break in the nature of technical change....

The widespread faith that we are much cleverer and more energetic in development than were our ancestors rests not upon any measured assessment of results but upon two critical assumptions. First, that because there are more persons equipped with technical knowledge employed in this task, results are about to emerge. This seems to deny the possibilities of diminishing returns. The second assumption is that by employing technicians in larger, organised groups, the effectiveness of their work will be enhanced, that in a community with a given number of technicians it is better to have a few large groups than a larger number of small groups, or a mixture of groups of varying size. Here we are confronted with such a tangle of intricate economic and administrative factors that it may be sheer futility to try to find answers. . . .

In the twentieth century, apart from the special awards to war-time inventors, the emphasis of public endeavour has been upon institutions and not upon incentives. Governments have sought to encourage invention by setting up research organisations under their own control, or have subsidised research groups set up otherwise. Firms themselves have established industrial research laboratories and, at least in the United States, private enterprise has created specialised research organisations working for profit. The underlying principle, rarely formulated precisely but ever present, has been that originality can be organised; that, provided more people can be equipped with technical knowledge and brought together in larger groups, more new ideas must emerge; that mass production will produce originality just as it can produce sausages.

Under the influence of this doctrine the process of discovery and invention is becoming progressively institutionalised. The disposition of individuals to pursue their own ways with their own resources is weakened in many ways. High taxation (which in part is high because governments collect resources from the citizen for the purpose of stimulating public scientific and technical research) makes difficult the accumulation of private means. The lure of adequate equipment, congenial intellectual society and a secure livelihood provided by the institution is strong. In turn, institutions will naturally place emphasis upon the formal training and academic qualifications of those they employ: they will therefore become increasingly staffed by men who have been subject to common moulding influences. There is a possibility of inbreeding from which the more eccentric strains of native originality may be excluded.

Even as between the various types of research institutions the scales seem to be weighted in favour of those which, in the nature of things, will tend to rely upon advanced organisation and planning and which are, in consequence, less able to give rein to the autonomist bent of genius. Thus, measured by the scale of effort, though not necessarily by results, the universities and the independent non-profit-seeking research organisations lose ground to the government research organisations, the industrial research laboratories and the profit-seeking research institutions. Room for independence is gradually narrowed down. This, of course, is not to say that in what has been happening in recent years there has been any deliberate intention of stifling individual originality. The purpose presumably has been to supplement, to enlarge, to make fuller use of innovating ability. It may, therefore, come as a surprise to find how completely, at some points, the institution has ousted the individual and in what unlikely quarters, and in what extreme terms, the doctrine is now promulgated that industrial society can get along without individual independence and generate its necessary innovations by mass effort. . . .

It is opportune to ask whether Western societies are fully conscious of what they have been doing to themselves recently; whether, even although there now may be more people with the knowledge and training that may be a prerequisite for invention, they are used in ways which make it less likely that they will invent; whether there is truth in the idea that, although the institution is a powerful force in accumulating, preserving, discriminating, rejecting, it will normally be weak in its power to originate and will, therefore, carry within itself the seeds of stagnation unless the powers and opportunities of individuals to compete with, resist, challenge, defy and, if necessary, overtopple the institutions can be preserved.

To the majority of people the suggestion that technical progress is likely to be endangered by the very means now so commonly employed to stimulate it may appear nothing short of fantastic pessimism. For if present-day technical advance is so swift, how can it be supposed that brakes are being applied? One possible rejoinder is that perhaps our age is not unique, that the belief in our technical virtuosity is at least partly an hallucination born of the self-generating clamour of popular writing. Be that as it may, it is incontestable that if the current trend for institutionalising research is not bringing ill effects, then we are succeeding in a most difficult task: that of organising originality without diminishing its power, a task in which in the past there are many more failures to record than successes.

There is always a serious danger that institutions will tend with time to grow stiff in outline and rigid in outlook; their early successes may enhance the dangers, for their very confidence may prove their undoing. It is in the nature of an institution that it must be organised in some degree, organisation involves centralisation in some measure and centralisation is the process of reducing the number of points of autonomous thought and action. When centralisation is complete, or approaches that point, then completely irrational actions may follow. It is known, for example, that the direct and over-riding intervention of Hitler blocked technical innovation at certain crucial periods and may have lost him the war. . . .

It is natural enough to look upon these as cases of the barbaric mishandling of knowledge arising from the over-concentration of political power and to deny their relevance in assessing the consequences of changes in the organisation of technology in the Western countries. On the other hand, it cannot be denied that in every hierarchical organisation, in proportion as it gradually accumulates traditions and respect for precedence and authority, originality begins to fight a losing battle with the forces of ossification. The evidence is overwhelming and need not be quoted in full. Military organisations have been peculiarly susceptible to resistance to ideas. Social classes united to break one set of privileges have become defenders of another. Religious orders created to establish freedom of worship have become the agents of intolerance; universities have impeded the free play of thought; associations founded for discovery have become centres of resistance to change. Society has to find ways both of recognising and fostering talent and of leaving itself open to the acceptance of genius. The encouragement of mere talent is a task which might be extremely well conducted in a highly institutionalised society: for all this goes on within the accepted rules. But genius breaks the rules and thereby confronts the institutions with challenges which may at first sight appear to constitute a danger to their very existence.

It is, therefore, not wholly perverse to pose the following question. *If present trends continue for any length of time, it is not improbable that all technical research will be carried on by men with high university degrees in institutions where a measure of organisation is necessarily imposed on their work; that such institutions will be looked upon as the sole source of technical ideas; that emphasis will be laid on the need for, indeed, the duty of, research workers to submit themselves to team work; that the activities of the groups will be planned with a view to eliminating overlapping efforts and "filling obvious gaps."* Is this the kind of system most conducive to innovation?

FRIEDRICH ENGELS

Freedom Through Socialism

The question of whether changes in society resulting from technology have, on the whole, benefited or harmed mankind has long been debated. For Friedrich Engels, who with Karl Marx developed the central ideas of scientific socialism, the effects were clearly negative. Writing in 1877, he describes how the rise of capitalism and factory production transformed independent artisans into machine attendants. Under capitalism, wealthy industrialists exploited proletarian workers, some of whom were continually unemployed and most of whom lived in misery and ignorance. The only way to end this degradation, Engels argued, was for the proletariat to seize the means of production—the mines and factories—and to erect a socialist system. Only under socialism, he concludes, can man be truly free.

The materialist conception of history starts from the principle that production, and with production the exchange of its products, is the basis of every social order; that in every society which has appeared in history the distribution of the products, and with it the division of society into classes or estates, is determined by what is produced and how it is produced, and how the product is exchanged. According to this conception, the ultimate causes of all social changes and political revolutions are to be sought, not in the minds of men, in their increasing insight into eternal truth and justice, but in changes in the mode of production and exchange; they are to be sought not in the

Friedrich Engels, *Herr Eugen Dühring's Revolution in Science (Anti-Dühring)*. International Publishers Co., Inc. 1939 and 1976. Reprinted by permission.

philosophy but in the *economics* of the epoch concerned. The growing realisation that existing social institutions are irrational and unjust, that reason has become nonsense and good deeds a scourge is only a sign that changes have been taking place quietly in the methods of production and forms of exchange with which the social order, adapted to previous economic conditions, is no longer in accord. This also involves that the means through which the abuses that have been revealed can be got rid of must likewise be present, in more or less developed form, in the altered conditions of productions. These means are not to be *invented* by the mind, but *discovered* by means of the mind in the existing material facts of production.

Where, then, on this basis, does modern socialism stand?

The existing social order, as is now fairly generally admitted, is the creation of the present ruling class, the bourgoisie. The mode of production peculiar to the bourgeoisie—called, since Marx, the capitalist mode of production—was incompatible with the local privileges and privileges of birth as well as with the reciprocal personal ties of the feudal system; the bourgeoisie shattered the feudal system, and on its ruins established the bourgeois social order, the realm of free competition, freedom of movement, equal rights for commodity owners, and all the other bourgeois glories. The capitalist mode of production could now develop freely. From the time when steam and the new tool-making machinery had begun to transform the former manufacture into large-scale industry, the productive forces evolved under bourgeois direction developed at a pace that was previously unknown and to an unprecedented degree. But just as manufacture, and the handicraft industry which had been further developed under its influence, had previously come into conflict with the feudal fetters of the guilds, so large-scale industry, as it develops more fully, comes into conflict with the barriers within which the capitalist mode of production holds it confined. The new forces of production have already outgrown the bourgeois form of using them; and this conflict between productive forces and mode of production is not a conflict which has risen in men's heads, as for example the conflict between original sin and divine justice; but it exists in the facts, objectively, outside of us, independently of the will or purpose even of the men who brought it about. Modern socialism is nothing but the reflex in thought of this actual conflict, its ideal reflection in the minds first of the class which is directly suffering under it—the working class.

In what, then, does this conflict consist?

Previous to capitalist production, that is to say, in the Middle Ages, small-scale production was general, on the basis of the private ownership by the workers of their means of production; the agricultural industry of the small peasant, freeman or serf, and the handicraft industry of the towns. The instruments of labour—land, agricultural implements, the workshop and tools—were the instruments of labour of individuals, intended only for individual use, and therefore necessarily puny, dwarfish, restricted. But just because of this they belonged, as a rule, to the producer himself. To concentrate and enlarge these scattered, limited means of production, to transform them into the mighty levers of production of the present day, was precisely the historic role of the capitalist mode of production and of its representative, the bourgeoisie. In Part IV of *Capital* Marx gives a detailed account of how, since the fifteenth century, this process has developed historically through the three stages of simple co-operation, manufacture and large-scale industry. But as Marx also points out, the bourgeoisie was unable to transform those limited means of production into mighty productive forces except by transforming them from individual means of production into *social* means of production, which could be used only *by a body of men as a whole.* The spinning wheel, the hand loom and the blacksmith's hammer were replaced by the spinning machine, the mechanical loom and the steam hammer; and the factory, making the co-operation of hundreds and thousands of workers necessary, took the place of the individual work-room. And, like the means of production, production itself changed from a series of individual operations into a series of social acts, and the products from the products of individuals into social products. The yarn, the cloth and the metal goods which now came from the factory were the common product of many workers through whose hands it had to pass successively before it was ready. No individual can say of such products: I made it, that is *my* product. . . .

The first capitalists found, as we have said, the form of wage labour already in existence; but wage labour as the exception, as an auxiliary occupation, as a supplementary, as a transitory phase. The agricultural labourer who occasionally went to work as a day labourer had a few acres of his own land, from which if necessary he could get his livelihood. The regulations of the guilds ensured that the journeyman of today became the master craftsman of tomorrow. But as soon as the means of production had become social and were concentrated in the hands of capitalists, this situation changed. Both the means of production and the products of the small, individual producer lost more and more of their value; there was nothing left for him to do but to go to the capitalist, and work for wages. Wage labour, hitherto an exception and subsidiary, became the rule and the basic form of all production; hitherto an auxiliary occupation, it now became the labourer's exclusive activity. The occasional wage worker became the wage worker for life. The number of lifelong wage workers was also increased to a colossal extent by the simultaneous disintegration of the feudal system, the dispersal of the retainers of the feudal lords, the eviction of peasants from their homesteads, etc. The separation between the means of production concentrated in the hands of the capitalists, on the one side, and the producers now possessing nothing but their labour power, on the other, was made complete. *The contradiction between social production and capitalist appropriation became manifest as the antagonism between proletariat and bourgeoisie.*

We saw that the capitalist mode of production thrust itself into a society of commodity producers, individual producers, whose social cohesion resulted from the exchange of their products. But every society based on commodity production has the peculiarity that in it the producers have lost control of their own social relationships. Each produces for himself, with the means of production which happen to be at his disposal and in order to satisfy his individual needs through the medium of exchange. No one knows how much of the article he produces is coming onto the market, or how much demand there is for it; no one knows whether his individual product will meet a real need, whether he will cover his costs or even be able to sell it at all. Anarchy reigns in social production. But commodity production, like all other forms of production, has its own laws, which are inherent in and inseparable from it; and these laws assert themselves in spite of anarchy, in and through anarchy. These laws are manifested in the sole form of social relationship which continues to exist, in exchange, and enforce themselves on the individual producers as compulsory laws of competition. At first therefore, they are unknown even to these producers, and have to be discovered by them gradually only through long experience. They assert themselves therefore apart from the producers and against the producers, as the natural laws of their form of production, working blindly. The product dominates the producers.

In mediaeval society, especially in the earlier centuries, production was essentially for the producer's own use; for the most part its aim was to satisfy only the needs of the producer and his family. Where, as in the countryside, personal relations of dependence existed, it also contributed towards satisfying the needs of the feudal lord. No exchange was involved, and consequently the products did not assume the character of commodities. The peasant family produced almost everything it required—utensils and clothing as well as food. It was only when it succeeded in producing a surplus beyond its own needs and the payments in kind due to the feudal lord—it was only at this stage that it also began to produce commodities; these surplus products, thrown into social exchange, offered for sale, became commodities. The town artisans, it is true, had to produce for exchange from the very beginning. But even they supplied the greatest part of their own needs themselves; they had gardens and small fields; they sent their cattle out into the communal woodland, which also provided them with timber and firewood; the women spun flax, wool, etc. Production for the purpose of exchange, the production of commodities, was only in its infancy. Hence, restricted exchange, restricted market, stable methods of production, local isolation from the outside world, and local unity within: the Mark in the countryside, the guild in the town.

With the extension of commodity production, however, and especially with the emergence of the capitalist mode of production, the laws of commodity production, previously latent, also began to operate more openly and more potently. The old bonds were loosened, the old dividing barriers broken through, the producers more and more transformed into independent, isolated commodity producers. The anarchy of social production became obvious, and was carried to further and further extremes. But the chief means through which the capitalist mode of production accentuated this anarchy in social production was the direct opposite of anarchy: the increasing organisation of production on a social basis in each individual productive establishment. This was the lever with which it put an end to the former peaceful stability. In whatever branch of industry it was introduced, it could suffer no older method of production to exist alongside it; where it laid hold of a handicraft, that handicraft was wiped out. The field of labour became a field of battle. The great geographical discoveries and the colonisation which followed on them multiplied markets and hastened on the transformation of handicraft into manufacture. The struggle broke out not only between the individual local producers; the local struggles developed into national struggles, the trade wars of the seventeenth and eighteenth centuries. Finally, large-scale industry and the creation of the world market have made the struggle universal and at the same time given it an unparalleled intensity. Between individual capitalists, as between whole industries and whole countries, advantages in natural or artificial conditions of production decide life or death. The vanquished are relentlessly cast aside. It is the Darwinian struggle for individual existence, transferred from Nature to society with intensified fury. The standpoint of the animal in Nature appears as the last word in human development. The contradiction between social production and capitalist appropriation reproduces itself as *the antithesis between the organisation of production in the individual factory and the anarchy of production in society as a whole....*

It is the driving force of the social anarchy of production which transforms the immense majority of men more and more into proletarians, and it is in turn the proletarian masses who will ultimately put an end to the anarchy of production. It is the driving force of the social anarchy of production which transforms the infinite perfectibility of the machine in large-scale industry into a compulsory commandment for each individual industrial capitalist to make his machinery more and more perfect, under penalty of ruin. But the perfecting of machinery means rendering human labour superfluous. If the introduction and increase of machinery meant the displacement of millions of hand workers by a few machine workers, the improvement of machinery means the displacement of larger and larger numbers of the machine workers themselves, and ultimately the creation of a mass of available wage workers exceeding the average requirements of capital for labour—a complete industrial reserve army, as I called it as long ago as 1845—a reserve that would be available at periods when industry was working at high pressure, but would be thrown out onto the streets by the crash inevitably following the boom; a reserve that would at all times be like a leaden weight on the feet of the working class in their fight for existence against capital, a regulator to keep wages down to the low level which suits the needs of capital. Thus it comes about that machinery, to use Marx's phrase, becomes the most powerful weapon in the war of capital against the working class, that the instruments of labour constantly tear the means of subsistence out of the hands of the labourer, that the very product of the labourer is turned into an instrument for his subjection. Thus it comes about that the economising of the instruments of labour becomes from the outset a simultaneous and absolutely reckless waste of labour power and robbery of the normal conditions necessary for the labour function; that machinery, "the most powerful instrument for shortening labour time, becomes the most unfailing means for placing every moment of the labourer's time and that of his family at the disposal of the capitalist for the purpose of expanding the value of his capital."

Thus it comes about that the excessive labour of some becomes the necessary condition for the lack of employment of others, and that large-scale industry, which hunts all over the world for new consumers, restricts the consumption of the masses at home to a starvation minimum and thereby undermines its own internal market. "The law that always equilibrates the relative surplus population, or industrial reserve army, to the extent and energy of accumulation, this law rivets the labourer to capital more firmly than the wedges of Vulcan did Prometheus to the rock. It establishes an accumulation of misery, corresponding with accumulation of capital. Accumulation of wealth at one pole is, therefore, at the same time accumulation of misery, agony of toil, slavery, ignorance, brutality, mental degradation, at the opposite pole.". . .

But if, on these grounds, the division into classes has a certain historical justification, it has this only for a given period of time, for given social conditions. It was based on the insufficiency of production; it will be swept away by the full development of the modern productive forces. And in fact the abolition of social classes has as its presupposition a stage of historical development at which the existence not merely of some particular ruling class or other but of any ruling class at all, that is to say, of class difference itself, has become an anachronism, is out of date. It therefore presupposes that the development of production has reached a level at which the appropriation of means of production and of products, and with these, of political supremacy, the monopoly of education and intellectual leadership by a special class of society, has become not only superfluous but also economically, politically and intellectually a hindrance to development.

This point has now been reached. Their political and intellectual bankruptcy is hardly still a secret to the bourgeoisie themselves, and their economic bankruptcy recurs regularly every ten years. In each crisis society is smothered under the weight of its own productive forces and products of which it can make no use, and stands helpless in face of the absurd contradiction that the producers have nothing to consume because there are no consumers. The expanding force of the means of production bursts asunder the bonds imposed upon them by the capitalist mode of production. Their release from these bonds is the sole condition necessary for an unbroken and constantly more rapidly progressing development of the productive forces, and therewith of a practically limitless growth of production itself. Nor is this all. The appropriation by society of the means of production puts an end not only to the artificial restraints on production which exist today, but also to the positive waste and destruction of productive forces and products which is now the inevitable acompaniment of production and reaches its zenith in crises. Further, it sets free for society as a whole a mass of means of production and products by putting an end to the senseless luxury and extravagance of the present ruling class and its political representatives. The possibility of securing for every member of society, through social production, an existence which is not only fully sufficient from a material standpoint and becoming richer from day to day, but also guarantees to them the completely unrestricted development and exercise of their physical and mental faculties— this possibility now exists for the first time, but it *does exist*.

The seizure of the means of production by society puts an end to commodity production, and therewith to the domination of the product over the producer. Anarchy in social production is replaced by conscious organisation on a planned basis. The struggle for individual existence comes to an end. And at this point, in a certain sense, man finally cuts himself off from the animal world, leaves the conditions of animal existence behind him and enters conditions which are really human. The conditions of existence forming man's environment, which up to now have dominated man, at this point pass under the dominion and control of man, who now for the first time becomes the real conscious master of Nature, because and in so far as he has become master of his own social organisation. The laws of his own social activity, which have hitherto confronted him as external, dominating laws of Nature, will then be applied by man with complete understanding, and hence will be dominated by man. Men's own social organisation which has hitherto stood in opposition to them as if arbitrarily decreed by Nature and history, will then become the voluntary act of men themselves. The objective, external forces which have hitherto dominated history, will then pass under the control of men themselves. It is only from this point that men, with full consciousness, will fashion their own history; it is only from this point that the social causes set in motion by men will have, predominantly and in constantly increasing measure, the effects willed by men. It is humanity's leap from the realm of necessity into the realm of freedom.

To carry through his world-emancipating act is the historical mission of the modern proletariat. And it is the task of scientific socialism, the theoretical expression of the proletarian movement, to establish the historical conditions and, with these, the nature of this act, and thus to bring to the consciousness of the now oppressed class the conditions and nature of the act which it is its destiny to accomplish.

CLARENCE E. AYRES

The Industrial Way of Life

Had institutional patterns been impossible to change, economist Clarence Ayres maintains, then the revolution leading to the overthrow of capitalism, predicted by Marx and Engels, might indeed have occurred. However, there was flexibility with the result that the lot of common people has continued to improve. The technology of mass production, which is a central feature of the industrial way of life, Ayres asserts, gives abundance, and abundance is a necessary condition for the attainment of freedom.

Modern man has more knowledge—clearer and more certain knowledge—of what his existence means than any previous generation has had, and for obvious reasons. Man has come a long way, far longer than any previous generation has realized, since his ancestors began tending "sacred" fires, and his achievements have been so great as to suggest the exercise of magical powers; and even so, we now stand only on the threshold of what one prophet has called the era of "neotechnics." In doing so, to be sure, we have left behind many of the fancies of our ancestors. It is those meanings, and only those, which we have "lost." Better than any previous generation we know what we are doing, and what is good for us.

Moreover, we have firmer ground for hope than any previous generation has ever had, precisely because of the technological threshold on which we now clearly see that we are standing. Men have always sought security, or as

C. E. Ayres, "The Industrial Way of Life" in *Toward a Reasonable Society*. University of Texas Press, Austin and London, 1961 and 1978. Reprinted by permission.

Dewey called it, certainty. They have always dreamed of better things; and if they could see no hope of better things in this world—for themselves, but even more for their children and their children's children—they have dreamed of other worlds beyond pearl-studded gates, where the streets would be paved with gold, and milk and honey would flow in the gutters. It is only since the effects of the great technological revolution of modern times began to be apparent that boldly imaginative men began to catch sight of the possibility of general abundance here and now. Beginning only a little more than a century and a half ago a whole series of utopian prophecies and schemes began to appear, to the great alarm of the more sober-minded members of the community.

For it has always been quite apparent that in order for abundance to be generally shared substantial modification of the prevailing institutional patterns would be necessary. Indeed, the dreams and prophecies of the late eighteenth and early nineteenth centuries were inspired not only by the steam engine and power-driven machinery but also by the American and French revolutions. Thus long before the time of Karl Marx "the conventional wisdom"... turned its face firmly away from all such utopianism. Not that such reactionaries as Malthus and John Stuart Mill were hard-hearted characters: quite the contrary was the case. But such men reacted against the new utopianism because, far more than they realized, the conventional wisdom with which their minds were learnedly saturated was a reflection of the prevailing institutional order.

That is still the case.... Our conventional wisdom still assumes that scarcity is the natural condition of mankind, to which Adam was condemned on the occasion of his expulsion from the Garden of Eden. Indeed, not only that: present-day exponents of the conventional wisdom still maintain that economic growth is possible at all only by virtue of the "involuntary saving" of the poor—that such affluence as we can achieve necessarily rests on the foundation of poverty.

Nevertheless, times change. Marx argued that the condition of "the working class" must inevitably grow progressively worse until the time came for a cataclysmic upheaval to destroy "the capitalistic system"; and that might indeed have been the case if the institutional patterns of Western society had proved obdurately unamenable to change. But such has not proved to be the case. The institutional flexibility which had already made it possible for the institution of property to supplant that of feudal fief continued to manifest itself, with the result that the condition of the common people has steadily improved.

Indeed, the whole attitude of the community has changed in this regard. One of the most important discoveries of modern times is the discovery that mass production requires a mass market, and that the adequacy of the mass market depends on the ability of the masses to buy the products of mass production. This idea, so obvious that it now seems almost axiomatic, has crept into the thinking of the modern community as gradually and anonymously as the idea of scarcity did in earlier times. Many eminent thinkers have given expression to it, but none more effectively than Gilbert Seldes in a single sentence which I have been quoting ever since: "The one luxury the rich cannot afford is the poverty of the poor." In order for the economy to operate at capacity, the steel industry must operate at capacity; and in order for the steel industry to operate at capacity, its principal customer, the automobile industry, must operate at capacity; and in order for the automobile industry to sell five or six million cars a year there must be each year five or six million people who are willing and able to buy new cars; and in order for them to do so there must be millions more who are willing and able to buy used cars of various ages. In short, our affluence rests not on poverty but on participation by the whole community in the benefits of industrial production. The conventional wisdom still persists in official pronouncements and academic lucubrations. But the axiom on which we actually operate is that of general participation.

And what is true of abundance is true of all real values, which indeed are indissociable from abundance as they are indissociable from each other. Freedom is a necessary condition to the attainment of abundance, and abundance is a necessary condition to the attainment of freedom. Freedom is possible only among equals, and equality is possible only when men are free from arbitrary social distinctions. These values are attainable only when men have achieved a measure of security; and real security is possible only when men have achieved a measure of abundance and a reasonable prospect of its continuance. But it is also possible only for those who enjoy freedom among equals; for any other condition implies a threat of insecurity. And the same is true of the highest achievements of the human spirit. The golden ages of mankind have invariably been periods of relative prosperity enjoyed by communities which had in some measure thrown off the shackles of their institutional past. Only free men can know excellence, and only affluent societies can afford to indulge in such pursuits. But only through excellence can societies become affluent.

Such is the industrial way of life. It is a way of life to which modern man has dedicated himself because it is the epitome of the real values which take their meaning from the life process of mankind. And its supreme value is hope—a hope, warranted by past achievements—of a far better life next year for ourselves, in the next century for our children's children, and in the next millennium for all mankind.

DAVID E. LILIENTHAL

Democracy
At the Grass Roots

*David Lilienthal, the original director of the Tennessee Valley Authority from
1933 to 1941, describes how this vast project of damming the rivers of the
region developed local resources and replenished farm lands. The most important aspect of TVA, he states, was that it fostered individualism and personal satisfaction. The participation of average citizens in the activities and
plans of TVA, he writes, was a fine example of democracy in action, and
without this personal activity the task would have been impossible to accomplish. The world of science and technology, Lilienthal says, is still a world of
men, and democratic action is necessary to prevent dehumanization.*

The river now is changed. It does its work. But it is on the land that the
people live. Millions of acres of the valley's land had lost its vitality. The
people had to make it strong again and fruitful if they themselves were to be
strong. For here in this valley more people depend for a living upon each acre
of farm land than in any other area in America. The farms are usually small,
an average of seventy-five acres. Farm families are large, and the birthrate is
the highest in the United States. Many people living on impoverished land—
that was the picture ten years ago. If the moral purpose of resource development—the greatest benefit to human beings—was to be achieved, TVA had
to see to it that the land changed as well as the river.

Abridged from pages 25–26 and 75–77 in *TVA: Democracy on the March* by David E. Lilienthal. Copyright 1944, 1953, by David E. Lilienthal. Reprinted by permission of Harper &
Row, Publishers, Inc.

And the land is changing. It is a slow job. Engineering a river with large-scale modern machinery and rebuilding soil that for generations has been losing its vitality are tasks of a different tempo. But even after these few years you can see the difference everywhere. The gullies are being healed. The scars of erosion are on the mend, slowly but steadily. The many wounds yet to be healed are by their contrast eloquent evidence of what a decade's work in restoration has accomplished. The cover of dark green, the pasture and deep meadow and upstanding fields of oats and rye, the marks of fertility and productiveness are on every hand. Matting and sloping, seeding and sodding have given protection to eroded banks on scores of thousands of acres. Ditches to divert the water and little dams to check it, hundreds of thousands of them, help control the course of the water on the land, hold it there till it can soak down and feed the roots of newly planted trees and grasses. A hundred and fifty million seedling trees have been planted on hundreds of thousands of acres of land from TVA nursery stock alone.

The farmers have built terraces on a million acres and more; their graceful design, following the contour, makes a new kind of landscape, one that led Jefferson, observing the effect upon the face of his own Monticello acres, to exclaim that in "point of beauty nothing can exceed" contour plowing with its "waving lines and rows winding along the face of the hills and valleys."

And on 20,000 individual farms embracing a total area of nearly 3,000,000 acres, actual farmers selected by their neighbors are carrying on a demonstration of modern farming, sponsored by TVA, built around a more scientific use by farmers of the almost magic mineral, phosphate, and the use of power and the machine. These 20,000 farm families have been willing, at considerable individual risk, to undertake a changed way of farming. . . .

*   *   *

People are the most important fact in resource development. Not only is the welfare and happiness of individuals its true purpose, but they are the means by which that development is accomplished; their genius, their energies and spirit are the instruments; it is not only "for the people" but "by the people."

The purpose of resource development must be more than the mere physical welfare of the greatest number of human beings. It is true that we cannot be starving and cold and still be happy. But an abundance of food, the satisfaction of elementary physical needs alone, is not enough. A man wants to feel that he is important. He wants to be able not only to express his opinion freely, but to know that it carries some weight; to know that there are some things that he decides, or has a part in deciding, and that he is a needed and useful part of something far bigger than he is.

This hankering to be an *individual* is probably greater today than ever before. Huge factories, assembly lines, mysterious mechanisms, standardization—these underline the smallness of the individual, because they are so fatally impersonal. If the intensive development of resources, the central fact in the immediate future of the world, could be made personal to the life of most men; if they could see themselves, because it was true, as actual participants in that development in their own communities, on their own land, at their own jobs and businesses—there would be an opportunity for this kind of individual satisfaction, and there would be something to tie to. Men would not only have more things; they would be stronger and happier men.

Resource development need not be held fast by the de-humanizing forces of modern life that whittle down the importance of the individual. Surely it should be freed of their grip, for they are the very negation of democracy. ". . . nothing is good to me now that ignores individuals."

It is the unique strength of democratic methods that they provide a way of stimulating and releasing the individual resourcefulness and inventiveness, the pride of workmanship, the creative genius of human beings whatever their station or function. A world of science and great machines is still a world of men; our modern task is more difficult, but the opportunity for democratic methods is greater even than in the days of the ax and the hand loom.

A method of organizing the modern task of resource development that not only will be based upon the principle of unity but can draw in the average man and make him a part of the great job of our time, in the day-to-day work in the fields and factories and the offices of business, will tap riches of human talent that are beyond the reach of any highly centralized, dictatorial, and impersonal system of development based upon remote control in the hands of a business, a technical, or a political elite.

It is just such widespread and intimate participation of the people in the development of their valley that has gone on here in these ten years past.

The spiritual yield of democratic methods, a renewed sense that the individual counts, would be justification enough. But there is yet another reason, a practical one, for seeking at every turn to bring people actively into the task of building a region's resources; there is, I think, really no other way in which the job can be done. The task of harmonizing and from time to time adjusting the intricate, detailed maze of pieces that make up the unified development of resources in a world of technology is something that simply cannot be done effectively from some remote government or business headquarters.

The people must be in on that job. The necessities of management make it mandatory. Efficiency, in the barest operating sense, requires it. There is nothing in my experience more heartening than this: that devices of management which give a lift to the human spirit turn out so often to be the most "efficient" methods. Viewed in any perspective there is no other way. No code of laws or regulations can possibly be detailed enough to direct the precise course of resource development. No district attorney or gestapo could, for long, hope to enforce such a regime. No blueprints or plans can ever be comprehensive enough, or sufficiently flexible, as a matter of management, for so ever-changing an enterprise. It is the people or nothing.

From the outset of the TVA undertaking it has been evident to me, as to many others, that a valley development envisioned in its entirety could become a reality if and only if the people of the region did much of the planning, and participated in most of the decisions. To a considerable degree this is what is happening. Each year, almost each month, one can see the participation of the people, as a fundamental practice, grow more vigorous, and, although it suffers occasional setbacks, it is becoming part of the thinking and the mechanics of the development of the Tennessee Valley.

ARTHUR E. MORGAN

The Garrison Dam Disaster

The late Arthur E. Morgan was one of the nation's foremost civil engineers, and he served as Chairman of the Board of the Tennessee Valley Authority from 1933 to 1937. His account of the planning for the construction of North Dakota's Garrison Dam by the U.S. Army Corps of Engineers contrasts with Lilienthal's narrative of the TVA. The bureaucratic and insensitive approach of the Corps of Engineers, Morgan asserts, essentially destroyed the civilization and culture of the Arikara Indians.

There is another phase of conclusive engineering analysis... which enters into the making of cost estimates and project justification. Results which actually increase hardships for individuals and communities affected by a project often are ignored by the Corps in estimating the cost of a project. Some of such costs and losses relate to entire cultures large and small, which may have existed for centuries and may sum up to massive hardship and unnecessary tragedy and injustice.

Conclusive analysis of a large public project may commonly make the difference between results which excellently serve their purpose and results which are irreparably disappointing. And this conclusive analysis should include not only a few of the major aims of the undertaking, but every element of vital human concern. In this respect the training of West Point and the long

Arthur E. Morgan, *Dams and Other Disasters: A Century of the Army Corps of Engineers in Civil Works,* Porter Sargent Publishers, Inc., 11 Beacon Street, Boston, MA 02108, 1971, pp. 40–51.

time tradition of the Corps of Engineers has been unfortunate not in a few particular cases, but habitually. This has been the case in many areas of human contact and especially in the taking of property for public works.

When a massive reservoir project is under way the public may see the large features, but may not clearly realize that, especially in the taking of property, the project is largely made up of many small economic and human relations, many of which may be major life issues to the persons directly involved. It has been characteristic of the Corps of Engineers to be deficient in human sensitivity in this class of relationships. The cases here mentioned are significant as representative of a type of attitude for which the typical patterns of West Point seem to be largely responsible.

In March, 1945, the Corps of Engineers began preliminary work on building the Garrison Dam. This dam was to flood the lands of the Three Tribes of the Fort Berthold Reservation in North Dakota. I do not have a clear judgment as to whether the Garrison Dam should have been built. It is my unconfirmed opinion that an adequately conclusive analysis might have disclosed very superior opportunities without the many losses which resulted from the building of the Garrison Dam. However, we shall assume here that the dam should have been built.

The Arikara Indians migrated about a thousand years ago from Asia across the Bering Strait, down to the tip of South America, and up the Mississippi and Missouri Rivers. They were searching for a place to settle where the environment would be hospitable and where they would be free from invasion from hostile tribes. Winding among the treeless, semi-arid and windswept prairies of North Dakota, the deep wooded valley of the Missouri River offered an almost ideal situation. Below the prairies and in the woods the driving, sub-zero winter winds of the prairies were tamed and more gentle. Free from tropical diseases and parasites, from which the Arikaras must have previously suffered, on their long, long search and remote from other human settlements, the valley abounded in game and provided shelter in the substantial forests that grew along the river margin. Along the lower slopes of that deep valley were perennial springs. The valley had fertile land for agriculture, wild fruits, and good rivers for fishing. The hillsides provided lignite for fuel. There was grass, water and shelter for cattle. The values of this location seemed to justify the long, long pilgrimage around two continents.

Just beyond the Missouri River bottoms were the bare prairies, without water except for wells 100 to 300 feet deep which were too alkaline to be used for drinking water. There was no fuel and no timber for building. The Missouri River bottom was a garden spot in a forbidding country and it was here that the Arikaras joined the Mandan and Hidatsa Indians. At their

North Dakota home, the Three Tribes set up an unusual culture marked by considerable fellowship and good will and skill in agriculture and other arts.

The tribes had a remarkable variety of vegetables, apparently accumulated from Asia and both continents of the Western Hemisphere. They had potatoes, varieties of beans, numerous varieties of squash, flint corn, dent corn, sweet corn, pop corn, and many other garden crops. Over the centuries they bred these vegetable species to survive the cold springs and short summers of the north and of the high altitude.

Their social relations were, in general, intimate and humane. The family pattern was extended to include grandparents, uncles and aunts. To a large extent the rearing of children was a community project.

The Three Tribes were largely a self-sufficient people. In their economy, money did not play a large part. The people exchanged their products and their labor and their interest with each other as neighbors.

> Their way of life rested upon a "gardening-gathering-hunting economy," made possible by the resources of the Missouri River and Lowlands.... Annual cash income was low because there was little need for it; "Grandmother River" provided the necessities of life.... before the flooding, the Ft. Berthold people had one of the lowest rates of Welfare in the United States.

On September 17, 1851 the U.S. Government approached the Three Tribes with the Treaty of Fort Laramie. In exchange for surrendering vast areas of their traditional domain, the Government made a solemn promise that their choice home tract—the river valley, as well as additional uplands, in all 12,500,000 acres—should be theirs in perpetuity. This treaty was not a generous gift from the federal government, but the purchase price for a vast ancestral domain. However, by successive executive orders and Acts of Congress the Three Tribes' 12.5 million acres were reduced to 643,368 acres, and often without the knowledge and consent of the Indians.

In the late 1940s and the 1950s the life and harmony of the Three Tribes were disrupted by the U.S. Corps of Engineers. The Corps sought flood control legislation calling for the construction of the Garrison Dam, which would flood the best lands of the Ft. Berthold Indians. And it was the Corps which was to be in charge of the Dam's construction. When built, the dam would be the largest earth dam in the world.

The Garrison Dam would uproot and destroy the economy and social organization of the Ft. Berthold Indians. The flooding of this valley would break up the existing balance between range, shelter, water and shade and disrupt the agricultural and livestock enterprises of the Indians which had come to provide 70% of their earned net income.

The Business Council of the Three Affiliated Tribes of the Fort Berthold Reservation adopted a resolution opposing the building of the Garrison Dam.

Part of the introduction to this resolution sums up the effects of the Garrison Dam on the Indians:

The construction of the Garrison Dam will have the following results to the Indian people of Fort Berthold Reservation:

1. All of the bottom lands, and all of the bench lands on this Reservation will be flooded, most of it will be under water to a depth of 100 feet or more.

2. The homes and lands of 349 families, comprising 1544 individuals will be covered with deep water.

3. The lands which will be flooded are practically all the lands on our Reservation which are of any use or value to produce feed for stock or winter shelter.

4. We are stockmen and our living depends on our production of cattle.

5. All of the area of this Reservation which will not be flooded will be of little or no value to us if the bottom and bench lands are lost.

6. There are over 2,000 individual members of the Three Affiliated Tribes of this Reservation and it is now proposed by Acts of Congress to remove 1544 of us to some other unknown location, leaving at least 456 of our people permanently separated from the others of the Tribes by the proposed removal.

7. All of our people have lived where we now are for more than 100 years. Our people have lived on and cultivated the bottom lands along the Missouri River for many hundreds of years. We were here before the first white man stepped foot on this land. We have always kept the peace. We have kept our side of all treaties. We have been, and now are, as nearly self-supporting as the average white community.

8. We recognize the value to our white neighbors, and to the people down stream, of the plan to control the River and to make use of the great surplus of flood waters; but we cannot agree that we should be destroyed, drowned out, removed and divided for the public benefit while all other white communities are protected and safe-guarded by the same River development plan which now threatens us with destruction.

9. We see on the plans and maps of the proposed Missouri River development, that five great dams are to be built across the River. Four of those dams are carefully located above the white communities of Yankton, Chamberlain and Pierre in South Dakota, and Bismarck in North Dakota. We also know that the Garrison Dam was first planned to hold water at the level of 1,850 feet above sea level but when it was shown to the Congress that water at that height would flood some of the streets of Williston, North Dakota, that plan was promptly changed to 1,830 feet to save Williston.

10. Our Indian community of 2,000 individuals is larger than some of the cities which have been so carefully safeguarded by the original plan but we are as much entitled to protection and consideration as is anyone or all of the cities along the River.

When it was decided to build the dam and to take away the river valley land they had prized through the centuries, the Three Tribes faced their desperate loss, and sought peaceful and friendly settlement. In such circumstances a decent course would have been to help them meet these difficult circumstances and to do all that would be reasonably possible to reduce their loss and tragedy. The effort would have been worthwhile even if chiefly to preserve this rare small culture as one would preserve a precious work of art or an outstanding work of nature such as the Everglades. As might be expected after West Point conditioning, this sensitivity was not in evidence in the Corps.

Negotiations with the Corps were formally initiated by the Tribes in accordance with law. A meeting was arranged and negotiations were begun in a spirit of friendly examination. The Indians requested that from the electric power generated by the dam, they be given a small amount—20,000 kilowatt hours per year—in order to light their houses and to pump their water when they were removed from the ready sources of water and were relocated on the dry prairies. They asked for the privilege of pasturing their cattle along the margin of the reservoir where grass would be available during the dryer season. They asked for the privilege of using the timber, which grew only in the narrow valley of the river, for building houses, fences, and so forth. They asked for a bridge across a narrow part of the reservoir, so that the different sections of the tribe could have communication with each other, without going 500 miles around the border of the reservoir. They also wanted access to the water of the reservoir for their cattle.

Since the U.S. Government, by formally enacted treaty, had unreservedly affirmed the full and perpetual ownership of this land to the Three Tribes, some of whom had owned and occupied this choice land for up to 900 years, it seems that the requests of the Indians were very moderate.

The negotiations were proceeding in apparently a friendly spirit, with acceptance by the Three Tribes of very moderate terms, when an incident occurred which ended the entire program of negotiation. This had the effect of putting the Indians again under arbitrary servitude to the Corps.

A small group of Indians, led by a man called Crow Flies High who had long considered themselves enemies of the Tribes and of the government, were opposed to negotiations and endeavored to get support for this position. They were able to get a petition signed by about 10% of the membership of the Three Tribes. When the negotiations were underway a few members of the Crow Flies High group appeared at the negotiations, dressed in ceremonial feathers. The leader of the Crow Flies High group pointed at General Pick, who was in charge of the negotiations for the Corps of Engineers, referred to Pick disrespectfully and condemned the negotiations. General Pick became enraged and said that he would remember that insult as long as he lived.

Without attempting to understand the situation, General Pick abruptly interrupted the negotiation, and despite the desire of the Three Tribes to continue, stated that he would have nothing to do with the negotiations. He

and his staff went to Washington and refused to visit the reservation for further negotiations. As stated by several of the Indians who were active at the time and were present at the negotiations: "He threw the negotiations into the waste basket" and repudiated all the elements of agreement which had been reached. On a recent visit to North Dakota I talked with members of the Tribes. I found a nearly unanimous opinion that the Corps welcomed the attack of the Crow Flies High group because it provided a semblance of justification for ignoring the clear terms of the law passed by Congress and for interrupting the negotiations, on the ground that negotiation was impossible with the Three Tribes. The Indians believed that the Corps did not want to negotiate as the law required, but wanted an excuse to negate the law in order to dictate. Rev. H. W. Case, a Congregationalist who worked with the Indians for 40 years, said:

> ... My own observation was that the Govt. had sent a man out, who knew so little of Indian History and people. One could see this when he said in his approach "I want to *show you* where we will place you people."

The Garrison Dam had been authorized by the Flood Control Act of 1944. The Bureau of Reclamation had disapproved of the Garrison Dam as unnecessary and as an undesirable waste of fertile land. General Lewis Pick of the Corps of Engineers was responsible for having the Garrison Dam included in this Act, which was part of the compromise with the Sloan plan of the Bureau of Reclamation known as the Pick-Sloan plan. It became the duty of the Secretary of the Interior to approve alternative land sites providing they were "comparable in quality and sufficient in area to compensate the said tribes for the land on the Fort Berthold Reservation." The lieu lands offered by the War Department were rejected by the Secretary of the Interior, J. A. Krug, as they did not meet these requirements.

Because of the need to conclude settlement with the Indians, and because there was no land comparable in quality to the valley that had not already been settled by white men, a campaign culminated with the passage of a bill providing for a payment to the Indians of $5,105,625 in exchange for the entire value of the taken lands—both above and below the surface.

The first bill drafted for the taking of the Indian lands seemed to have been carefully and fairly designed by the Bureau of Indian Affairs, and to give evidence of sensitiveness to the desperate adjustment presented to the Three Tribes. This bill left the administration of the change in the hands of the Bureau of Indian Affairs in the Department of the Interior.

Then, when the control of the proceedings was transferred from the Indian Bureau to the U.S. Engineer Corps, the bill was largely rewritten by the Committee on Interior and Insular Affairs to suit the Corps. The terms of the bill were greatly changed, probably by General Pick. Various provisions for protecting the rights and interests of the Three Affiliated Tribes were eliminated and the administration was put into the hands of the Corps of Engineers. It was adopted as Public Law 437 on October 29, 1949. The following extracts from the earlier law illustrate elements of protection which were included in the Indian Bureau draft *but were eliminated* from the final draft as desired by the Corps:

> Section 1. The tribes and the members thereof may salvage, remove, reuse, sell, or otherwise dispose of all or any part of their improvements within the Taking Area...
>
> Section 2. The tribes and the members thereof shall have the privilege of cutting timber and all forest products and removing sand and gravel, and may use, sell, or otherwise dispose of the same until at least October 1, 1950...
>
> Section 3. The tribes and the members thereof may remove, sell, or otherwise dispose of lignite until such date as the District Engineer, Garrison District, fixes for the impoundment of waters.

> If, in the future, sub surface values are discovered within the Taking Area, which if known at this time would increase the value of said area, and said values are reduced to money, then the tribes shall be entitled to have paid to them a royalty of one-eighth of the money received for the oil and gas extracted after the ratification of this agreement.

According to the Act passed by Congress, the Indians were not allowed the privilege to fish or to graze their cattle along the river, nor could they bring their cattle to drink at the river. Their mineral rights were denied. Their hunting and trapping rights were denied. Their right to some royalty in case oil or gas should be discovered was denied. Non-taxation of future land purchases within the boundaries of the remaining reservation was denied them. Twenty thousand KWH of electricity from the dam, at cost, was also refused. The irrigation facilities of the dam and reservoir were not made available to the Indians. As it was finally drafted, the Act forbade the Indians to use the funds provided in it to hire attorneys or agents to represent them. There was no assurance that the road system would be built, nor that schools would be moved or rebuilt, or that the Indian Agency would be reestablished. It was provided that the Tribes should receive interest from the date the Act is accepted, thus pressuring the Indians toward acceptance. A more reasonable provision would have the interest begin on the date of Presidential approval. Furthermore, their money was "placed to their credit" but remained under the absolute control of Congress.

Except in one case the Indians were not allowed to cut the timber before flooding. In this case when the Indians were permitted to cut timber, they were refused permission to take it away. Now the trees are dead and half submerged in water where they are useless, unsightly and a barrier to navigation and recreation.

The Indians were ordered to come to Washington, where, under pressure and in a strange environment and under threat that the alternatives were to sign or be entirely without protection, some of them were brought unwillingly to sign the agreement, which left decisions in the hands of the Engineer Corps. The Indians apparently did not understand its terms at the time it was being drawn up. This legislation was forced on the Indians and whether they understood it or not, they did not at any time favor it. Later Congress approved an additional $7,500,000 which gave the Indians approximately $12.5 million. There is a photograph of the signing of the agreement in Washington. The official head of the Three Affiliated Tribes is holding his hands over his face, showing his grief and despair over the course of the proceeding. After signing the agreement the Tribes were largely at the mercy of the Corps.

WIDE WORLD PHOTO

With heavy hearts: signing the Garrison Dam agreement. George Gillette, left foreground, chairman of the Three Affiliated Tribes, weeps in the office of Interior Secretary J. A. Krug. Gillette stated that the members of the tribal council "sign this contract with heavy hearts."

According to law, construction of the dam proper had to wait until agreement was reached with the Indians. Yet preliminary work on the project was done during 1945 and 1946. Representative D'Ewart said "The wrong in this method is that negotiations with the Indians were not started until after construction was actually begun on this project." Representative Lemke said "I do not consider it a just or moral settlement." And according to Representative Francis Case of South Dakota, there was $60,000,000 already invested in the Garrison project and a money solution was the only solution to be considered at this late stage.

Without even token concessions, the Affiliated Tribes had to leave their precious river bottom land and home for the treeless, waterless, relatively barren prairie, where temperatures ranged from 40 below zero to more than 100 above. The Tribes left behind them the natural values and their centuries of development, living and culture. Families were largely scattered across the prairies on patches of land which were assigned them. Almost no attempt was made to maintain their high degree of community life, and the fundamental basis of their culture was destroyed. All of this added much to the emotional injury over and above the economic.

JAMES C. WALLACE

Freedom and Direction

James C. Wallace, a professor of social studies, believes that for too long society has been the pawn of a "free" engineering ethic. There are critical situations, he warns, wherein society should "direct" engineering. Just because our advanced technology gives engineers the capability of accomplishing some type of change, it does not follow that change is desirable from society's point of view.

Nearly everyone agrees that engineers need more and better education in the humanities and the social sciences. I have no quarrel with such a proposition as long as its purpose is to make engineers more sensitive to the values of our society, values which are often grounded in considerations lying wholly, or in part, beyond the reach of technology. My objections arise when such additional preparation is justified, as it is, more often than not, in terms of subverting those values rather than in affirming them.

Therefore, I have deep misgivings about the ubiquitous rationale which is ascendant, even regnant, today—a point of view which is characterized by an insistence upon the need for the humanities, the social sciences, morals, ethics, religion, and a hundred other things, to catch up with the accomplishments of science and technology, in order that the benefits and understanding therefrom might be distributed throughout the population.

James C. Wallace, "The Engineering Use of Human Beings," from *Technology and Culture* 10 (1969). Published by the University of Chicago Press, Chicago 60637. © 1969 by the Society for the History of Technology. All rights reserved. Reprinted by permission.

Implicit in this whole mode of thought is the very nub of environmental determinism; and, while it is clear that environmental factors have played a major part in the human story, I am not persuaded that they are a sufficient explanation of the plight in which we presently find ourselves. Yet there is the theme which runs throughout most of the articles on the engineer-as-humanist: If only we could dress up the engineer in the essential accouterments of economics, sociology, ethics, and theology, we might be able to budge society from its backward ways and into the paths of rationalization through technology.

This is a simple-minded proposition. Stripped of the aura of "progress" which current journalism lards into such a notion, its real nature is clear: Its very heart is the thought that science and technology, knowledge and know-how, are the essential arbiters of moral and ethical behavior for the human race.

Such a presumption must be emphatically rejected. I have grave doubts that our increasing knowledge of mechanism implies a general and mandatory alteration of all principles of right conduct. It is true, I realize, that certain technological possibility generates certain moral necessity. Thus, the sixty-five-year-old pauper, on the hospital steps with an operable brain tumor, is cared for at public expense, because advancing medical technology has created a moral question where none existed before, and because we have become sensitive to this new situation.

But does it follow that *all* technological possibility—including H-bombs, germ warfare, total automation, and all the rest—must be allowed to come to pass, and must be accompanied by "necessary" adjustments within the society? I think not. While it is true that our developing technology will continue to pose moral questions, it is not true that society is a helpless spectator, forever doomed to seek accommodation to a world of aimless change.

Society should not be forced into such a demeaning role. We must disabuse ourselves of the notion that technological perfection, continuously accomplished, is the eternal *independent* variable of the piece, producing the need for society to conform to the new circumstances—a society which plays the part, always, of the *dependent* variable, being operated *upon* by mechanical change.

Before we usher in the Age of the Engineer, we should ask several fundamental questions. Must technology always alter society? Are there not points at which society might alter technology? Are there not many critical junctures at which a "free" society and a "directed" engineering might be preferable to a "free" engineering and a "directed" society?

This last question is the essence of my concern. Is it enough to prepare engineers who are capable of greasing the skids toward that ever-receding

Nirvana which lies in the tomorrows yet to come? I do not think so. This is merely the case of breeding a technical functionary, who, convinced of the rectitude of his occupation, seeks to convert society into a cocoon for his works in the interest of compatibility.

If this analysis is correct, it is clear that much of the complaining on the part of engineers, and others, is due to their view that there is an inadequacy of lubrication in the process, a kind of social sluggishness. That their outlook is deficient has not crossed their minds. It has not occurred to many of them that their view of what is right, just, and proper—a view based on a juiced-up application of eighteenth-century mechanism—is little, if any, better than the tangled verities of society which they decry.

Sooner or later, we must come to grips with the essential problem posed for society by our rampant technology. We must decide whether capability must always be identified with desirability. We must decide if the fundamental interests of society are, at critical points, antithetical to the unimpeded ongoing of a technology which is motivated by a syndrome of growth and increasing efficiency.

As an example, I cite the question of whether the fact that a plant employing 500 people can be automated, reducing the employees to 50 people, necessarily means that it should be done. Shall other factors, lying beyond the single undoubted fact of increased efficiency, be considered before the 450 workers lose their employment? Is the best engineering in this case actually *no* engineering? And who shall decide the question? Does the technological possibility of achieving a higher level of mechanical efficiency (and profits) *require* that such efficiency be forthwith achieved?

In my judgment, not at all. While technological possibility might imply moral necessity in many cases, it is certainly not a proposition that is general in its application. The possibility that one might, through a knowledge of atomic processes, destroy the whole world, surely does not thereby connote the necessity to do so. Conversely, the possibility of healing the sick, on a scale never before contemplated, might well imply a command to do so. In short, the individual, human decisions in matters of morality remain with us, even in these days of science. So it is that many of the precepts which come to us from the most ancient times are no less valid today than when they first were uttered.

I would argue, therefore, that an engineer whose career is grounded in the assumption that the health and merit of society depend on the speed with which adaptations are made to the changes wrought by an undirected technology is short sighted indeed. Such a person will fail to see that society, per se, has anything whatever to offer. He will demand *change* and call it progress, and he will refuse to believe that society, in its resistance to change, might possess a legitimacy greater than his own.

And there is, lamentably, *nothing* in the many articles now appearing in the journals—articles usually titled "Technology for Man," or words to that effect—to suggest that an elongated program for engineers would yield any significant criticism of engineering: criticism of the persistent misuse of its magnificent powers and of its frequent debasement in the achievement of unworthy objects. The implication of such articles, whether intended or not, is that engineers graduated from such a program, and armed with a new social expertise, will be better prepared to accommodate the society to an engineering whose ultimate thrust will yet remain unchallenged and unspecified.

In sum, I would say that if we can educate the engineer to the critical need for directing the technology in the interests of preserving and enhancing certain values within the society, and if we can discourage him from altering the society at the expense of those values, then we will have accomplished a great thing. And we will have moved to a higher plateau of human endeavor.

However, if by further exposure of the engineer to the humanities and the social sciences we merely empower the facile technician to dampen or to stimulate various processes of the society in an effort to reduce its multifactorial complexity to a type of social catechism, we will have raised up an ignoramus, of colossal size and major consequence, and we will eventually suffer because of it.

KENNETH KENISTON

Technology
And Human Nature

Psychologist Kenneth Keniston suggests that advancing technology may be changing human nature. Technology gives the opportunity to many of a more advanced education, and thereby permits the development of previously latent human talents. On the other hand, Keniston warns, there are signs that a "technological man" may be emerging, an individual whose ethical commitments are subordinated to his belief in professional competence and technical skill. These qualities, he concludes, must be guided by an ethic that values human life; otherwise the future of man on Earth will be uncertain.

Man is learning to understand the inner processes of nature, to intervene in them and to use his understanding for his own purposes, both destructive and benign. Increasingly, the old reins are off, and the limits (if any) of the future remain to be defined.

But what of man himself? Do the constraints of human nature still apply? How will the change in man's relationship to nature change man? No one can answer these questions with assurance, for the future of humanity is not predestined but created by human folly and wisdom. Yet what is already happening to modern men can provide some insight into our human future.

If there is any one fact that today unites all men in the world, it is adaptation to revolutionary change in every aspect of life—in society, in values, in technology, in politics and even in the shape of the physical world. In the underdeveloped world, just as in the industrialized nations, change has encroached upon every stable pattern of life, on all tribal and traditional values, on the structure and functions of the family and on the relations between the generations.

Furthermore, every index suggests that the rate of change will increase up to the as yet untested limits of human adaptability. Thus, man's relationship to his individual and collective past will increasingly be one of dislocation, of that peculiar mixture of freedom and loss that inevitably accompanies massive and relentless change.

As the relevance of the past decreases, the present—all that can be known and experienced in the here-and-now—will assume even greater importance. Similarly, the gap between the generations will grow, and each new generation will feel itself compelled to define anew what is meaningful, true, beautiful and relevant, instead of simply accepting the solutions of the past. Already today, the young cannot simply emulate the parental generation; tomorrow, they will feel even more obliged to criticize, analyze, and to reject, even as they attempt to re-create.

A second characteristic of modern man is the prolongation of psychological development. The burgeoning technology of the highly industrialized nations has enormously increased opportunities for education, has prolonged the postponement of adult responsibilities and has made possible an extraordinary continuation of emotional, intellectual and ethical growth for millions of children and adolescents.

In earlier eras, most men and women assumed adult responsibilities in childhood or at puberty.

Today, in the advanced nations, mass education continues through the teens and for many, into the twenties. The extension of education, the postponement of adulthood, opens new possibilities to millions of young men and women for the development of a degree of emotional maturity, ethical commitment and intellectual sophistication that was once open only to a tiny minority. And in the future, as education is extended and prolonged, an ever larger part of the world population will have what Santayana praised as the advantages of a "prolonged childhood."

This will have two consequences. First, youth, disengaged from the adult world and allowed to question and challenge, can be counted on to provide an increasingly vociferous commentary on existing societies, their institutions and their values. Youthful unrest will be a continuing feature of the future.

Second, because of greater independence of thought, emotional maturity and ethical commitment, men and women will be more complex, more finely differentiated and psychologically integrated, more subtly attuned to their environments, more developed as people. Perhaps the greatest human accomplishment of the technological revolution will be the unfolding of human potentials heretofore suppressed.

Finally, in today's developed nations we see the emergence of new life styles and outlooks that can be summarized in the concept of technological man. Perhaps here the astronauts provide a portent of the future. Studies of the men who man the space capsules speak not of their valor, their dreams, or their ethical commitments, but of their "professionalism and feeling of craftsmanship," their concern "with the application of thought to problems solvable in terms of technical knowledge and professional experience," and their "respect for technical competence."

The ascendancy of technological man is of course bitterly resisted. The technological revolution creates technological man but it also creates two powerful reactions against the technological life style. On the one hand it creates, especially in youth, new humanist countercultures devoted to all that technological man minimizes: feeling, intensity of personal relationships, fantasy, the exploration and expansion of consciousness, the radical reform of existing institutions, the furtherance of human as opposed to purely technical values.

On the other hand, technological change creates reactionary counterforces among those whose skills, life styles and values have been made obsolete. In the future, the struggle between these three orientations—technological, humanistic and reactionary—will inevitably continue. Technological man, like the technology he serves, is ethically neutral. The struggle for the social and political future will therefore be waged between those who seek to rehumanize technology and those who seek to return to a romanticized, pre-technological path.

Much of what will happen to men and women in the future is good—or if not good, then at least necessary. Yet it may not be good enough. The revolution over nature has already given men the capacity to destroy tenfold all of mankind, and that capacity will be vastly multiplied in the future. And many of the likely future characteristics of men—openness, fluidity, adaptability, professional competence, technical skill and the absence of passion—are essentially soulless qualities. They can equally be applied to committing genocide, to feeding the starving, to conquering space or to waging thermonuclear war.

Such qualities are truly virtuous only if guided by an ethic that makes central the preservation and unfolding of human life and that defines "man" as any citizen of this spinning globe. So far, the technological revolution has neither activated nor extended such an ethic.

Indeed, I sometimes feel that we detect no life on any of the myriad planets of other suns in distant galaxies for just this reason. I sometimes fear that creatures on other planets, having achieved control of nature but lacking an overriding devotion to life, ended by using their control of nature to destroy their life.

In this regard, the future of man remains profoundly uncertain.

PETER F. DRUCKER

The Futility
And Dangers
Of Technology
Assessment

In 1972, Congress established the Office of Technology Assessment in an attempt to curb unwanted effects of technological innovation. Peter Drucker believes that any such attempt to predict the impact and side effects of new technologies before they are put to use is perilous and futile. He cites a number of examples to show that experts in the past have been unable to foresee the uses to which inventions have been put. Technology assessment, he warns, may result in the prohibition of important and beneficial technologies because of the presumed harmful side effects, which may never occur.

There is great interest in "technology assessment" these days. This means anticipating the impact and side effects of new technology *before* going ahead with it.

Congress has actually set up an Office of Technology Assessment. This new agency is expected to be able to predict which new technologies are likely to become important and what long-range effects they are likely to have. It is then expected to advise the Government which new technologies it should encourage and which new technologies it should discourage, if not forbid altogether.

This, one can say with certainty, is going to be a fiasco. Not only is "technology assessment" of this kind impossible but also it is likely to lead to encouragement of the wrong technologies and to discouragement of the technologies we need.

The future impact of new technology is almost always beyond anybody's imagination.

DDT is an example. This pesticide was synthesized during World War II to protect American soldiers against disease-carrying insects, especially in the tropics. Some of the scientists then envisaged the use of the new chemical to protect civilian populations as well.

But not one of the many men who worked on DDT thought of applying the new pesticide to control insects infesting crops, forests or livestock. If DDT had been restricted to the use for which it was developed, the protection of humans, it would never have become an environmental hazard.

Use of DDT for this original purpose accounted for no more than 5 or 10 per cent of the total at the pesticide's peak, in the mid-sixties. Farmers and foresters, without much help from the scientists, saw that what killed bugs on men would also kill bugs on plants, and they turned DDT into a massive assault on the environment.

Another example is the population explosion in the developing countries. DDT and other pesticides were a factor in that. So were the new antibiotics. Yet the two technologies were developed quite independently of each other, and no one "assessing" either technology could have foreseen their convergence. Indeed, no one did.

But more important as causative factors in the sharp drop in infant mortality, which set off the population explosion, were two very old "technologies" to which no one paid any attention.

One was the elementary public-health measures of keeping latrine and well apart. This precaution was known to Macedonians before the time of Alexander the Great.

The other was the wire-mesh screen for doors and windows, invented by an unknown American about 1860.

Both "technologies" were suddenly adopted even by backward tropical villages after World War II. Together they were probably the main cause of the population explosion.

At the same time, a technology impact that the "experts" foresee almost never actually occurs. One example is the "private flying boom" widely predicted during and shortly after World War II.

The private plane, owner-piloted, would become as common, we were told, as the Model T became after World War I. Indeed, "experts" among city planners, engineers and architects advised New York City not to go ahead with the second tube for the Lincoln Tunnel or with the second deck on the George Washington Bridge but to build instead a number of small airports along the west bank of the Hudson River.

It would have taken only elementary mathematics to disprove this particular "technology assessment." There just is not enough airspace for commuter traffic by plane. But this did not occur to the "experts." No one then realized how finite airspace is.

At the same time, almost no "experts" foresaw the expansion of commercial air traffic or anticipated, at the time the jet plane was first developed, that the jet would lead to mass transportation by air.

To be sure, trans-Atlantic travel was expected to grow fast, but of course people would go by ship. These were the years in which all the governments along the North Atlantic heavily subsidized the building of new super-luxury liners, just at the time the passengers deserted the liners and switched to the new jet plane.

A few years later we were told by everybody that automation would have tremendous economic and social impacts, and it has had practically none. And the computer is an even more bizarre story.

In the late nineteen-forties nobody predicted that the computer would be used by business and government. While the computer was a "major scientific revolution," everybody "knew" that its main use would be in science and in warfare.

Indeed, the most extensive market research study undertaken at that time reached the conclusion that the world computer market would, at most, be able to absorb 1,000 computers by the year 2000. Now, only 25 years later, there are some 150,000 computers installed in the world, most of them doing the most mundane bookkeeping work.

A few years later, however, when it became apparent that business was buying computers for payroll or billing, everybody predicted that the computer would revolutionize business.

The "experts" predicted that the computer would displace middle management, so that there would be nobody left between the chief executive officer and the foreman.

"Is middle management obsolete?" asked a widely quoted *Harvard Business Review* article of the early nineteen-fifties. And it answered this rhetorical question with a resounding "Yes." At that moment the tremendous expansion of middle management jobs began.

In every developed country, over the last 20 years, middle management

DRAWING BY JAN ADKINS; REPRINTED WITH ARTIST'S PERMISSION

jobs, in business as well as in government, have grown about three times as fast as total employment. And that growth has been directly correlated with the growth of computer usage.

Anyone depending on "technology assessment" in the early nineteen-fifties would have abolished the graduate business schools as likely to produce graduates who could not possibly find jobs. Fortunately, the kids knew better and flocked to the graduate business schools to get the good jobs that the computer was creating instead of obsoleting.

At the same time, no one predicted the real revolution in business in the fifties and sixties: the merger wave and the conglomerates.

It is not only that man has the gift of prophecy no more in respect to technology than in respect to anything else. Technology is actually far more difficult to predict, especially in its impact, than most other developments.

In the first place, as the population explosion shows, social and economic impact is almost always the result of the convergence of a number of factors, not all of them technological. And each of these factors has its own origin, its own development, its own dynamics and its own experts.

"The expert" in one field, epidemiology, has never heard of agricultural pests and would never think of them. The expert on antibiotics is concerned with the treatment of disease. Whereas the actual explosion of the birth rate largely resulted from elementary and long-known public health measures.

But equally important, which technology is likely to become important and have an impact and which technology either will fizzle out (like the "flying Model T") or will have minimal social or economic impact (like automation) is simply impossible to predict.

The most successful prophet of technology is a good example. Jules Verne predicted a great deal of twentieth-century technology a hundred years ago, though few scientists of that time took him seriously. But, while Jules Verne foresaw revolutionary new technology, he anticipated no social impact but an unchanged Victorian society and economy.

Economic and social prophets, in turn, have the most dismal record as predictors of technology. The only thing an Office of Technology Assessment is likely to produce, therefore, would be not "prophecies" but fifth-rate trashy fiction.

However, this is not the major danger. The major danger is that the delusion that we can foresee the impact of new technology will lead us to slight the really important task.

For technology does have impacts, and serious ones, beneficial as well as detrimental. These do not require prophecy. They require careful monitoring of the actual impact of a technology once it has become effective.

In 1948 practically no one correctly saw the impact of the computer. Five or six years later one could and did know. Then one could say, "Whatever the technological impact, socially and economically, this is not a major threat."

In 1943 no one could predict the impact of DDT. Ten years later it was already accomplished fact that DDT had primarily become a worldwide tool of farmer, forester and livestock breeder and, with that, a major ecological factor.

Then it was time to think through what action to take, to work on the development of pesticides without the major environmental impact of DDT and, failing this development, to think through the trade-offs between food production and environmental damage—which neither the unlimited use nor the present complete ban on DDT sufficiently considers.

Technology monitoring is a serious, an important, indeed a vital task. But it is not prophecy. The only thing possible, in respect to a new technology, is speculation with about one chance in a hundred of being right. And there is a much better chance of doing harm by encouraging the wrong new technology or discouraging the most beneficial one.

What needs to be watched is "young technology," one that has already had a substantial impact, enough to be judged, to be measured, to be evaluated.

And technology monitoring, in respect to the social as well as technological impact of a "young technology," is, above all, a managerial responsibility.

LANGDON WINNER

Technology
As Legislation

Langdon Winner, a political scientist, sees two quite different kinds of current concern about technology. The first considers technology as essentially neutral, and subject to direction by technology assessment. But technology assessment, Winner argues, is merely another method of regulation; it is a political process in which trade-offs are made, and which will be of little lasting value. The second takes the position that technology shapes the fundamental pattern and content of human activity. To ameliorate and enhance human life, then, it is necessary to change the entire structure and direction of current technology. It is clear that Winner is an adherent of the second point of view.

The ecology movement, consumerism, future studies, the technology assessors, students of innovation and social change, and what remains of the "counterculture" all have something to say about the ways in which technology presents difficulties for the modern world. Since the reader is no doubt familiar with the debates now raging over these issues, I will not review the details. But in their orientations toward politics and their conceptions of how a better state of affairs might be achieved, the issue areas sort themselves into roughly two categories.

In the first domain, far and away the most prominent, the focus comes to rest on matters of risk and safeguard, cost and benefit, distribution, and the familiar interest-centered style of politics. Technology is seen as a cause of

"Frankenstein's Problem," reprinted from *Autonomous Technology* by Langdon Winner, 1977, by permission of the MIT Press, Cambridge, Massachusetts.

certain problematic effects. All of the questions raised... would be interpreted as "risks taken" and "prices paid" in the course of technological
advance. Once this is appreciated, the important tasks become those of
(1) accurate prediction and anticipation to alleviate risk, (2) adequate evaluation of the costs that are or might be incurred, (3) equitable distribution of
the costs and risks so that one portion of the populace neither gains nor
suffers excessively as compared to others, and (4) shrewd evaluation of the
political realities bearing upon social decisions about technology.

Under this model the business of prediction is usually meted out to the
natural and social sciences. Occasionally, some hope is raised that a new art
or science—futurism or something of the sort—will be developed to improve
the social capacity of foresight. The essential task is to devise more intelligent
ways of viewing technological changes and their possible consequences in
nature and society. Ideal here would be the ability to forecast the full range of
significant consequences in advance. One would then have a precise way of
assigning the risk of proceeding in one way rather than another.

The matter of determining costs is left to orthodox economic analysis. In
areas in which "negative externalities" are experienced as the result of
technological practice, the loss can be given a dollar value. The price paid for
the undesirable "side effects" can then be compared to the benefit gained. An
exception to this mode of evaluation can be seen in some environmental and
sociological arguments in which nondollar value costs are given some weight.
On the whole, however, considerations of cost follow the form Leibniz
suggested for the solution of all rational disputes: "Let us calculate." Taking
this approach one tends to ask questions of the sort: How much are you
prepared to pay for pollution-free automobiles? What is the public prepared
to tax itself for clean rivers? What are the trade-offs between having wilderness and open space as opposed to adequate roads and housing? Are the costs
of jet airport noise enough to offset the advantage of having airports in the
middle of town? Such questions are answered at the cash register, although
the computer shows a great deal more style.

Once the risks have been assigned, the safeguards evaluated, and the costs
calculated, one is then prepared to worry about distribution. Who will enjoy
how much of the benefit? Who will bear the burden of the uncertainty or the
price tag of the costs? Here is where normal politics—pressure groups, social
and economic power, private and public interests, bargaining, and so forth—
enters. We expect that those most aware, best supplied, and most active will
manage to steer a larger proportion of the advantages of technological productivity their way while avoiding most of the disadvantages. But for those
who have raised technology as a political problem under this conception,
reforms are needed in this distributive process. Even persons who have no

quarrel with the inequities of wealth and privilege in liberal society now step forth with the most trenchant criticisms of the ways in which technological "impacts" are distributed through the social system. A certain radicalism is smuggled in through the back door. The humble ideal of those who see things in this light is that risks and costs be allotted more equitably than in the past. Those who stand to gain from a particular innovation should be able to account for its consequences beforehand. They should also shoulder the major brunt of the costs of undesirable side effects. This in turn should eliminate some of the problems of gross irresponsibility in technological innovation and application of previous times. Since equalization and responsibility are to be induced through a new set of laws, regulations, penalties, and encouragements, the attention of this approach also aims at a better understanding of the facts of practical political decision making.

Most of the work with any true influence in the field of technology studies at present has its basis in this viewpoint. The ecology movement, Naderism, technology assessment, and public-interest science each have somewhat different substantive concerns, but their notions of politics and rational conduct all fit within this frame. There is little new in it. What one finds here is the utilitarian-pluralist model refined and aimed at new targets. In this form it is sufficiently young to offer spark to tired arguments, sufficiently critical of the status quo to seem almost risqué. But since it accepts the major premises and disposition of traditional liberal politics, it is entirely safe. The approach has already influenced major pieces of legislation in environmental policy and consumer protection. It promises to have a bright future in both the academic and the political realms, opening new vistas for "research," "policy analysis," and, of course, "consulting."

On the whole, the questions I have emphasized here are not those now on the agendas of persons working in the first domain. But for those following this approach I have one more point to add. It is now commonly thought that what must be studied are not the technologies but their implementing and regulating systems. One must pay attention to various institutions and means of control—corporations, government agencies, public policies, laws, and so forth—to see how they influence the course our technologies follow. Fine. I would not deny that there are any number of factors that go into the original and continued employment of these technical ensembles. Obviously the "implementing" systems have a great deal to do with the eventual outcome. My question is, however, In what technological context do such systems themselves operate and what imperatives do they feel obliged to obey? ... The hope for some "alternative implementation" is largely misguided. *That* one employs something at all far outweighs (and often obliterates) the matter of *how* one employs it. This is not sufficiently appreciated by those working within the utilitarian-pluralist framework. We may firmly believe that we are developing ways of regulating technology. But is it perhaps more likely that the effort will merely succeed in putting a more elegant administrative façade on old layers of reverse adapted rules, regulations, and practices?

The second domain of issues is less easily defined, for it contains a collection of widely scattered views and spokesmen. At its center is the belief that technology is problematic not so much because it is the origin of certain undesirable side effects but rather because it enters into and becomes part of the fabric of human life and activity. The maladies technology brings—and this is not to say that it brings only maladies—derive from its tendency to structure and incorporate that which it touches. The problems of interest, therefore, do not arrive by chain reaction from some distant force. They are present and immediate, built into the everyday lives of individuals and institutions. Analyses that focus only upon risk/safeguard, cost/benefit, and distribution simply do not reveal problems of this sort. They require a much more extraordinary, deep-seeking response than the utilitarian-pluralist program can ever provide.

What, then, are the issues of this second domain? Some of the most basic of them are mirrored in . . . the theory of technological politics. This model represents the critical phase of a movement of thought, the attempt to do social and political analysis with technics as its primary focus. But these thoughts so far have given little care to matters of amelioration. . . . It is my experience that inquiries pointing to broad, easy solutions soon become cheap merchandise in the commercial or academic marketplace. They become props for the very thing criticized.

For better or worse, however, most of the thinking in the second domain at present is highly specific, solution oriented, and programmatic. The school of humanist psychology, writers and activists of the counterculture, utopian and communal living experiments, the free schools, proponents of encounter groups and sensual reawakening, the hip catalogers, the peace movement, pioneers of radical software and new media, the founders and designers of alternative institutions, alternative architecture, and "appropriate" or "intermediate" technology—all of these have tackled the practical side of one or more of the issues raised in this essay.

Much of the work has begun with a sobering recognition of the psychological disorders associated with life in the technological society. The world of advanced technics is still one that makes excessive demands on human performance while offering shallow, incomplete rewards. The level of stress, repression, and psychological punishment that rational-productive systems extract from their human members is not matched by the opportunity for personal fulfillment. Men and women find their lives cut into parcels, spread out, and dissociated. While the neuroses generated are often found to be normal and productive in the sociotechnical network, there has been a strong revolt against the continuation of such sick virtues. Both professionals and amateurs in psychology have come together in a host of widely differing attempts to find the origins of these maladies and to eliminate them.

Other enterprises of this kind have their roots in a pervasive sense of personal, social, and political powerlessness. Confronted with the major forces and institutions that determine the quality of life, many persons have begun to notice that they have little real voice in most important arrangements affecting their activities. Their intelligent, creative participation is neither necessary nor expected. Even those who consider themselves "well served" have cause to wonder at decisions, policies, and programs affecting them directly, over which they exercise no effective influence. In the normal state of affairs, one must simply join the "consensus." One consents to a myriad of choices made, things built, procedures followed, services rendered, in much the same way that one consents to let the eucalyptus trees continue growing in Australia. There are some, however, who have begun to question this submissive, compliant way of life. In a select few areas, some people have attempted to reclaim influence over activities they had previously let slip from their grasp. The free schools, food conspiracies and organic food stores, new arts and crafts movement, urban and rural communes, and experiments in alternative technology have all—in the beginning at least—pointed in this direction. With mixed success they have sought to overcome the powerlessness that comes from meting out the responsibility for one's daily existence to remote large-scale systems.

A closely related set of projects stems from an awareness of the ways organized institutions in society tend to frustrate rather than serve human needs. The scandal of productivity has reached astounding proportions. More and more is expended on the useless, demeaning commodities idealized in the consumer ethos (for example, vaginal deodorants), while basic social and personal needs for health, shelter, nutrition, and education fall into neglect. The working structures of social institutions that provide goods and services seem themselves badly designed. Rather than elicit the best qualities of the persons they employ or serve, they systematically evoke the smallest, the least creative, least trusting, least loving, and least lovable traits in everyone. Why and how this is so has become a topic of widespread interest. A number of attempts to build human-centered and responsive institutions, more reasonable environments for social intercourse, work, and enjoyment, are now in the hands of those who found it simply impossible to continue the old patterns.

Finally, there is a set of concerns, evident in the aftermath of Vietnam,

Watergate, and revelations about the CIA, which aims at restoring the element of responsibility to situations that have tended to exclude responsible conduct. There is a point, after all, where compliance becomes complicity. The twentieth century has made it possible for a person to commit the most ghastly of domestic and foreign crimes by simply living in suburbia and doing a job. The pleas of Lieutenant Calley and Adolf Eichmann—"I just work here"—become the excuse of everyman. Yet for those who perceive the responsibility, when distant deeds are done and the casualties counted, the burdens are gigantic. As Stanley Cavell and Nadezhda Mandelstam have observed, there is a sense in which one comes to feel responsible for literally everything. Evils perpetrated and the good left undone all weigh heavily on one's shoulders. Like Kafka's K. at the door of the castle, the concerned begin a search for someone or something that can be held accountable.

I admit that I have no special name for this collection of projects. *Humanist technology* has been suggested to me, but that seems wide of the mark. At a time in which the industrialization of literature demands catchy paperback titles for things soon forgotten, perhaps it is just as well to leave something truly important unnamed.

The fundamental difference between the two domains, however, can be stated: a difference in insight and commitment. The first, the utilitarian-pluralist approach, sees that technology is problematic in the sense that it now *requires legislation.* An ever-increasing array of rules, regulations, and administrative personnel is needed to maximize the benefits of technological practice while limiting its unwanted maladies. Politics is seen as the process in representative government and interest group interplay whereby such legislation takes shape.

The second approach, disjointed and feeble though it still may be, begins with the crucial awareness that technology in a true sense *is legislation.* It recognizes that technical forms do, to a large extent, shape the basic pattern and content of human activity in our time. Thus, politics becomes (among other things) an active encounter with the specific forms and processes contained in technology. . . .

The idea central to all thinking in the second domain [is] that *technology is itself a political phenomenon.* A crucial turning point comes when one is able to acknowledge that modern technics, much more than politics as conventionally understood, now legislates the conditions of human existence. New technologies are institutional structures within an evolving constitution that gives shape to a new polity, the technopolis in which we do increasingly live. For the most part, this constitution still evolves with little public scrutiny or debate. Shielded by the conviction that technology is neutral and tool-like, a whole new order is built—piecemeal, step by step, with the parts and pieces linked together in novel ways—without the slightest public awareness or opportunity to dispute the character of the changes underway. It is somnambulism (rather than determinism) that characterizes technological politics—on the left, right, and center equally. Silence is its distinctive mode of speech. If the founding fathers had slept through the convention in Philadelphia in 1787 and never uttered a word, their response to constitutional questions before them would have been similar to our own.

Indeed, there is no denying that technological politics as I have described it is, in the main, a set of pathologies. To explain them is to give a diagnosis of how things have gone wrong. But there is no reason why the recognition of technology's intrinsic political aspect should wed us permanently to the ills of the present order. On the contrary, projects now chosen in the second domain bear a common bond with attempts made to redefine an authentic politics and reinvent conditions under which it might be practiced. As a concern for political theory this work has been admirably carried forward by such writers as Hannah Arendt, Sheldon Wolin, and Carole Pateman. In the realm of historical studies it appears as a renewed interest in a variety of attempts— the Paris Communes of 1793 and 1871, nineteenth-century utopian experiments, twentieth-century Spanish anarchism, the founding of worker and community councils in a number of modern revolutions—to create decentralist democratic politics. In contemporary practice it can be seen in the increasingly common efforts to establish worker self-management in factories and bureaucracies, to build self-sufficient communities in both urban and rural settings, and to experiment with modes of direct democracy in places where hierarchy and managerialism had previously ruled.

Taken in this light, it is possible to see technology as legislation and then follow that insight in hopeful directions. An important step comes when one recognizes the validity of a simple yet long overlooked principle: *Different ideas of social and political life entail different technologies for their realization.* One can create systems of production, energy, transportation, information handling, and so forth that are compatible with the growth of autonomous, self-determining individuals in a democratic polity. Or one can build, perhaps unwittingly, technical forms that are incompatible with this end and then wonder how things went strangely wrong. The possibilities for matching political ideas with technological configurations appropriate to them are, it would seem, almost endless. If, for example, some perverse spirit set out deliberately to design a collection of systems to increase the general feeling of powerlessness, enhance the prospects for the dominance of technical elites, create the belief that politics is nothing more than a remote spectacle to be experienced vicariously, and thereby diminish the chance that anyone would take democratic citizenship seriously, what better plan to suggest than that we simply keep the systems we already have? There is, of course, hope that we may decide to do better than that. The challenge of trying to do so now looms as a project open to political science and engineering equally. But the notion that technical forms are merely neutral and that "one size fits all" is a myth that no longer merits the least respect.

DOROTHY NELKIN

The Technological Imperative versus Public Interests

Unlike Winner, who dismisses governmental decisions involving controversial technologies as merely a process of political trade-offs, political scientist Dorothy Nelkin sees decision-making as an exciting challenge. We must recognize the demands for increased public participation as legitimate, she argues, but at the same time we must ensure that the opinion of technical experts is heard so that decisions can be informed. Decisions involving technological innovations, she concludes, constitute a major dilemma of democracy today.

Nearly every technological development, be it the siting of a large-scale facility such as a nuclear power plant or hospital, the implementation of a new biomedical or agricultural innovation, or a technology-based program such as fluoridation or genetic screening, is surrounded by controversy. Decisions once defined as technical are increasingly forced into the political arena by people who are skeptical about the value of technological progress, who perceive a gap between technology and human needs, or who mistrust the concentration of authority in bureaucracies responsible for technological change.

How do such technological controversies differ from other political events?

Are there really new issues in the politics of technology? Are there special organizational and political manifestations that follow from the characteristics of technology? Political conflict over technological innovations offers a natural opportunity to explore such questions, providing insight into the politics of technical decisions, the role of experts in policymaking, and the implications of technical complexity and expert decision making for the increased public participation demanded by protest groups.

It is useful to think of the politics of technological innovations as a dialectic between efficiency and democratic ideology. The tendency toward efficiency, as well as the intrinsic complexity of technology, calls for defining problems as technical (i.e., requiring solution by experts). Yet the democratic ideal requires participation in decision making by affected interests. The planning and development of technology is the responsibility of public or private agencies whose autonomy is maintained by the esoteric nature of the technology they deal with and legitimized by the authority of the technical advice available to them. The complexity of technology and the specialized character of technological planning limits public choice, giving special authority to those people who have the required knowledge; and this authority is reinforced by the tendency of bureaucracies to claim absolute expertise in order to maintain unfettered control.

Concerned with efficiency, developers of any new technology prefer to define decisions as technical rather than political. As long as the problem remains one of performing a narrowly defined task there is no need to weigh conflicting interests, but only the relative effectiveness of various technical alternatives. This situation is surely more comfortable and efficient than the negotiation and compromise required in the political arena.

There are, however, many factors provoking political conflict. A generalized mistrust of authority has compounded sensitivity to the social and environmental impacts of technology. Technologies seem to perpetuate themselves often irrespective of human needs. Moreover, many technological decisions involve nonvoluntary compliance; the risks of nuclear power, the neighborhood disruption from a new technical facility, the economic disruption from an agricultural innovation, cannot be avoided by those people who oppose a technology.

Until recently it was assumed that progress has inevitable costs and that people ought to adapt or leave (one is often reminded of Lord Coke's famous statement, made some three hundred years ago, that one ought not to have so delicate a nose that one cannot bear the smell of hogs). However, it has become increasingly difficult to avoid the burdens of technology; and as the options decrease, the tendencies to protest increase. Indeed, for those people affected by technical decisions there are powerful incentives to try to influence them.

Participation as an ideology in American society seems to be of growing importance just when technical complexity threatens to limit effective political choice. The actual scope of citizen influence on technological development depends on many of the usual factors that affect any political decision: leadership, community organization, access to the media, the visibility and urgency of the issue. However, other factors assume special importance in the case of decisions concerning technology: the availability of information and the distribution of expertise. To influence technological innovation, to take issue with the validity of forecasts that are used to justify new technologies, requires coping with complex and often uncertain technical material.

It is characteristic of conflicts over technological innovation that there is seldom full and conclusive understanding of the technical issues that can serve as a definitive guide to policy. Information can be mustered to support either side of a debate, and power hinges on the ability to manipulate knowledge and control uncertainty. Technical expertise, therefore, is a crucial political resource in the politics of technology. And the key questions focus on the relationship between policymakers and their experts, on the ways in which decisions about innovation deal with uncertainty concerning social costs, and on the dilemma of democracy in this increasingly complex and professionalized policy arena.

HARVEY BROOKS

Technology Assessment In Retrospect

Harvey Brooks, professor of technology and public policy at Harvard University, was the chairman of the National Academy of Sciences committee that in 1967 studied the concept of technology assessment and recommended to the Congress that it establish an Office of Technology Assessment. Brooks notes in this selection that the two major criticisms of technology assessment are first, that it is impossible to foresee the effects of technological innovation, and second, that technology assessment is merely a Band-Aid to avoid more fundamental criticisms of our technological society. He thinks that both criticisms fail to take into account the realities of the assessment mechanism. Merely revealing the discrepancies in the process, he argues, does not provide a sound argument for abandoning the ideal. After discussing the problem areas in the technology assessment process, Brooks concludes that it still has considerable promise as a method of preventing unwanted effects of future technological innovation.

Introduction

Since about 1966 there has been a rapid growth of public, political and scholarly interest in the secondary consequences of technological progress and the

From the *Newsletter on Science, Technology and Human Values* (Harvard University) Number 17, October 1976. Copyright © by the President and Fellows of Harvard University. Reprinted by permission.

applications of technology. This has been accompanied by a rapid proliferation of new legislation to regulate technology and to create new bureaucracies to refine and enforce the regulations. . . .

There has been a parallel growth of the environmental movement and the technology assessment movement. . . . Although parallel in purpose the movements have developed rather separately, with different bureaucracies, and distinct client groups and operating philosophies. Of the two, environmental assessment came first, and has been politically more influential, with a larger social impact. Yet the technology assessment and environmental movements cannot be treated entirely separately; they are too closely related, and any assessment of technology assessment in retrospect will have to consider both environmental and technological aspects.

The Magna Carta of the environmental movement was NEPA, the National Environmental Policy Act of 1969, and particularly Section 102, which required environmental impact statements for "all federal actions significantly affecting the environment." The corresponding charter for technology assessment was the Technology Assessment Act of 1972, which created the Office of Technology Assessment in the Congress. But both environmental and technology assessment have been in effect the subject of hundreds of pieces of legislation too numerous to list. One has only to mention auto safety, consumer product safety, pesticide regulation, the clean air amendments, the water pollution control act, the occupational health and safety act, the creation of the Nuclear Regulatory Agency, and so on down the line. All of these pieces of legislation require what amounts to more or less elaborate technology assessments prior to any positive action to permit the application of technology, either in general or with respect to a specific project, such as a dam or a nuclear power plant, or even a specific regulatory action.

It is difficult to distinguish between environmental and technology assessment substantively, since on the one hand the environment has tended to be interpreted more and more broadly to include personal health and safety, depletion of resources and even social effects, while on the other, it is impossible to carry out an assessment of a particular technology without thorough consideration of environmental effects. A true technology assessment may in fact require the comparative environmental assessment of several alternate technologies designed for similar purposes, e.g., coal fired vs. nuclear power plants. One distinction is that environmental and technology assessment have appealed to different professional and political constituencies. By and large, environmental concerns and regulation have become the province of lawyers and biologists (biomedical with respect to public health aspects, or ecological with respect to the natural environment). In contrast, technology assessment (TA) has become the province primarily of engineers and economists. As a result TA is somewhat more positive in its stance towards technology, especially new technology. Whereas environmentalists are concerned almost exclusively with the control of technology and with its possible negative consequences or "externalities," the TA constituency is also concerned with the benefits of technology and with the identification of new or underdeveloped technologies, which might have social benefits or positive "externalities" not wholly realizable within the incentive structure of the private market. In fact the TA Act of 1972 refers explicitly to the potential benefits of neglected technologies.

All this has resulted in an enormous shift in the burden of proof with respect to the introduction of new technology or the expansion of the application of old technology. Whereas the burden of proof used to be on those advocating the slowdown or halting of particular technological developments or applications, it is now on those seeking to advance technological innovations or particular projects. This is due in part to legislation which gives sweeping new powers to new regulatory bureaucracies such as EPA or the Consumer Product Safety Commission or the Occupational Health and Safety Administration. But also, encouraged by legislation, public interest groups have acquired new standing to sue in the courts for the halting of certain kinds of technological developments or projects. Furthermore, the courts have interpreted such legislative mandates as exist in the broadest possible terms. . . .

The concept of liability has been extended by judicial interpretation so as to embrace "strict liability," i.e., the notion that the manufacturer is liable for injury caused by his product or activity even in the absence of a showing of negligence on his part. Moreover, this concept of liability has been extended well back along the chain of suppliers, like the house that Jack built.

The interpretation of Section 102 of NEPA by the courts has also made it necessary to prepare a full environmental or technological assessment for a technology even to obtain authorization to build a prototype for test or demonstration purposes. For example, it has been difficult to obtain a variance for emission standards for an experimental coal gasification plant, even though such standards would only be relevant when that particular type of plant was deployed commercially on a large scale. . . .

Lessons from Recent History

What, then, can we say about what we have learned? I would like to discuss this under several headings, which I will now summarize.

1. Is technology or environmental assessment really technically feasible, and under what conditions? Can we foresee the consequences of technology sufficiently well to make rational decisions which will not be overtaken by

subsequent events? Can we forestall or modify adverse consequences by foreseeing them? Here we must look at both the technical feasibility of TA and the political feasibility of implementing its conclusions.

2. What is the appropriate role of the public in TA? Is so-called "participatory technology" workable in practice? What are the ultimate implications of public participation and of the growing strength of political groups which claim to act as surrogates for the public interest?

3. What standards of evidence and proof should be required in TA? Is it possible to define "reasonable" standards of proof such that the burden of proof placed on one side or the other of a controversy is not totally unrealistic? Can opposing sides be induced to accept the same criteria for the acceptability of technical evidence? What is the appropriate role of scientists and engineers relative to other interest groups and the general public in the technology assessment process?

4. What is the impact, actual and potential, of TA on the process of technological innovation, both in the private and public sectors? What is the impact on economic growth and on the net growth of individual and social welfare, taking into account "externalities" as well as direct benefits? Are the secondary and unforeseen consequences of TA and of the regulation of technology tending to price technological innovation risks out of the market and thus to deprive the public of benefits which might outweigh the benefits of the protection they receive from the regulations? What about the synergistic interactions between different and separate regulatory actions?

5. Conversely, to what extent can it be said that TA and regulation are actually stimulating innovation in new and socially beneficial directions which are more important than the perhaps minor product improvements to which much industrial and public innovative activity was previously directed? For example, are not emission controls of greater social benefit than riding comfort or automatic window controls?

6. Will regulation and TA ultimately become captive to the technological momentum of the area being regulated and to the institutions and professions that advocate or promote the technologies in question? If so, to what extent is this necessarily bad for society?

7. What has been the effect of TA and its attendant processes on the health of science as distinct from technology? What will be its long range impact on the demand for scientists, what scientists do, and the status and role of scientific institutions?

Feasibility of TA

About a year ago, Peter Drucker, in one of his usually provocative articles, ridiculed the whole notion of technological assessment, pointing out that it was impossible to foresee the consequences of technology and that, in fact, the whole TA movement was nothing but a manifestation of scientific *hubris*— a new intellectual promotion of a piece with other fads such as systems analysis or space exploration. At the same time a number of political scientists took up the cudgels on the opposite side, with the view that TA was just a palliative to avoid more fundamental criticism of the basic assumptions underlying our technological society and the cult of technological progress. Neither critic touched much on the realities of TA as it might be carried out in practice—who will do it, how it will be financed, how and when the general public will have an input, in what form the conclusions will be presented and how the results will be implemented by various institutional decision makers.

Ideally the concept of TA is that it should forecast, at least on a probabilistic basis, the full spectrum of possible consequences of technological advance, leaving to the political process the actual choice among the alternative policies in the light of the best available knowledge of their likely consequences. Such an idealization implies the possibility of a greater separation of value judgments and technical judgments than many people with practical experience in TA would consider feasible. Nevertheless, by a process of iteration or dialogue involving experts and decision-makers (including the public), the ideal might be approached by successive approximations. In this respect the situation would be no different from other political processes, for which there are large differences between the ideal prescription and reality. Such discrepancies are not necessarily an argument for abandoning the ideal.

A more fundamental difficulty is that the evidence to conduct a confident assessment is seldom available at the time when important decisions may have to be taken in view of the pressures emanating from the political process or the sequential nature of important decisions. TA really has to be an iterative learning process, with the first assessment often doing little more than identifying areas where more research is needed. But in many cases definitive research results cannot be available before *some* decision has to be made. Changing the course of a development, or changing regulations after a technology is partly deployed, can be costly and disruptive, and the unpredictability and risk which it introduces from the standpoint of the developer may be a strong deterrent to enterprising action. . . .

In the end, I think the question can only be answered by saying some knowledge is better than none. It is better to proceed with incomplete or inadequate information than with none and some risks will have to be taken. It is the "no risk" assumption sometimes underlying implementation of TA rather than the TA process itself that is at fault. Too much regulation is predicated on an assumption of technological determinism, i.e., that anything that is researched or developed will eventually be deployed. This is not true historically and the TA process cannot realistically be based on such an assumption. Where high risks are involved to the innovator, some sort of insurance may be desirable and necessary so that the psychological and financial pressures resulting from sunk costs will not distort the decision process, either through the suppression of adverse information or through failure to begin a project because of technical uncertainties that can only be resolved by practical experience.

The record on implementation of TA has not been particularly happy. The outcome, whether negative or positive, tends to be more determined by political momentum and bureaucratic balance of power than by a rational process. Despite the implausibility of the assumption of complete technological determinism, it has frequently been difficult for rational analysis to change a technological trend whose directions have been well established. The SST development continued despite many unfavorable TA's and then was cancelled for reasons which were extremely shaky—at least at the time the decision was made by Congress, although subsequent research did confirm what was only a speculative suspicion. The history of auto emission legislation is explained better by political dynamics than by a rational evolution of choices based on improved technology assessments. But perhaps there is some convergence when viewed in a larger perspective. Given the fact that the acquisition of new scientific knowledge is itself a disorderly and halting process, scientists should perhaps allow the political process equal margin for learning by trial and error.

Public Input

This is one of the most controversial questions, and one on which I tend to be more conservative than is currently fashionable. Clearly we are passing through a phase of severe reaction against a purely technocratic mode of decision-making exemplified by the career of Robert Moses of the Port of New York Authority, in which he rode rough-shod over the views of local communities in the interests of a bold technological vision (although this vision was probably one that was supported by the majority of New York State voters as an abstract proposition). At the same time I doubt if it is possible to have wide public participation in every technological decision without a virtual paralysis of all decision-making, and without the complete disappearance of any coherent plan or vision of the future. The problem of public participation is that unless there are some incentives for consensus on an overall strategy embracing more than the particular decision in question, a few people who believe themselves adversely affected or whose values are offended can often dominate the decision process as against more diffuse benefitted interests. In the past, of course, the reverse was often true, and it was economic interests that tended to dominate the decision process in this way; today it is more likely to be political groups claiming to represent the public interest, though often having their own axes to grind. It is one thing to give affected interests an adequate hearing, but quite another to allow particular interests a *de facto* veto or the ability to ride rough-shod over other interests. In the Jamaica Bay study of the National Academy of Sciences it was pointed out that some thirteen different agencies had the power to impose an absolute veto on the extension of the runways at Kennedy Airport, regardless of any overall assessment of the public merits. The problem may be less serious when it comes to the assessment of a generic technology, such as the fast breeder program, than in the case of specific projects where the selection of a particular reactor site is in question. In the latter case the adversely affected interests are much more concentrated and hence easy to mobilize, though in a minority compared with those who might ultimately benefit. Nevertheless, unless there is a forum where all interests can be finally balanced and a decision reached, public participation may be merely a prescription for paralysis. It is interesting to note that Congress finally intervened to waive the application of NEPA in the case of the Alaska pipeline after litigation threatened to delay a decision indefinitely in the face of the energy crisis.

There probably needs to be priority with respect to what problems require public participation and what the nature and process of this participation should be. There is no assurance that the interests and viewpoints actually represented in the participation process will include all those that *should* be represented. . . . The politically active public interest groups or the affected industrial interests are not necessary the best sampling of the public interest, which is itself rather poorly defined. More importantly, public participation as it has been practiced in the last decade has been very costly in terms of time and money; in the jargon of economists, the "transaction costs" have been very high. The only "success stories" have been those in which technological developments have been stopped by a public process where they might have gone forward in its absence. Thus, in practice, public participation appears to be primarily a strategy for stopping technology. . . . I can think of no instance in which an important generic technology or even a specific technological project has been *advanced* by a participatory process. Perhaps solar power will prove to be such an example, but that can only be judged in the future.

Experience with public participation is very recent. Thus participatory technological decision-making is still subject to considerable social learning. It may be premature to write it off as counterproductive, as I am sure many scientists and technologists who have been involved in the process are inclined to do. Certainly if a process evolves which is not too costly in time and money, it has a high positive value in legitimizing public decisions about technology and insuring better public understanding and acceptance once a consensus is reached. The negative side of this is that if a decision outcome is thought to be the result of manipulation by a minority with a special interest or a special viewpoint, it will not long remain legitimate in the eyes of the public. Popular referenda on technological questions, a tactic followed by opponents in the case of fluoridation and of nuclear power, is a form of participation which is especially prone to manipulation by extremists and by public relations gimmicks. I predict that the public will once again return to decision by experts if it comes to feel that the participatory processes are being used not to better define the public interest, but rather to further special interests or political ideologies out of the mainstream. The benefits in better decisions must be seen over the long run to outweigh the costs of a lengthy and uncertain process.

On balance, then, it is my opinion that public participation has so far been the least satisfactory aspect of TA and that unless a more rapidly convergent process can be devised, the political process with either reject participatory decision-making or will show symptoms of political frustration such as recourse to demagogues or a search for scapegoats.

Standards of Proof

Most of the battles between the proponents and critics of particular technologies boil down to disagreement over what standard of evidence or argument is to be applied. At least this is true when the disagreements appear to be technical rather than explicitly involving a strong difference in ethical or political values. Battles over nuclear power safety and waste disposal, the controversy regarding auto emission standards, most controversies over dam siting or highway impact, etc. are perceived by the public as technical, even though value preferences may be embedded in the outcomes.

For example, in cases of safety the critics demand a high level of proof that a product or activity is safe, while the proponents want proof that the product is unsafe before any action against it is taken. This may be true even when the two sides agree on the definition of acceptable safety. . . .

[For example] in the controversy over the effects of freon on stratospheric ozone, the manufacturers dismissed the work of the scientists as "abstract speculation" with inadequate observation and experiment to make a case for regulatory action. Many environmentalists, on the other hand, were prepared to ban freon production regardless of economic consequences, on the basis of evidence which was suggestive of deleterious effects, but far from certain. In the hearings before Congressional committees there was virtually no confrontation of opposing *technical* arguments; the battle was all over the implications of degrees of confidence in the calculations. . . .

Consensus between critics and proponents might be helped if the two sides could agree on acceptable standards of proof prior to collection of the evidence. This might be possible in a sufficiently well-ordered participatory process, analogous to stipulation in judicial proceedings, especially if guided by experts trusted by both sides. The problem is to secure consensus on standards of proof before the issue becomes heavily polarized by political advocacy. In many controversies it would be interesting to challenge each side to state beforehand what kind of evidence and argument it would need to induce it to reverse or alter its initial position. . . .

It would also be helpful to force the adversaries in a technical or quasi-technical controversy to be explicit about the value assumptions and judgment criteria underlying their conclusions. It would also be helpful to force them to be clearer in their estimates of uncertainties and the scientific confidence with which their conclusions can be affirmed.

Impact on Innovation

When the NAS committee wrote its report *Technology: Processes of Assessment and Choice* in 1967, a report which in some ways launched the TA movement, one member wrote an appendix expressing reservations about the possible adverse impact of TA on technological progress and innovation. He pointed to the sensitivity of the innovation process to small changes in the perception of entrepreneurial risk.

What has been the impact of TA, consumerism and environmentalism on technological progress? The question is hard to answer because of so many other environmental factors that have changed—inflation, recession, shortages of capital, radical increases in energy and material costs, and uncertainties about future prices. It is difficult to point to any major innovation that has failed primarily because of TA. The SST was close to being abandoned by its potential clients, the airlines, and would probably have died without assistance from the skin cancer scare, although it is true that the adverse economic assessment by the airlines was largely the result of the banning of overland flights because of the sonic boom. LWR's have been delayed, and their capital costs have far exceeded estimates, largely owing to regulatory delays and increasing regulatory conservatism under pressure from public interest groups. But there is not yet a nuclear moratorium. . . .

Environmental regulations have impacted small businesses and older, marginal manufacturing facilities much more than they have the leaders of an industry who are usually the innovators.

There have been some studies to show that the U.S. has lagged behind other countries in drug innovation and that this was the result of the complexity and cost of procedures required for FDA approval, but there are other studies which raise doubts about this conclusion. So the matter must still be considered as moot.

During the last 15 years the comparative U.S. position in high-technology internationally-traded capital goods has slipped, but this trend began before environmental assessments could have been an important factor. If the U.S.' standards and the rigor of its assessments continue, there may be a future effect on international trade, but as indicated in the next section, this effect could just as likely prove positive as negative.

Talks with industrial research leaders reveal many plans for retrenchment in research and development in industry, and a growing tendency to concentrate on short-term evolutionary product improvements, with abandonment of projects aimed at more fundamental innovations and brand new technologies. Again, however, it is hard to correlate this with external technology assessments or with the fear of regulation or environmental controls. There has been some trend towards migration of industries with high emissions or hazardous processes out of this country in order to avoid the high costs of doing business under such constraints, but this does not yet appear to be an unmistakable trend, and in each instance factors other than environmental controls or other assessments are involved, e.g., the dangers of expropriation.

In short, one cannot make a strong argument that the application or prospect of TA has yet been a major negative factor in the innovativeness of U.S. industry or even in the introduction of innovations by the government. When all industry faces the same regulatory environment, the effect on innovativeness does not appear to be major. On the other hand, it would be wrong to assert there has been no effect, especially since one would expect it to be cumulative and not to be readily detectable at the beginning of a period of tighter controls. At the moment one can only say that there are grounds neither for complacency nor excessive alarm over the effects of environmental regulations or technology assessment on the innovation process itself.

Stimulation of Innovation

As I have pointed out elsewhere, TA and regulation can be a stimulus as well as an inhibition to innovation. Auto emission standards *have* led to major progress in the technology of emission controls, much more than many experts anticipated. On the other hand, concentration on meeting early deadline dates for the standards has probably seriously inhibited work on new types of power plants that might meet standards at lower cost with higher reliability while reducing fuel consumption.

It could well turn out that because of more rigorous standards U.S. industry will pioneer in abatement and environmental monitoring technologies and will find itself in an advantageous international position as other industrialized countries begin to adopt U.S. practices, out of imitation or necessity. This would apply both to specific abatement technologies and to environmental monitoring instrumentation, as well as to wholly new alternate manufacturing processes which are less polluting or less dangerous.

TA in the field of energy efficiency may well serve to stimulate new technologies. The energy crisis has stimulated a lot of hard scholarly thinking in this domain, previously an almost neglected topic of engineering and physics. . . .

Currently U.S. industry appears to be under greater pressure to meet environmental, energy conservation and occupational health standards than some of its foreign competitors, and in the long run this may turn out to give it a new kind of competitive edge.

Who Will Capture Assessment?

The history of regulation has been that after a while the regulated industry tends to dominate the perspective of the regulatory agency. Will this also happen eventually with EPA, OTA and other similar agencies? It seems less likely because it is not industry-specific like ICC, FPC, FCC, FDA, NRC and other regulation and assessment bodies. In the case of the older industry-specific regulatory agencies the interests of the public are diffuse and scattered, while the interests of the regulated industry are coherent and focused, hence easy to mobilize. In the newer broad agencies such as EPA almost the opposite is the case. The affected industries form a broad spectrum whose interests are too diverse to be brought together in terms of a few simple arguments, while environmental interests are increasingly well-organized and also in a position to present their case in the over-simplified terms that can garner public support. Despite this, there is evidence that as time goes on the point of view of industry tends to sink in as it is presented more consistently and persistently and with growing technical depth and sophistication to public agencies. EPA has receded from a number of recent borderline positions, at least partly under industry influence. In saying this I am not making a value judgment. EPA may have receded because in fact industry's technical arguments were more persuasive and corresponded with the conclusions of EPA's own scientists. In practice it may be very difficult to draw the line between technical persuasiveness and improper influence.

Organizations such as EPA and OTA have more incentive to develop common standards of analysis and evaluation across many different technologies. This is their great advantage over old-line regulatory agencies. It is conceivable, however, that they may be more subject to capture by a particular political interest, such as environmental advocates with a strong bias against private enterprise or against economic growth.

Also, it could be argued that environmental assessment has been too much captured by the lawyers and the public health professions, who tend to view health as an absolute good not to be traded off against any economic values. This is, of course, partly a professional bias, partly a political judgment, except that treating health as an absolute is never really practical and hence results in glaring inconsistencies between standards applied to different technologies, e.g., smoking vs. mercury in swordfish, auto accidents vs. radiation safety, auto emissions vs. stationary source emissions. More recently there has been a trend towards modulation of absolute positions on safety and the admission of economic costs as a legitimate consideration in the setting of standards.

Conversely, OTA could become the sounding board for technological promoters, the advocates of government funding for particular technologies not viewed with sufficient favor by executive agencies.

Health of Science

... There is no question that TA has provided a new and fascinating domain for scientists and engineers, and a role that enhances the relative importance of science politically. TA is bound to reveal the glaring lack of basic knowledge in many areas which are of vital importance to assessment, and also increasingly to demonstrate the inefficiency of filling the gaps in an *ad hoc* way for each new assessment as it comes along.

TA has also given many scientists a taste of a more holistic approach to problems. Witness the work of the APS groups on energy conservation and reactor safety. It has helped many academic scientists better understand the relevance of their own disciplines to national issues, and this may be a healthy thing for science in the long run. TA may prove to be for this generation of scientists what the war effort was for my generation—especially if TA is viewed in its full scope of identifying new technological possibilities as well as side effects.

APPENDIX A

Notes About the Authors

T. S. Ashton served as emeritus Professor of Economic History at the University of London from 1960 until his death in 1968. His works include *An Economic History of the Eighteenth Century* and *The Industrial Revolution.*

Clarence E. Ayres taught economics at Reed College, Brown University, and the University of Texas for almost fifty years. He is the author of *The Theory of Economic Progress* (1944) and *Toward a Reasonable Society: The Values of Industrial Civilization* (1961).

Jacob Bronowski was born in Poland and educated in England. A mathematician by training, he was Deputy Director of the Salk Institute for Biological Studies (San Diego) on his death in 1974. A popularizer of scientific ideas, he is best known for *Science and Human Values* (1958) and *The Ascent of Man* (1973).

Harvey Brooks was trained as a physicist and is currently Benjamin Peirce Professor of Technology and Public Policy at Harvard University. He served as Chairman of the National Academy of Sciences Panel, which in 1969 issued the report *Technology: Process of Assessment and Choice.*

Gary Brynner was President of United Auto Workers Local 1112 at the Chevrolet plant in Lordstown, Ohio, where they produced Vegas.

John G. Burke is Professor of History at the University of California, Los Angeles. He is author of *Origins of the Science of Crystals* (1966) and editor of *The New Technology and Human Values* (1966, 1972).

Carlo M. Cipolla is Professor of Economic History at the University of Pavia and the University of California, Berkeley. He has written more than a half dozen books, including *Guns, Sails, and Empires* (1965) and *Clocks and Culture: 1300–1700* (1967).

Dan Clark was an assembler on the line of the Chevrolet Vega plant in Lordstown, Ohio.

Wilson Clark is the author of *Energy for Survival: The Alternative to Extinction* (1974).

Barry Commoner received his training as a biologist and is Director of the Center for the Biology of Natural Systems at Washington University, St. Louis. He is the author of *Science and Survival* (1966) and *The Closing Circle* (1972).

Ruth Schwartz Cowan is Associate Professor of History at State University of New York, Stony Brook. She has published articles in professional journals on the history of biology and the history of technology.

George H. Daniels is a historian of American science and technology. He is the author of *American Science in the Age of Jackson* (1968) and *Science in American Society: A Social History* (1971).

Kingsley Davis is Emeritus Professor of Sociology, University of California, Berkeley, and Distinguished Professor of Sociology at the Population Research Laboratory at the University of Southern California. He is the author of *Human Society* (1949), *World Urbanization* (1969–1972), and the editor of *Cities: Their Origin, Growth, and Human Impact* (1973).

Giulio Douhet was an Italian army general and one of the first military men to grasp the potential of long-range strategic bombing as a weapon. His best-known and most controversial work, published in 1921, was translated into English with the title *The Command of the Air* (1942).

Peter F. Drucker is Clarke Professor of Social Sciences at Claremont Graduate School, Claremont, California. He is a management expert and is the author of more than a dozen books, including *The Age of Discontinuity* (1969), *Technology, Management, and Society* (1970), and *Men, Ideas, and Politics* (1971).

René Dubos was trained as a bacteriologist and is Emeritus Professor at Rockefeller University, New York. He is the author of many books, including *So Human an Animal* (1968), *Reason Awake* (1970), and *Only One Earth* (1972).

Albert Einstein was born in Germany in 1879. He revolutionized modern physics with the statement of the theories of special and general relativity in the early years of the twentieth century. He was awarded the Nobel Prize in 1922 and was a member of the Institute for Advanced Study at Princeton from 1933 until his death in 1955.

Jacques Ellul is a Professor in the Faculty of Law at the University of Bordeaux, France. His most famous work, translated into English as *The Technological Society* (1964) was a best seller in the United States and Europe.

Friedrich Engels was born in Germany in 1820 and collaborated with Karl Marx in founding modern socialism. Marx and Engels were co-authors of *The Communist Manifesto*. Engels clarified the bases of Marxism in a work published in 1878, which was translated into English as *Herr Eugen Dühring's Revolution in Science (Anti-Dühring)* (1934).

Sir William Fairbairn was one of England's most noted nineteenth-century civil engineers and inventors. He was a pioneer in the use of wrought iron in shipbuilding and in bridges, introduced a more efficient type of steam boiler, and collaborated in the design of the Britannia bridge, spanning the Menai Strait in Wales.

Eugene S. Ferguson is a Professor of History at the University of Delaware and Curator of Technology at the Hagley Museum, Greenville, Wilmington. Among his publications is his *Bibliography of the History of Technology* (1968).

Victor C. Ferkiss is a Professor of Government at Georgetown University. He is the author of *Technological Man* (1968) and *The Future of Technological Civilization* (1974).

Samuel C. Florman is a consulting engineer who also holds a Master's degree in English literature. His publications include *Engineering and the Liberal Arts* (1968) and *The Existential Pleasures of Engineering* (1976).

Siegfried Giedion was born in Switzerland in 1894 and played a leading role in the promotion of modern art and architecture. His publications include *Space, Time, and Architecture* (1941) and *Mechanization Takes Command* (1948).

Jean Gimpel is a medieval scholar and social historian who currently lives in London. He is the author of *The Cathedral Builders* (1961) and *The Medieval Machine* (1976).

Clarence J. Glacken is Emeritus Professor of Geography at the University of California, Berkeley. He is the author of *Traces on the Rhodian Shore* (1967).

G. Allen Greb is a Research Historian in the Program in Science, Technology, and Public Affairs at the University of California, San Diego. He is co-author of "Military Research and Development: A Postwar History," *Bulletin of the Atomic Scientists* (1977), and other articles.

Robert H. Guest is Professor of Organizational Behavior at Dartmouth College. He is co-author of *The Man on the Assembly Line* (1952) and *Organizational Change through Effective Leadership* (1977).

Robert L. Heilbroner is Norman Thomas Professor of Economics at the New School for Social Research, New York. Among his publications are *The Great Ascent* (1963), *Between Capitalism and Socialism* (1970), and *An Inquiry into the Human Prospect* (1974).

Carter Henderson is Co-director of the Princeton Center for Alternative Futures, Inc., an independent research institute. With Albert C. Lasher, Henderson co-authored *Twenty Million Careless Capitalists* (1967).

Walter R. Hibbard, Jr. was Director of the U.S. Bureau of Mines from 1965 and 1968 and is presently a Professor of Engineering at Virginia Polytechnic Institute. He has contributed many articles to professional journals and is co-editor of *Progress in Very High Pressure Research* (1961).

John A. Hostetler was raised in an Amish community and is now a Professor of Sociology at the University of Alberta. Among his publications are *The Sociology of Mennonite Evangelism* (1954) and *Amish Society* (1963).

Neal F. Jensen is Liberty Hyde Bailey Professor of Plant Breeding in the New York State College of Agriculture and Life Sciences at Cornell University. He has contributed many articles to professional journals.

John Jewkes is Emeritus Fellow at Oxford University, where he taught economic organization from 1948 to 1969. Among his many published works are *The Sources of Invention* (1958) and *Delusions of Dominance* (1977).

Kenneth Keniston is Professor of Psychiatry and Director of the Behavioral Sciences Study Center at Yale Medical School. Among his publications are *The Uncommitted: Alienated Youth in American Society* (1965) and *Radicals and Militants* (1973).

Sanford A. Lakoff is Professor of Political Science at the University of California, San Diego. He is the author of *Equality in Political Philosophy* (1964) and editor/contributor of *Knowledge and Power: Essays on Science and Government* (1966).

David E. Lilienthal was trained in law and was Director of the Tennessee Valley Authority from 1933 to 1941, and Chairman of the Atomic Energy Commission from 1946 to 1950. His publications include *TVA: Democracy on the March* (1944) and *Change, Hope, and the Bomb* (1963).

Donald E. Marlowe is Executive Director of the American Society for Engineering Education. He was formerly Dean of the Engineering School of the Catholic University of America, and served in 1969 as President of the American Society of Mechanical Engineers.

Marshall McLuhan is Director of the Center for Culture and Technology at the University of Toronto. His books include *The Gutenberg Galaxy* (1962) and *Understanding Media* (1964).

Hugo A. Meier is Associate Professor of History at Pennsylvania State University. He has written articles in professional journals on technology and American social history.

Arthur E. Morgan was a consulting engineer and an expert in flood control. He was Chairman of the Board of the Tennessee Valley Authority from 1933 to 1937. He was a prolific writer, and published *Dams and Other Disasters* in 1971, a few years before his death.

Lewis Mumford has taught at many universities and lectured widely on the history of technology. Over the last half century he has written a score of books, of which the best known are *Technics and Civilization* (1934), *The City in History* (1961), and *The Myth of the Machine* (1967–1970).

Ralph Nader was trained as a lawyer and has become a noted consumer advocate, lecturer, and author. He is the author of *Unsafe at Any Speed* (1965) and has co-authored many books, including *The Consumer and Corporate Accountability* (1973).

Dorothy Nelkin is Associate Professor in the Program on Science, Technology, and Society in the Department of City and Regional Planning at Cornell University. She is the author of a half-dozen books, including *Nuclear Power and Its Critics* (1971), *The University and Military Research* (1972), *Jetport* (1974), and *Science Textbook Controversies and the Politics of Equal Time* (1977).

William F. Ogburn was, before his death in 1959, a Professor of Sociology at Columbia University and at the University of Chicago. Among his writings are *Social Change* (1922, 1950), *You and Machines* (1935), and *Technology and the Changing Family* (1953).

Derek J. de Solla Price is Avalon Professor of the History of Science at Yale University. He is the author of a half-dozen books, including *Science Since Babylon* (1961), *Little Science, Big Science* (1965), and *Gears from the Greeks: The Antikythera Mechanism* (1975).

Carroll W. Pursell, Jr. is Professor of History at the University of California, Santa Barbara. He is the author of *Early Stationary Steam Engines in America* (1969) and the editor of *The Military-Industrial Complex* (1972).

Simon Ramo was trained in science and engineering, and was chief scientist of the U.S. intercontinental ballistic missile program from 1954 to 1958. He is one of the founders of TRW, one of the world's largest communications companies. He has written many technical books and articles, and is the author of *Cure for Chaos* (1969).

Jerome R. Ravetz was trained as a mathematician and specializes in the history and philosophy of science. He holds the position of Reader at the University of Leeds in England and is the author of *Scientific Knowledge and Its Social Problems* (1971).

Nathan Rosenberg is Professor of Economics at Stanford University. He is the author of *Technology and American Economic Growth* (1972) and *Perspectives on Technology* (1976).

Theodore Roszak is Professor of History at California State University, Hayward. One of the best known and most articulate critics of modern technology, his works include *The Making of a Counterculture* (1969) and *Where the Wasteland Ends* (1972).

Andrei D. Sakharov is a Russian nuclear physicist who aided the development of the Soviet hydrogen bomb. He is the author of *Progress, Peaceful Coexistence, and Intellectual Freedom* (1968) and *Sakharov Speaks* (1974). He was awarded the Nobel Peace Prize in 1975.

David Sawers has been since 1977 the Undersecretary of the Departments of Environment and Transport of Great Britain. He is the co-author of *The Sources of Invention* (1958) and wrote, with Ronald Miller, *The Technical Development of Modern Aviation*.

Jacob Schmookler taught economic history at the University of Minnesota and was the author of *Invention and Economic Growth* (1966) and *Patents, Invention, and Economic Change* (1972).

Herbert A. Simon is Richard King Mellon University Professor of Computer Sciences and Psychology at Carnegie-Mellon University. He was a member of the President's Science Advisory Council from 1968 to 1971 and is a member of the National Academy of Science. The recipient of the Nobel Prize for Economics, he is the author or co-author of many books, including *The Shape of Automation* (1965), *Human Problem Solving* (1972), and *Models of Discovery* (1977).

Richard Stillerman is the co-author of *The Sources of Invention* (1958).

Thorstein Veblen was an economist and social scientist who taught at several colleges before his death in 1929. He gained fame in 1899 when his *Theory of the Leisure Class* was published, and was the author of *The Engineers and the Price System* (1921).

J. Alan Wagar teaches forestry at the University of Washington and is an environmental researcher for the United States Forest Service.

James C. Wallace teaches social studies and history of science at North Carolina State University.

Charles Weiner is Professor of History of Science and Technology at the Massachusetts Institute of Technology. He has edited *Exploring the History of Nuclear Physics* (1972) and *Topics in the History of Twentieth Century Physics* (1974).

Lynn White, Jr. is Emeritus University Professor of History, University of California, Los Angeles. A specialist in medieval history, his best-known works are *Medieval Technology and Social Change* (1962), *Machina ex Deo: Essays in the Dynamism of Western Culture* (1968), and *Medieval Religion and Technology* (1978).

Reynold M. Wik is Professor of American History at Mills College. He is the author of *Steam Power on the American Farm* (1953) and *Henry Ford and Grass-roots America* (1972).

Langdon Winner is Assistant Professor of Political Science and Technology at Massachusetts Institute of Technology. He is the author of *Autonomous Technology: Technics-out-of-Control as a Theme in Political Thought* (1977).

Quincy Wright was Professor of Political Science and International Law at the University of Chicago until his death in 1970. He was the author of many books, including *The Control of American Foreign Relations* (1922), *International Law and the United Nations* (1960), and *A Study of War* (1942, 1965).

Herbert F. York is Professor of Physics and Director of the Program in Science, Technology, and Public Affairs at the University of California, San Diego. He was Director of Defense Research and Engineering in the Office of the Secretary of Defense from 1958 to 1961. He is the author of *Race to Oblivion* (1970), *Arms Control* (1973), and *The Advisors: Oppenheimer, Teller, and the Superbomb* (1976).

S. Husain Zaheer was educated in England and Germany and received his doctorate in science from Heidelberg University in 1930. He taught chemistry at Lucknow University (India) from 1930 to 1946, and was Director of the Hyderabad Regional Research Laboratory from 1948 to 1962. At that time he became Director General of the Council for Science and Industrial Research of India.

APPENDIX B

Suggestions for Further Reading

Part One: Technology on Trial

Commoner, Barry. "Healthy Environment: The Life-or-Death Challenge to Technology," *Popular Science* 200 (May 1972), 100–101. A good summary of Commoner's view on the causes of the environmental crisis. He discusses the primary technologies "with high biological impact."

Ellul, Jacques. *The Technological Society* (trans. John Wilkinson) (New York: Knopf, 1964). The classic anti-technology statement, which had a profound influence on the anti-technology movement in the United States during the sixties.

Florman, Samuel C. *The Existential Pleasures of Engineering* (New York: St. Martin's Press, 1976). Well-written and articulate examination of the role of the engineer in contemporary society, which confronts the arguments of the anti-technologists.

Harris, Neil. "We and Our Machines: 200 Years of Love and Hate," *New Republic* 171 (November 23, 1974), 24–33. Traces the evolution of popular attitudes about machines over the course of our nation's history.

Juenger, Friedrich G. *The Failure of Technology* (Chicago: Henry Regnery Company, 1956). A pessimistic portrayal of the evils of technology.

Lawless, Edward B. *Technology and Social Shock* (New Brunswick, N.J.: Rutgers University Press, 1977). Documents 45 case histories of controversial problems engendered by science and technology.

Roszak, Theodore. *Where the Wasteland Ends: Politics and Transcendence in Postindustrial Society* (Garden City, N.Y.: Doubleday & Company, Inc., 1972). This sequel to *The Making of a Counterculture* amplifies Roszak's earlier rejection of the scientific world view.

Schwartz, Eugene S. *Overkill: The Decline of Technology in Modern Civilization* (Chicago: Quadrangle Books, 1971). An analysis of science and technology, which concludes that they cannot help to solve the world's problems since they are the prime contributors to the problems.

Taviss, Irene. "A Survey of Popular Attitudes toward Technology," *Technology and Culture* 13 (October 1972), 606–621. Samples opinions concerning technology of Americans from differing social, educational, and financial backgrounds, and indicates that people do not generally feel alienated or dominated by technology.

Winner, Langdon. *Autonomous Technology: Technics-out-of-Control as a Theme in Political Thought* (Cambridge, Mass.: MIT Press, 1977). An attempt to explore the consequences of holding a belief that "somehow technology has gotten out of control and follows its own course, independent of human direction."

Part Two: Technology's Effects

Technology and Social Organization

Ashton, T. S. *The Industrial Revolution, 1769–1830* (London: Oxford University Press, 1948). Describes the changes in technology, industry, and society during the period of the Industrial Revolution in England.

Bacon, Francis. *New Atlantis* (New York: P. F. Collier & Son, 1901). Originally published in 1627, this book pictures a utopia having the goal of improving society by means of a rational technology based on science.

Boorstin, Daniel J. *The Republic of Technology: Reflections on Our Future Community* (New York: Harper & Row, 1978). Reflects on the importance of technology in forming the American national character, and in making America democratic, literate, and free.

Burnham, James. *The Managerial Revolution: What Is Happening in the World* (New York: John Day, 1941). Foreshadows the counterculture theories of the technocratic society in predicting control by a managerial elite.

Drucker, Peter F. *The Age of Discontinuity* (New York: Harper & Row, 1968). Treats four areas of "discontinuity": explosive technology, rapid economic change, the rise of pluralistic institutions, and mass education.

Drucker, Peter F. *Technology, Management, and Society* (New York: Harper & Row, 1970). Twelve provocative essays on technology and its influence on work, management, and society.

Mumford, Lewis. *Technics and Civilization* (New York: Harcourt, Brace & Company, 1934). Examines the development of technology from the tenth century forward, and seeks to discover the conditions necessary for the "machine" to be directed toward more humane purposes.

White, Lynn T., Jr. *Medieval Technology and Social Change* (New York: Oxford University Press, 1962). Chapter 1. Describes the introduction of the stirrup into Europe during the eighth century and demonstrates its influence on the creation of feudalism.

Technology and Complexity

Elsner, Henry, Jr. *The Technocrats: Prophets of Automation* (Syracuse, N.Y.: Syracuse University Press, 1967). A history of the 1930s movement, which sought to promote a society run by scientific and engineering experts.

Gotlieb, C. C., and A. Borodin. *Social Issues in Computing* (New York: Academic Press, 1973),. Treats computer problem areas: technical, managerial, economic, legal, social, and ethical.

Meynaud, Jean. *Technocracy* (trans. Paul Barnes) (London: Faber and Faber, 1968). A social and political analysis of the meaning and characteristics of technocracy.

Price, Derek J. de Solla. "Automata and the Origins of Mechanism and Mechanistic Philosophy," *Technology and Culture* 5 (Winter 1964), 9–23. Describes automata constructed in the ancient and early modern periods, and postulates that the increasing ability to simulate nature by mechanical means had a profound influence on thought.

Ramo, Simon. *Century of Mismatch* (New York: David McKay, 1970). A prediction of how science and technology will be able to solve the problems of mankind.

Raphael, Bertram. *The Thinking Computer: Mind Inside Matter* (San Francisco: W. H. Freeman, 1976). Fine general introduction for the layman on the workings, possibilities, and applications of computers.

Snow, C. P. *The Two Cultures: And a Second Look* (Cambridge: Cambridge University Press, 1964). Expounds the belief that society is increasingly separating into two different cultures: one versed in the humanities and the other in the sciences, each becoming less capable of understanding the other.

Weizenbaum, Joseph. *Computer Power and Human Reason* (San Francisco: W. H. Freeman, 1976). One of the best treatments of the problems posed by computers.

Technology and Work

Bendix, Reinhard. *Work and Authority in Industry: Ideologies of Management in the Course of Industrialization* (New York: Harper & Row, 1963). Describes how entrepreneurs and managers developed ideologies during the Industrial Revolution to impose work discipline, legitimize their authority, and justify their positions.

Gies, Joseph, and Frances Gies. "A City Working Woman: Agnes Li Patiniere of Couai; Women and Guilds," in *Women in the Middle Ages* (New York: Thomas Y. Crowell, Inc., 1978). Describes the putting-out system of the Middle Ages and its effect on workers.

Gies, Joseph, and Frances Gies. "Sam Slater Discovers America," "Eli Whitney: The Cotton Gin and Interchangeable Parts," "Francis Cabot Lowell, Boston Pirate," "John Hall and Sam Colt: The Birth of the 'American System,'" in *The Ingenious Yankees* (New York: Thomas Y. Crowell, Inc., 1976). Describes the birth of mass production in nineteenth-century American industry.

Gutman, Herbert G. "Work, Culture, and Society in Industrializing America," *American Historical Review* 78 (June 1973), 537–581. An account of the impact of mass-production work roles on workers in late nineteenth-century America.

Haber, Samuel. *Efficiency and Uplift: Scientific Management in the Progressive Era, 1890–1920* (Chicago: University of Chicago Press, 1964). Describes Frederick W. Taylor's program of industrial management in which efficiency was the key concept.

Kranzberg, Melvin, and Joseph Gies. *By the Sweat of Thy Brow* (New York: Putnam, 1975). A brief history of the organization of work in the western world from classical times to the present.

Thompson, Edward P. *The Making of the English Working Class* (New York: Vintage Books, 1963). An account of the creation of the industrial working class in England in the period 1792 to 1832.

U.S. Department of Health, Education, and Welfare. *Work in America* (Cambridge, Mass.: MIT Press, 1972). A large-scale study of the effects on workers of workplace conditions.

Walker, Charles R. "The Social Effects of Mass Production," in Melvin Kranzberg and Carroll W. Pursell, Jr., eds., *Technology in Western Civilization,* vol. 2 (New York: Oxford University Press, 1967). An account of the effects of the assembly line and other modern industrial methods.

Part Three: Conditions of Technological Development

Nature and Technology

Bates, Marston. *The Forest and the Sea* (New York: Random House, 1960). Nature and natural resources are discussed from an ecological point of view.

Braudel, Fernand. *The Mediterranean and the Mediterranean World in the Age of Philip II* (New York: Harper & Row, 1972), vol. 1, part 1. A minutely detailed analysis of the interrelationships of the natural environment, resources, technology, cities, and communications in the Mediterranean world of the sixteenth century.

Carson, Rachel. *Silent Spring* (Boston: Houghton Mifflin, 1962). One of the first and most lucid critiques of the use of chemical pesticides.

Dickens, Charles. *Hard Times* (New York: Doubleday, n.d.), Book 1, Chapter 5. Famous description of Coketown—its monotony, pollution, and depressing practicality in the early days of the Industrial Revolution.

Glacken, Clarence J. *Traces on the Rhodian Shore: Nature and Culture from Ancient Times to the End of the Eighteenth Century* (Berkeley and Los Angeles: University of California Press, 1967). A history of the ideas of a harmoniously designed earth, of the influence of the environment on culture, and of the modifications of the earth by human agency.

Klingender, Francis D. *Art and the Industrial Revolution* (London: Evelyn, Adams & Mackay, 1968). Penetrating discussions and many fine illustrations of industrial art in the English landscape in a period generally called Romantic, whose artists were thought hostile to the machine.

Malone, Thomas F. "Weather: Man Will Control Rain, Fog, Storms, and Even Possibly the Climate," in Foreign Policy Association, *Toward the Year 2018* (New York: Cowles Education Corporation, 1968), 61–74. Traces recent successful efforts to control weather, and is optimistic about the unlimited possibilities for the use of technology to manipulate the natural world.

Marsh, George P. *Man and Nature, or Physical Geography as Modified by Human Action* (Cambridge, Mass.: The Belknap Press of Harvard University Press, 1965). First published in 1864; Marsh was the most perceptive writer of the nineteenth century on the effects of long and continued human use of the environment.

Thomas, William L., Jr., ed. *Man's Role in Changing the Face of the Earth* (Chicago: University of Chicago Press, 1956). The best single volume on the effects of human agency in modifying the natural environment.

Thoreau, Henry David. "The Wild," in Charles R. Anderson, ed., *Thoreau's Vision: The Major Essays* (Englewood Cliffs, N.J.: Prentice-Hall, 1973), 131–158. A succinct exposition of Thoreau's philosophy of nature and human participation in it.

Societal Values and Technology

Daniels, George H., John G. Burke, and Edwin Layton. "Symposium: The Historiography of American Technology," *Technology and Culture* 11 (January 1970), 1–35. A wide-ranging discussion of the relationship of technology and social change.

Flink, James T. *America Adopts the Automobile, 1895–1910* (Cambridge, Mass.: MIT Press, 1970). A social history of the automobile in America from its introduction to the beginning of the mass production of Ford's Model T, in which the author is concerned with the automobile in American thought and values.

Kasson, John F. *Civilizing the Machine: Technology and Republican Values in America, 1776–1900* (New York: Grossman Publishers, 1976). An excellent description of the clash of values accompanying industrialization in nineteenth-century America.

Kouwenhoven, John A. *The Arts in Modern American Civilization* (New York: W. W. Norton, Inc., 1967). An analysis of how technology and art in the United States were shaped by the American experience.

Morison, Elting E. *Men, Machines, and Modern Times* (Cambridge, Mass.: MIT Press, 1966). Case studies of resistance to change engendered by technological innovation.

Ogburn, William F. *Social Change with Respect to Culture and Original Nature* (New York: Viking Press, Inc., 1923). Presents the theory of "social lag," which postulates that society only gradually adapts to changes created by technological innovations.

Sanford, Charles L. "The Intellectual Origins and New-Worldliness of American Industry," *Journal of Economic History* 18 (March 1958), 1–16. Theorizes that the Puritan work ethic had a profound influence upon the establishment of industrialism in America.

Smith, Cyril Stanley. "Art, Technology, and Science: Notes on Their Historical Interaction," *Technology and Culture* 11 (October 1970), 493–549. Attacks idea that the evolution of technology can be explained by reference to narrowly utilitarian, economic, or scientific influences, and demonstrates by many examples that purely aesthetic considerations have been responsible for many important technological innovations.

Technology, Population, and Resources

Boserup, Esther. *The Conditions of Agricultural Growth: The Economics of Agrarian Change under Population Pressure* (London: George Allen & Unwin, 1965). Expounds the thesis that increasing population has been the cause of technological change in agriculture.

Callahan, Daniel, ed. *The American Population Debate* (Garden City, N.Y.: Doubleday, 1971). A collection of essays on population representing just about every viewpoint and every possible solution.

Davis, Kingsley, ed. *Cities: Their Origin, Growth, and Human Impact* (San Francisco: W. H. Freeman, 1973). A collection of articles from *Scientific American,* including sections on urban environments in less developed countries, and the effects of urban living on health.

Ehrlich, Paul R. *The Population Bomb* (New York: Ballantine, 1968). Argues that the root of the environmental crisis and the world food problem is too many people, and calls for radical measures to limit population growth.

Lappé, Frances Moore, and Joseph Collins, with Cary Fowler. *Food First: Beyond the Myth of Scarcity* (Boston: Houghton Mifflin, 1977). Argues that there is and will be enough food for the world's people if only the large corporate interests return control of food production to the people and ensure that it is properly distributed on a world-wide basis.

Lazlo, Erwin, et al. *Goals for Mankind: A Report to the Club of Rome on the New Horizons of Global Community* (New York: E. P. Dutton, 1977). Focuses on the different cultural attitudes and values of First, Second, and Third World nations, and attempts to set goals for various countries.

Mamdani, Mahmood. *The Myth of Population Control: Family, Caste, and Class in an Indian Village* (New York: Monthly Review Press, 1973). An analysis of why an elaborate and costly effort by Americans to reduce the birth rate by introducing "family planning" to an area of India failed utterly.

National Academy of Sciences. *Rapid Population Growth: Consequences and Policy Implications* (Baltimore: Johns Hopkins Press, 1971). A comprehensive symposium on various aspects of contemporary population changes and their significance.

Poleman, Thomas T., and Donald K. Freebairn, eds. *Food, Population, and Employment: The Impact of the Green Revolution* (New York: Praeger Publishers, 1973). Contains original analyses of the economic, political, and demographic effects of the "Green Revolution."

Scientific American (September 1974). The entire issue is devoted to various aspects of the human population.

Part Four: Sources of Technological Change

Economic Growth and Technology

Bury, J. B. *The Idea of Progress* (New York: Dover Publications, 1955). A historical description of the genesis and evolution of the idea of progress with an analysis of the influence on it of technological achievements.

Chandler, Alfred D., Jr. *The Visible Hand: The Managerial Revolution in American Business* (Cambridge, Mass.: The Belknap Press of Harvard University Press, 1977). Theorizes that modern business enterprise replaced market mechanisms in the coordination of the activities of the economy and in the allocation of resources.

Cole, H. S. D., et al., eds. *Models of Doom: A Critique of "Limits to Growth"* (New York: Universe Books, 1973). Criticizes the Meadows Club of Rome study, arguing that the analysis is faulty.

Eckhaus, Richard. *Appropriate Technologies for Developing Countries* (Washington, D.C.: National Research Council, 1977). Investigates the problem of appropriate technology from a perspective which views growth and efficiency as the logical aims of any production system.

Meadows, Donella H., et al. *The Limits to Growth* (New York: Universe Books, 1972). First report to the Club of Rome; examines five factors which the authors believe will ultimately limit growth: population, agricultural production, resources, industrial output, and pollution.

Miles, Rufus E., Jr. *Awakening from the American Dream: The Social and Political Limits to Growth* (New York: Universe Books, 1976). A critique of America's affluence and its impending demise, in which the author seeks to make readers aware of the need to create a new society based upon the limits-to-growth approach.

Noble, David F. *America by Design: Science, Technology, and the Rise of Corporate Capitalism* (New York: Alfred A. Knopf, Inc., 1977). Advances the thesis that an alliance between scientific and engineering educational institutions and corporations has produced a managerial elite that has taken control of the American economy.

Rosenberg, Nathan. *Technology and American Economic Growth* (New York: Harper & Row Publishers, 1972). An exploration of the relationship between technological change and the long-term growth of the American economy.

Schumacher, E. F. *Small Is Beautiful: Economics as if People Mattered* (New York: Harper & Row Publishers, 1973). Argues that large-scale technology, destructive of nature and humans, lies at the root of three crises: pollution, resource scarcity, and human discontent; advocates smaller and more appropriate technologies.

Science and Technology

Beer, John J. *The Emergence of the German Dye Industry* (Urbana, Ill.: University of Illinois Press, 1959). A wide-ranging study of the influence of chemical science in stimulating German industry in the nineteenth century.

Birr, Kendall. *Pioneering in Industrial Research: The Story of the General Electric Research Laboratory* (Washington, D.C.: Public Affairs Press, 1957). An excellent account of the establishment, management, and achievements of a major industrial research laboratory.

Hewlett, R. G., and O. E. Anderson. *The New World, 1939–1946* (University Park, Pa.: Pennsylvania State University Press, 1962). A history of the wartime Manhattan Project and the development of the atomic bomb.

Josephson, Matthew. *Edison: A Biograpy* (New York: McGraw-Hill, 1959). An outstanding biography which emphasizes Edison's economic motivations.

Multhauf, Robert P. "The Scientist and the 'Improver' of Technology," *Technology and Culture* 1 (Winter 1960), 38–47. Clarifies the distinction between science and technology by describing goals and tasks performed.

Passer, Harold C. *The Electrical Manufacturers, 1875–1900* (Cambridge, Mass.: Harvard University Press, 1953). A study of the rise of the U.S. electrical industry and its response to innovations and requirements in lighting, power, and railway traction.

Price, Derek J. de Solla. *Little Science, Big Science* (New York: Columbia University Press, 1963). An analysis of the general problems of the growth and size of the scientific enterprise, and its effect on contemporary society.

Ravetz, Jerome. *Scientific Knowledge and Its Social Problems* (Oxford: The Clarendon Press, 1971). Addresses current problems of science and society, and argues that scientific research and development must seek to protect humans and the natural environment.

Sinsheimer, Robert L. "The Presumptions of Science," *Daedalus* (Spring 1978), 23–26. Cites examples of scientific research which the author thinks are of dubious merit and should not be pursued.

Ziman, John. *The Force of Knowledge: The Scientific Dimension of Society* (Cambridge: The University Press, 1976). Explores the social relations of science and technology, introducing the reader to a wide range of issues.

Engineering and Technological Innovation

Ferguson, Eugene S. "Why Systems Go on the Blink," and Frank J. Regan, "The Tough Problems Are Non-Technical," *Machine Design* (February 23, 1978), 28–30. A criticism and a defense of modern engineering design.

Giedion, Siegfried. *Mechanization Takes Command* (New York: Oxford University Press, 1948). Traces the ways in which mechanization has come to pervade twentieth-century life, and reveals the hidden influences of the mechanical environment on humans.

Gille, Bertrand. *Engineers of the Renaissance* (Cambridge, Mass.: MIT Press, 1966). Describes the achievements of Renaissance engineers, and emphasizes the continuity of technology from antiquity to the Renaissance.

Hughes, Thomas P., ed. *Lives of the Engineers by Samuel Smiles* (Cambridge, Mass.: MIT Press, 1966). A contribution to the social history of technology in considering the milieu and accomplishments of prominent eighteenth- and nineteenth-century English engineers.

Layton, Edwin T., Jr. *The Revolt of the Engineers: Social Responsibility and the American Engineering Profession* (Cleveland: Case Western Reserve University Press, 1971). A critical survey of the American engineering profession; explores the reasons for the lack of development of a social ethic among engineers.

Nader, Ralph. *Unsafe at Any Speed* (New York: Grossman Publishers, 1965). Charges that auto manufacturers have failed to make cars safe even though the technical ability to do so has been available for years.

Odiorne, George S. "The Trouble with Engineers," *Harper's Magazine* 210 (January 1955), 41–46. Analyzes the difficulties faced by young engineers in adjusting to the "unscientific" real world in which they must work.

Pacey, Arnold. *The Maze of Ingenuity* (New York: Holmes & Meier Publishers, 1975). Describes how non-economic incentives such as religion have influenced technological innovations.

Rolt, L. T. C. *Isambard Kingdom Brunel* (Harmondsworth, Middlesex: Penguin Books, 1970). Chapters 1 and 11. Describes the characters and accomplishments of two prominent nineteenth-century engineers, Marc and Isambard Brunel, in which some of the imperatives that drive an engineer are clearly revealed.

Veblen, Thorstein. *The Engineers and the Price System* (New York: B. W. Huebsch, Inc., 1921). Argues that the engineer's logical and objective approach to problems qualified him to take command of industrial production.

War and Technological Change

Baxter, James Phinney. *Scientists Against Time* (Boston: Little, Brown and Company, 1946). A history of the activities and accomplishments of the Office of Scientific Research and Development in World War II.

Brodie, Bernard. *War and Politics* (New York: Macmillan Publishing Co., 1973). An analysis of the relations between politics and military affairs in the United States since World War II.

Cipolla, Carlo. *Guns, Sails, and Empires: Technological Innovation and the Early Phases of European Expansion, 1400–1700* (New York: Minerva, 1966). A history of the technological innovations that permitted European countries to conquer and colonize other continents.

Douhet, Giulio. *The Command of the Air* (New York: Coward McCann, Inc., 1942). The first statement of the potentialities of strategic bombing with aircraft.

Nef, John U. *War and Human Progress* (New York: W. W. Norton & Co., 1968). A history of the relation between war, technological advance, and intellectual changes in western civilization since the fifteenth century.

Wright, Quincy, et al. *A Study of War* (Chicago: University of Chicago Press, 1965). A multi-volume history and analysis of all aspects of warfare.

York, Herbert. *The Advisors: Oppenheimer, Teller, and the Superbomb* (San Francisco: W. H. Freeman, 1976). Describes the political process culminating in President Truman's decision to develop the hydrogen bomb.

York, Herbert. *Race to Oblivion: A Participant's View of the Arms Race* (New York: Simon and Schuster, 1970). An analysis of the course and direction of the arms race.

York, Herbert, and G. Allen Greb. "Military Research and Development: A Postwar History," *Bulletin of the Atomic Scientists* 33 (January 1977), 13–26. Describes post–World War II military research and development.

The Government and Technological Change

Bronk, Detlev W. "The National Science Foundation: Origins, Hopes, and Aspirations," *Science* 188 (May 2, 1975), 409–414. A retrospective look at the National Science Foundation by a former chairman of the National Science Board.

Dupree, A. Hunter. *Science in the Federal Government: A History of Policies and Activities to 1940* (Cambridge, Mass.: Harvard University Press, 1957). A comprehensive history of government involvement with and promotion of science and technology from the founding of the United States to the outbreak of World War II.

Greenberg, Daniel S. *The Politics of Pure Science* (New York: New American Library, 1967). An analysis of the relations between the federal government and the scientific community since World War II.

Hewlett, Richard, and Francis Duncan. *Nuclear Navy, 1946–1962* (Chicago: University of Chicago Press, 1974). A semi-official history of the development of nuclear propulsion systems in the U.S. Navy.

Killian, James R., Jr. *Sputnik, Scientists, and Eisenhower* (Cambridge, Mass.: MIT Press, 1977). The recollections of the first special assistant to the President for science and technology.

Parry, John H. *The Age of Reconnaissance* (Cleveland: World Publishing Co., 1963). Treats the role of governments in encouraging technology in early modern times.

Penick, James L., et al., eds. *The Politics of American Science, 1939 to the Present* (Chicago: Rand McNally & Company, 1965). Describes how major governmental problems of the post–World War II period helped to form the structure of the U.S. scientific community.

Price, Don K. *The Scientific Estate* (Cambridge, Mass.: The Belknap Press of Harvard University Press, 1965). Postulates that the rise of the federal science establishment has weakened traditional political processes.

Part Five: Retrospect and Prospect

The Process of Invention

Gilfillan, S. C. *The Sociology of Invention* (Cambridge, Mass.: MIT Press, 1970). A study of the nature and origins of invention, originally published in 1935.

Jewkes, John, et al. *The Sources of Invention* (London: Macmillan, 1958). Detailed study of the nature, origins, and history of inventions.

Schmookler, Jacob. *Invention and Economic Growth* (Cambridge, Mass.: Harvard University Press, 1966). An attempt to explain variations in the rate of inventions in the same industry over time and between different industries, using patent data and chronologies of inventions.

Utterback, James M. "Innovation in Industry and the Diffusion of Technology," *Science* 183 (February 15, 1974), 620–626. A survey of available knowledge concerning innovative processes in industry.

Vanderbilt, Byron M. *Inventing: How the Masters Did It* (Durham, N.C.: Moore Publishing Co., 1974). An analysis of the careers of six noted nineteenth-century inventors: Goodyear, Bell, Edison, Westinghouse, Acheson, and Nobel.

Wall Street Journal Staff. *The Innovators* (Princeton, N.J.: Dow Jones Books, 1968). An account of many aspects of the process of invention.

White, Lynn T., Jr. "Eilmer of Malmesbury, an Eleventh Century Aviator," *Technology and Culture* 2 (Spring 1961), 97–111. An interesting analysis of the first documented attempt to fly.

White, Lynn T., Jr. "The Invention of the Parachute," *Technology and Culture* 9 (October 1968), 462–467. Describes early modern concepts of the parachute.

White, Lynn T., Jr. "Technology and Invention in the Middle Ages," *Speculum* 15 (1940), 141–159. Documents numerous medieval inventions, all of which produced social complexity.

Technology and Ethical Dilemmas

Aeschylus. *The Prometheus Bound,* ed. George Thomson (Cambridge: The University Press, 1932). A classic Greek tragedy conveying several modern themes: the arrogance of man in transforming nature into artifacts; the indulgence of self-will to the exclusion of wise counsel; unrestrained power contrasted with the life-giving arts.

Ayres, Clarence E. *Toward a Reasonable Society: The Values of Industrial Civilization.* (Austin: University of Texas Press, 1961). Argues that anything promoting life-activities is good, and that in modern society science and technology fulfill this function.

Barbour, Ian G. *Science and Secularity: The Ethics of Technology* (New York: Harper & Row, 1970). Asserts that religion and religious values can be utilized to control technology and direct it to achieve human purposes.

Dewey, John. *Freedom and Culture* (New York: G. P. Putnam's Sons, 1939). Proposes the employment of the methods of science and technology to achieve ends of life that are more social and more free.

Eliade, Mircea. *The Forge and the Crucible* (trans. Stephen Corrin) (New York: Harper, 1962). Describes how primitive societies close to nature employed myth and folklore in their mining and metallurgical operations.

Ferkiss, Victor C. *Technological Man: The Myth and the Reality* (New York: George Braziller, 1969). Observes that profound changes have occurred because of technology, and argues that "technological man" must develop a new code of ethics.

Galbraith, John Kenneth. *The New Industrial State* (Boston: Houghton Mifflin, 1971). Views the new industrial system as a product of a changing world, utilizing an extended form of planning and exercising such power as is necessary to maintain itself.

Jonas, Hans. "Technology and Responsibility: Reflections on the New Task of Ethics," *Social Research* 40 (1973), 31–54. Contends that technological advance has created new possibilities for human action, which have made all standards and canons of traditional ethics obsolete.

Lilienthal, David. *TVA: Democracy on the March* (New York: Harper & Bros., 1944). An account of the development of the resources of the Tennessee River Valley and of the transformation of a disheartened people into one capable of managing their own affairs and living a life of fulfillment.

Marx, Karl. *Capital* (trans. Ben Brewster) (Leicester University, 1966). Vol. I, Part IV, Chapter 15, Section 1. Marx writes that under capitalism the alienation of humans reveals itself as a class society, internally in a state of contradiction, unstable, and incapable of maintaining itself.

Veblen, Thorstein. *The Instinct of Workmanship and the State of the Industrial Arts* (New York: The Macmillan Company, 1918). Claims that man in culture has an instinct to work in efficient and economical ways, but that under capitalism the ownership of productive capacity is a type of parasitism that harms this instinct.

Monitoring Technology

Baram, Michael. "Technology Assessment and Social Control," *Science* 180 (May 4, 1973), 465–473. Opposes adversary relationship in the discussion of technological impacts, and calls for negotiation and joint decision-making in dealing with problems arising from the introduction of a new technology.

Coates, Joseph F. "Technology Assessment," *Yearbook of Science and Technology* (New York: McGraw-Hill, 1974). Treats the methodology of technology assessment and presents the results of some assessment studies.

Illich, Ivan. *Tools for Conviviality* (New York: Perennial Library, 1973). A critique of modern industrial society, offering an alternative course for a future society, not dominated by industry, in which technology is harnessed and employed for the benefit of mankind.

Mesthene, Emmanuel G. "Some General Implications of the Research of the Harvard University Program on Technology and Society," *Technology and Culture* 10 (October 1969), 489–513. Notes that new technology has both positive and negative effects, and speaks of the need to develop new institutional methods to deal with the problems raised by technology.

National Academy of Sciences. *Technology: Processes of Assessment and Choice* (Washington, D.C.: U.S. Government Printing Office, 1969). The basic report to the Congress underlying its decision to establish an Office of Technology Assessment.

Starr, Chauncey. "Social Benefits versus Technological Risk," *Science* 165 (September 19, 1969), 1232–1238. An attempt to quantify social benefits and technological risks by providing a methodology to answer the question: "How safe is safe enough?"

Tamplin, Arthur, and John Goffman. "Towards an Adversary System of Scientific Enquiry," *The Ecologist* 1 (October 1971), 10–11. Points out the need for extensive public participation and the need for adversary assessment of new technologies.

White, Lynn T., Jr. "Technology Assessment from the Stance of a Medieval Historian," *The American Historical Review* 79 (1974), 1–13. Describes how it would have been difficult to predict the long-range social effects of some medieval inventions, and argues that imponderables should be considered in technology assessment, not just the measurable social elements.